Foundations of
Environmental Law and Policy

Interdisciplinary Readers In Law
ROBERTA ROMANO, *General Editor*

Foundations of Corporate Law
ROBERTA ROMANO

Foundations of Tort Law
SAUL LEVMORE

Foundations of Administrative Law
PETER SCHUCK

Foundations of Contract Law
RICHARD CRASWELL AND ALAN SCHWARTZ

Foundations of Environmental Law and Policy
RICHARD L. REVESZ

Foundations of Environmental Law and Policy

RICHARD L. REVESZ

New York   Oxford
Oxford University Press
1997

Oxford University Press

Oxford   New York
Athens   Auckland   Bangkok
Bogota   Bombay   Buenos Aires   Calcutta
Cape Town   Dar es Salaam   Delhi
Florence   Hong Kong   Istanbul   Karachi
Kuala Lumpur   Madras   Madrid   Melbourne
Mexico City   Nairobi   Paris   Singapore
Taipei   Tokyo   Toronto

and associated companies in
Berlin   Ibadan

Library of Congress Cataloging-in-Publication Data
Foundations of Environmental Law and Policy / [edited by] Richard L. Revesz
p. cm.—(Interdisciplinary readers in law)
Includes bibliographical references.
ISBN 0-19-509151-5
ISBN 0-19-509152-3 (pbk)
1. Environmental law—United States.   2. Environmental policy—
United States.   3. Environmental law, International. I. Revesz,
Richard L., 1958–   .   II. Series
KF3775.A7F68   1996
346.73'046—dc20
[347.30646]   96–23101

9 8 7 6 5 4 3 2 1

Printed in the United States of America
on acid-free paper

Preface

This book is designed to introduce students to the major theoretical approaches in the field of environmental law and policy. It can be used as a companion volume to the case materials used in a survey course on environmental law, or as a textbook for law school seminars on environmental law and policy, and for undergraduate and graduate seminars on environmental policy in a variety of disciplines, including government, public policy, forestry, and resources management. The book can also be used for self-study.

The readings are organized in a manner that is quite different from that of traditional environmental law case books. Except for some brief introductory materials, such case books generally devote the bulk of their attention to the major federal environmental law statutes, and discuss theoretical issues, such as the design of regulatory tools for environmental policy, primarily as they relate to problems that arise under these statutes. Thus, for example, marketable permit schemes are introduced in the section on the acid rain provisions of the Clean Air Act; effluent fees are presented in the chapter of the Clean Water Act in connection with taxation approaches in Germany; and the possibility of transmitting incentives through liability rules is raised in the chapter on Superfund. Such an organization, while readily understandable given the way that the field of environmental law developed, is poorly suited for a rigorous analysis of the range of policy instruments and of the factors affecting the choice among these instruments in particular environmental contexts. Similarly, theoretical issues relating to risk assessment, risk management, and federalism are treated in a disjointed manner. This reader, by contrast, attempts to provide a comprehensive treatment of these matters.

This book begins with eight foundational chapters dealing with issues that are central to the design of environmental policy. The next two chapters deal with case studies concerning the Clean Air Act and Superfund, which apply the foundational principles previously developed. The final two chapters deal with the problems of environmental regulation in an international community. The selections have been extensively edited to facilitate accessibility. Each chapter has an introduction that highlights the most important contributions of the readings for the purpose of efficiently directing the attention of students. The chapters end with an extensive set of notes and questions, designed to provide a deeper understanding of the readings, as well as to introduce and critique a broader set of perspectives.

In my four-credit environmental law survey course, I spend the first four and a half weeks on the eight foundational chapters (supplemented by a few relevant cases, such as *Industrial Union Department, AFL-CIO v. American Petroleum Institute*, 448 U.S. 607 (1980), and *Corrosion Proof Fittings v. EPA*, 947 F.2d 1201 (5th Cir. 1991)). Then, I use a case book for the discussion of the individual statutes (supplemented by the two chapters dealing with the Clean Air Act and Superfund). I end the course with one week on the two chapters dealing with international issues. I find that the perspectives acquired from foundational chapters of this reader make it possible to study the statutes at both a deeper level and a quicker pace.

In the case of a seminar, this reader can constitute the main text. Each of the twelve chapters is well suited for a two-hour discussion. For law school seminars, some of the chapters can perhaps be supplemented with one illustrative case or regulatory problem.

Vicki Been gave me important comments on an earlier draft; I also benefited from several conversations with Lewis Kornhauser. I am grateful for the able secretarial assistance of Evelyn Palmquist at the New York University School of Law, my home institution; Isabelle Girardi at the Graduate Institute for International Studies in Geneva, Switzerland, where I was a visiting professor during 1994 and 1995; and Thompson Potter at Harvard University Law School, where I am a visiting professor during 1995 and 1996. I am also indebted to the reference librarians at the New York University School of Law, who went well beyond the call of duty in locating the several hundred articles and books that I consulted in choosing the selections in this reader.

I dedicate this book to my children, Joshua, who at age four has made me think more deeply about the case for vegetarianism by repeatedly inquiring at the dinner table whether I am eating dead sheep or dead cow, and Sarah, who, since age one has taken weekly trips to a recycling center with her day care group.

R.L.R.

Cambridge, Mass.
June 1996

Contents

The Theoretical Foundations of Environmental Law

I *The Economic Perspective on Environmental Degradation, 3*

The Tragedy of the Commons, 5
Garrett Hardin

The Problem of Social Cost, 8
Ronald H. Coase

Notes and Questions, 13

II *Noneconomic Perspectives on Environmental Degradation, 18*

The Economy of the Earth: Philosophy, Law, and the Environment, 20
Mark Sagoff

Respect for Nature: A Theory of Environmental Ethics, 29
Paul W. Taylor

Notes and Questions, 39

III *The Scientific Predicate for Environmental Regulation: Risk Assessment, 45*

Risk, Science, and Democracy, 48
William D. Ruckelshaus

Legislating Acceptable Cancer Risk from Exposure to
Toxic Chemicals, 52
Alon Rosenthal, George M. Gray, and John D. Graham

In Search of Safety: Chemicals and Cancer Risk, 58
John D. Graham, Laura C. Green, and Marc J. Roberts

Breaking the Vicious Circle: Toward Effective Risk Regulation, 65
Stephen Breyer

Notes and Questions, 70

IV *The Objectives of Environmental Regulation:
Risk Management, 76*

Acceptable Risk, 80
Baruch Fischhoff, Sarah Lichtenstein, Paul Slovic, Stephen L. Derby,
and Ralph L. Keeney

The Strategy of Social Regulation: Decision Frameworks
for Policy, 84
Lester B. Lave

Cost-Benefit Analysis: An Ethical Critique, 93
Steven Kelman

Notes and Questions, 98

V *Distributional Consequences of Environmental Policy, 102*

Anatomy of Environmental Racism and the Environmental
Justice Movement, 104
Robert D. Bullard

Locally Undesirable Land Uses in Minority Neighborhoods:
Disproportionate Siting or Market Dynamics?, 107
Vicki Been

What's Fairness Got to Do With It? Environmental Justice
and the Siting of Locally Undesirable Land Uses, 112
Vicki Been

Environmental Policy and the Distribution of Benefits
and Costs, 117
Henry M. Peskin

Notes and Questions, 124

VI *The Choice of Regulatory Tools, 130*

Reforming Environmental Law, 133
Bruce A. Ackerman and Richard B. Stewart

The Theory of Environmental Policy, 139
William J. Baumol and Wallace E. Oates

Deposit-Refund Systems: Theory and Applications to Environmental,
Conservation, and Consumer Policy, 141
Peter Bohm

Liability for Harm Versus Regulation of Safety, 144
Steven Shavell

Informational Approaches to Regulation, 149
Wesley A. Magat and W. Kip Viscusi

Notes and Questions, 151

The Political Dimensions of Environmental Law

VII *Federalism and Environmental Regulation, 161*

Pyramids of Sacrifice? Problems of Federalism in
Mandating State Implementation of National
Environmental Policy, 163
Richard B. Stewart

Rehabilitating Interstate Competition: Rethinking the
"Race-to-the-Bottom" Rationale for Federal
Environmental Regulation, 169
Richard L. Revesz

Notes and Questions, 178

VIII *Environmental Law and Public Choice, 183*

Toward a Theory of Statutory Evolution:
The Federalization of Environmental Law, 186
E. Donald Elliott, Bruce A. Ackerman, and John C. Millian

Clean Coal/Dirty Air, 193
Bruce A. Ackerman and William T. Hassler

Environmental Regulation: Whose Self-Interests Are
Being Protected?, 201
B. Peter Pashigian

A Positive Theory of Environmental Quality Regulation, 205
Michael T. Maloney and Robert E. McCormick

Politics and Procedure in Environmental Law, 210
Daniel A. Farber

Notes and Questions, 215

Case Studies

IX *Control of Air Pollution, 223*

The Irrational National Air Quality Standards:
Macro- and Micro-Mistakes, 226
James E. Krier

The Confusion of Goals and Instruments: The Explicit
Consideration of Cost in Setting National Ambient
Air Quality Standards, 231
George Eads

Prevention of Significant Deterioration: Control-
Compelling Versus Site-Shifting, 236
Craig N. Oren

Notes and Questions, 241

X *Liability for the Cleanup of Hazardous Waste Sites, 247*

The Superfund Debate, 249
Richard L. Revesz and Richard B. Stewart

The Magnitude and Policy Implications of Health Risks
from Hazardous Waste Sites, 256
James T. Hamilton and W. Kip Viscusi

Evaluating the Effects of Alternative Superfund
Liability Rules, 264
Lewis A. Kornhauser and Richard L. Revesz

Notes and Questions, 271

Environmental Law in an International Community

XI *Environmental Regulation and International Trade, 279*

International Trade and Environment: Lessons from the
Federal Experience, 281
Richard B. Stewart

An Economic Analysis of Trade Measures to Protect the
Global Environment, 289
Howard F. Chang

Trade and the Environment: Does Environmental
Diversity Detract from the Case for Free Trade?, 295
Jagdish Bhagwati and T. N. Srinivasan

Notes and Questions, 301

XII *International Environmental Law, 307*

Intergenerational Equity: A Legal Framework for
Global Environmental Change, 309
Edith Brown Weiss

An Almost Practical Step Toward Sustainability, 312
Robert Solow

The Politics of International Regime Formation:
Managing Natural Resources and the Environment, 315
Oran R. Young

Subsistence Emissions and Luxury Emissions, 322
Henry Shue

Notes and Questions, 329

Foundations of
Environmental Law and Policy

The Theoretical Foundations
of Environmental Law

The Economic Perspective on Environmental Degradation

The readings in this chapter analyze the problem of environmental degradation from an economic perspective. This perspective can be defined by reference to normative, positive, and attitudinal characteristics.

The economic perspective's normative goal is to maximize social welfare—the sum of the private welfare of each individual in a society. Pollution and other forms of environmental degradation are generally a by-product of profitable economic activity. A reduction in pollution is socially advantageous only if it increases the welfare of the victims of pollution by more than it decreases the welfare of those who cause the pollution. Thus, under the economic perspective, there is a socially optimal amount of pollution, and less pollution can be as undesirable as more pollution.

The positive, or descriptive, characteristic of the economic perspective is that it explains the existence of excessive pollution by reference to a divergence between the polluter's private costs and the social costs imposed by its activity. For example, a steel manufacturer's private costs consist of the inputs, such as raw materials, electricity, and labor, that it must purchase. The manufacturer may use other kinds of goods in its production process as well—the air or water to dispose of the by-products of steel production, for example. If the manufacturer is not required to "purchase" these goods, others (either society as a whole or some subset of society, such as the plant's neighbors) will have to bear the costs of the use. The costs are therefore external to the manufacturer, or, in economic parlance, they are "externalities." An economically rational manufacturer makes its production decisions without regard to these social costs, seeking simply to maximize the difference between its private costs and the benefits that it accrues from selling its products.

Finally, the economic perspective's attitudinal characteristic is that it does not view pollution as the result of antisocial action worthy of moral opprobrium. Rather, it sees it as the natural response of rational individuals who seek to further their self-interest.

The readings differ principally in their prescriptions for the design of social mechanisms to control the undesirable aspects of environmental degradation. In particular, they have varying degrees of faith in the beneficial effects of governmental action.

Garrett Hardin's classic article analogizes the problem of pollution to that presented by an open pasture. The problem in this "commons" is that each herder has an incentive to add cattle to the pasture, though the aggregate effect is to render the land unproductive as a result of overgrazing. In a much quoted sentence, Hardin concludes that "[f]reedom in a commons brings ruin to all."

A firm contemplating the discharge of pollution faces the same calculus as the herder, receiving a benefit from adding pollution to an environmental commons, such as an airshed or a river or lake. The aggregate effect of such decisions, however, is to produce an excessive amount of pollution, harming society as a whole. Hardin advocates the use of the coercive powers of government to prevent excessive exploitation of a commons.

Ronald Coase, in an essay that in part earned him the Nobel Prize in Economics, makes four important claims. First, he argues that the problem of pollution is a reciprocal one, which arises because of the simultaneous presence of two parties, for example, a factory that emits fumes and a laundry that is harmed by the presence of these fumes; the problem is not caused solely by the factory. Protecting the laundry by enjoining the fumes imposes harm on the factory, just as protecting the factory by not enjoining its actions imposes harm on the laundry. The relative desirability of these alternative rules depends on a comparison of the harms to the laundry and the factory.

Coase then shows that when a polluter and a pollutee, such as the factory and the laundry, can bargain costlessly, they will reach socially desirable agreements, and that the resulting amount of pollution will be independent of the legal regime. So, if the legal regime enjoins the pollution but the harm to the factory is greater than the harm that the laundry would have suffered in the absence of such an injunction, the parties will enter into a contract under which, in return for a payment, the laundry will agree not to exercise its right to seek an injunction. Conversely, if the legal regime allows the pollution but the resulting harm to the laundry is greater than the harm that the injunction would impose on the factory, the parties will enter into contract under which, again in return for a payment, the factory agrees not to pollute. Thus regardless of the initial legal rule, bargaining will produce two results: (1) it will lead to the same amount of pollution (the invariance claim); and (2) it will lead to the maximization of social welfare (the efficiency claim).

Coase then shows that these results will not be attained if the costs of bargaining are sufficiently high. If such costs are greater than the benefit that a party can obtain from the bargain, no agreement will take place. Thus, there would be no contractual modification of a rule enjoining the fumes even if the resulting harm to the factory from the injunction were greater than the harm that the laundry would suffer in the absence of the injunction. Similarly, there would be no contractual modification of a rule allowing the fumes even if the resulting harm to the laundry were greater than the harm that the factory would suffer as a result of an injunction. When bargaining costs are high, the choice of legal rule affects both the amount of pollution and the level of social welfare.

The Tragedy of the Commons

GARRETT HARDIN

The tragedy of the commons develops in this way. Picture a pasture open to all. It is to be expected that each herdsman will try to keep as many cattle as possible on the commons. Such an arrangement may work reasonably satisfactorily for centuries because tribal wars, poaching, and disease keep the numbers of both man and beast well below the carrying capacity of the land. Finally, however, comes the day of reckoning, that is, the day when the long-desired goal of social stability becomes a reality. At this point, the inherent logic of the commons remorselessly generates tragedy.

As a rational being, each herdsman seeks to maximize his gain. Explicitly or implicitly, more or less consciously, he asks, "What is the utility *to me* of adding one more animal to my herd?" This utility has one negative and one positive component.

1. The positive component is a function of the increment of one animal. Since the herdsman receives all the proceeds from the sale of the additional animal, the positive utility is nearly +1.

2. The negative component is a function of the additional overgrazing created by one more animal. Since, however, the effects of overgrazing are shared by all the herdsmen, the negative utility for any particular decision-making herdsman is only a fraction of −1.

Adding together the component partial utilities, the rational herdsman concludes that the only sensible course for him to pursue is to add another animal to his herd. And another; and another. . . . But this is the conclusion reached by each and every rational herdsman sharing a commons. Therein is the tragedy. Each man is locked into a system that compels him to increase his heard without limit—in a world that is limited. Ruin is the destination toward which all men rush, each pursuing his own best interest in a society that believes in the freedom of the commons. Freedom in a commons brings ruin to all. . . .

In an approximate way, the logic of the commons has been understood for a long time, perhaps since the discovery of agriculture or the invention of private property in real estate. But it is understood mostly only in special cases which are not sufficiently generalized. Even at this late date, cattlemen leasing national land on the western ranges demonstrate no more than an ambivalent understanding, in constantly pressuring federal authorities to increase the head count to the point where overgrazing produces erosion and weed-dominance. Likewise, the oceans of the world continue to suffer from the survival of the philosophy of the commons. Maritime nations still respond automatically to the shibboleth of the "freedom of the

ing to believe in the "inexhaustible resources of the oceans," they
after species of fish and whales closer to extinction.

nal Parks present another instance of the working out of the tragedy of
u.. s. At present, they are open to all, without limit. The parks themselves
are limited in extent—there is only one Yosemite Valley—whereas population
seems to grow without limit. The values that visitors seek in the parks are steadily
eroded. Plainly, we must soon cease to treat the parks as commons or they will be of
no value to anyone.

What shall we do? We have several options. We might sell them off as private
property. We might keep them as public property, but allocate the right to enter
them. The allocation might be on the basis of wealth, by the use of an auction system. It might be on the basis of merit, as defined by some agreed-upon standards. It
might be by lottery. Or it might be on a first-come, first-served basis, administered
to long queues. These, I think, are all the reasonable possibilities. They are all objectionable. But we must choose—or acquiesce in the destruction of the commons that
we call our National Parks.

Pollution

In a reverse way, the tragedy of the commons reappears in problems of pollution.
Here it is not a question of taking something out of the commons, but of putting
something in—sewage, or chemical, radioactive, and heat wastes into water;
noxious and dangerous fumes into the air; and distracting and unpleasant advertising signs into the line of sight. The calculations of utility are much the same
as before. The rational man finds that his share of the cost of the wastes he discharges into the commons is less than the cost of purifying his wastes before
releasing them. Since this is true for everyone, we are locked into a system of
"fouling our own nest," so long as we behave only as independent, rational, free-
enterprisers.

The tragedy of the commons as a food basket is averted by private property, or
something formally like it. But the air and waters surrounding us cannot readily be
fenced, and so the tragedy of the commons as a cesspool must be prevented by different means, by coercive laws or taxing devices that make it cheaper for the polluter to treat his pollutants than to discharge them untreated. We have not progressed
as far with the solution of this problem as we have with the first. Indeed, our particular concept of private property, which deters us from exhausting the positive
resources of the earth, favors pollution. The owner of a factory on the bank of a
stream—whose property extends to the middle of the stream—often has difficulty
seeing why it is not his natural right to muddy the waters flowing past his door. The
law, always behind the times, requires elaborate stitching and fitting to adapt it to
this newly perceived aspect of the commons.

The pollution problem is a consequence of population. It did not much matter
how a lonely American frontiersman disposed of his waste. "Flowing water purifies
itself every 10 miles," my grandfather used to say, and the myth was near enough to

the truth when he was a boy, for there were not too many people. But as population became denser, the natural chemical and biological recycling processes became overloaded, calling for a redefinition of property rights.

How to Legislate Temperance?

Analysis of the pollution problem as a function of population density uncovers a not generally recognized principle of morality, namely: *the morality of an act is a function of the state of the system at the time it is performed.* Using the commons as a cesspool does not harm the general public under frontier conditions, because there is no public; the same behavior in a metropolis is unbearable. A hundred and fifty years ago a plainsman could kill an American bison, cut out only the tongue for his dinner, and discard the rest of the animal. He was not in any important sense being wasteful. Today, with only a few thousand bison left, we would be appalled at such behavior.

In passing, it is worth noting that the morality of an act cannot be determined from a photograph. One does not know whether a man killing an elephant or setting fire to the grassland is harming others until one knows the total system in which his act appears. "One picture is worth a thousand words," said an ancient Chinese; but it may take 10,000 words to validate it. It is as tempting to ecologists as it is to reformers in general to try to persuade others by way of the photographic shortcut. But the essence of an argument cannot be photographed: it must be presented rationally—in words.

That morality is system-sensitive escaped the attention of most codifiers of ethics in the past. "Thou shalt not . . ." is the form of traditional ethical directives which make no allowance for particular circumstances. The laws of our society follow the pattern of ancient ethics, and therefore are poorly suited to governing a complex, crowded, changeable world. Our epicyclic solution is to augment statutory law with administrative law. Since it is practically impossible to spell out all the conditions under which it is safe to burn trash in the back yard or to run an automobile without smog-control, by law we delegate the details to bureaus. The result is administrative law, which is rightly feared for an ancient reason—*Quis custodiet ipsos custodes*?—"Who shall watch the watchers themselves?" John Adams said that we must have "a government of laws and not men." Bureau administrators, trying to evaluate the morality of acts in the total system, are singularly liable to corruption, producing a government by men, not laws.

Prohibition is easy to legislate (though not necessarily to enforce); but how do we legislate temperance? Experience indicates that it can be accomplished best through the mediation of administrative law. We limit possibilities unnecessarily if we suppose that the sentiment of *Quis custodiet* denies us the use of administrative law. We should rather retain the phrase as a perpetual reminder of fearful dangers we cannot avoid. The great challenge facing us now is to invent the corrective feedbacks that are needed to keep custodians honest. We must find ways to legitimate the needed authority of both the custodians and the corrective feedbacks.

The Problem of Social Cost
RONALD H. COASE

The Problem to Be Examined

This paper is concerned with those actions of business firms which have harmful effects on others. The standard example is that of a factory the smoke from which has harmful effects on those occupying neighbouring properties. The economic analysis of such a situation has usually proceeded in terms of a divergence between the private and social product of the factory, in which economists have largely followed the treatment of Pigou in *The Economics of Welfare*. The conclusions to which this kind of analysis seems to have led most economists is that it would be desirable to make the owner of the factory liable for the damage caused to those injured by the smoke, or alternatively, to place a tax on the factory owner varying with the amount of smoke produced and equivalent in money terms to the damage it would cause, or finally, to exclude the factory from residential districts (and presumably from other areas in which the emission of smoke would have harmful effects on others). It is my contention that the suggested courses of action are inappropriate, in that they lead to results which are not necessarily, or even usually, desirable.

The Reciprocal Nature of the Problem

The traditional approach has tended to obscure the nature of the choice that has to be made. The question is commonly thought of as one in which A inflicts harm on B and what has to be decided is: how should we restrain A? But this is wrong. We are dealing with a problem of a reciprocal nature. To avoid the harm to B would inflict harm on A. The real question that has to be decided is: should A be allowed to harm B or should B be allowed to harm A? The problem is to avoid the more serious harm. I instanced in my previous article the case of a confectioner the noise and vibrations from whose machinery disturbed a doctor in his work. To avoid harming the doctor would inflict harm on the confectioner. The problem posed by this case was essentially whether it was worth while, as a result of restricting the methods of production which could be used by the confectioner, to secure more doctoring at the cost of a reduced supply of confectionery products. Another example is afforded by the problem of straying cattle which destroy crops on neighbouring land. If it is inevitable that some cattle will stray, an increase in the supply of meat can only be obtained at the expense of a decrease in the supply of crops. The nature of the choice is clear: meat or crops. What answer should be given is, of course, not clear unless we know the value of what is obtained as well as the value of what is sacrificed to obtain it. To give another example, Professor George J. Stigler instances the contamination of a stream. If we assume that the harmful effect of the pollution is that it kills the fish,

R. H. Coase, "The Problem of Social Cost," 3 *Journal of Law and Economics* 1 (1960) (various pages, edited). Reprinted by permission of the University of Chicago Law School *Journal of Law and Economics*.

the question to be decided is: is the value of the fish lost greater or less than the value of the product which the contamination of the stream makes possible. . . .

The Pricing System with Liability for Damage

I propose to start my analysis by examining a case in which most economists would presumably agree that the problem would be solved in a completely satisfactory manner: when the damaging business has to pay for all damage caused *and* the pricing system works smoothly (strictly this means that the operation of a pricing system is without cost).

A good example of the problem under discussion is afforded by the case of straying cattle which destroy crops growing on neighbouring land. Let us suppose that a farmer and a cattle-raiser are operating on neighbouring properties. Let us further suppose that, without any fencing between the properties, an increase in the size of the cattle-raiser's herd increases the total damage to the farmer's crops. . . .

To simplify the argument, I propose to use an arithmetical example. I shall assume that the annual cost of fencing the farmer's property is $9 and that the price of the crop is $1 per ton. Also, I assume that the relation between the number of cattle in the herd and the annual crop loss is as follows:

Number in Herd (Steers)	Annual Crop Loss (Tons)	Crop Loss per Additional Steer (Tons)
1	1	1
2	3	2
3	6	3
4	10	4

Given that the cattle-raiser is liable for the damage caused, the additional annual cost imposed on the cattle-raiser if he increased his herd from, say, 2 to 3 steers is $3 and in deciding on the size of the herd, he will take this into account along with his other costs. That is, he will not increase the size of the herd unless the value of the additional meat produced (assuming that the cattle-raiser slaughters the cattle), is greater than the additional costs that this will entail, including the value of the additional crops destroyed. Of course, if, by the employment of dogs, herdsmen, aeroplanes, mobile radio and other means, the amount of damage can be reduced, these means will be adopted when their cost is less than the value of the crop which they prevent being lost. Given that the annual cost of fencing is $9, the cattle-raiser who wished to have a herd with 4 steers or more would pay for fencing to be erected and maintained, assuming that other means of attaining the same end would not do so more cheaply. When the fence is erected, the marginal cost due to the liability for damage becomes zero, except to the extent that an increase in the size of the herd necessitates a stronger and therefore more expensive fence because more steers are liable to lean against it at the same time. But, of course, it may be cheaper for the cattle-raiser not to fence and to pay for the damaged crops, as in my arithmetical example, with 3 or fewer steers.

ght be thought that the fact that the cattle-raiser would pay for all crops aged would lead the farmer to increase his planting if a cattle-raiser came to occupy the neighbouring property. But this is not so. If the crop was previously sold in conditions of perfect competition, marginal cost was equal to price for the amount of planting undertaken and any expansion would have reduced the profits of the farmer. In the new situation, the existence of crop damage would mean that the farmer would sell less on the open market but his receipts for a given production would remain the same, since the cattle-raiser would pay the market price for any crop damaged. . . .

I have said that the occupation of a neighbouring property by a cattle-raiser would not cause the amount of production, or perhaps more exactly the amount of planting, by the farmer to increase. In fact, if the cattle-raising has any effect, it will be to decrease the amount of planting. The reason for this is that, for any given tract of land, if the value of the crop damaged is so great that the receipts from the sale of the undamaged crop are less than the total costs of cultivating that tract of land, it will be profitable for the farmer and the cattle-raiser to make a bargain whereby that tract of land is left uncultivated. This can be made clear by means of an arithmetical example. Assume initially that the value of the crop obtained from cultivating a given tract of land is $12 and that the cost incurred in cultivating this tract of land is $10, the net gain from cultivating the land being $2. I assume for purposes of simplicity that the farmer owns the land. Now assume that the cattle-raiser starts operations on the neighbouring property and that the value of the crops damaged is $1. In this case $11 is obtained by the farmer from sale on the market and $1 is obtained from the cattle-raiser for damage suffered and the net gain remains $2. Now suppose that the cattle-raiser finds it profitable to increase the size of his herd, even though the amount of damage rises to $3; which means that the value of the additional meat production is greater than the additional costs, including the additional $2 payment for damage. But the total payment for damages is now $3. The net gain to the farmer from cultivating the land is still $2. The cattle-raiser would be better off if the farmer would agree not to cultivate his land for any payment less than $3. The farmer would be agreeable to not cultivating the land for any payment greater than $2. There is clearly room for a mutually satisfactory bargain which would lead to the abandonment of cultivation. . . .

I think it is clear that if the cattle-raiser is liable for damage caused and the pricing system works smoothly, the reduction in the value of production elsewhere will be taken into account in computing the additional cost involved in increasing the size of the herd. This cost will be weighed against the value of the additional meat production and, given perfect competition in the cattle industry, the allocation of resources in cattle-raising will be optimal. What needs to be emphasized is that the fall in the value of production elsewhere which would be taken into account in the costs of the cattle-raiser may well be less than the damage which the cattle would cause to the crops in the ordinary course of events. This is because it is possible, as a result of market transactions, to discontinue cultivation of the land. This is desirable in all cases in which the damage that the cattle would cause, and for which the cattle-raiser would be willing to pay, exceeds the amount which the farmer would pay for use of the land. In conditions of perfect competition the amount which the farmer would pay for the use of the land is equal to the difference between the value of the total production when the factors are employed on this land and the value of the additional product yielded in their next best use (which would be what the farmer would have to pay for the factors). If damage exceeds the amount the farmer would pay for the use of the land, the value of the additional product of the factors employed elsewhere would exceed the value of the total product in this use after damage is taken into account. It follows that it would be desirable to abandon cultivation of the land and to release the factors employed for production elsewhere. A procedure which merely provided for payment for damage to the crop caused by the cattle but which did not allow for the possibility of cultivation being discontinued would result in too small an employment of factors of production in cattle-raising and too large an employment of factors in cultivation of the crop. But given the possibility of market transactions, a situation in which damage to crops exceeded the rent of the land would not endure. Whether the cattle-raiser pays the farmer to leave the land uncultivated or himself rents the land by paying the land-owner an amount slightly greater than the farmer would pay (if the farmer was himself renting the land), the final result would be the same and would maximise the value of production. Even when the farmer is induced to plant crops which it would not be profitable to cultivate for sale on the market, this will be a purely short-term phenomenon and may be expected to lead to an agreement under which the planting will cease. The cattle-raiser will remain in that location and the marginal cost of meat production will be the same as before, thus having no long-run effect on the allocation of resources.

The Pricing System with No Liability for Damage

I now turn to the case in which, although the pricing system is assumed to work smoothly (that is, costlessly), the damaging business is not liable for any of the damage which it causes. This business does not have to make a payment to those damaged by its actions. I propose to show that the allocation of resources will be the same in this case as it was when the damaging business was liable for damage caused. As I showed in the previous case that the allocation of resources was optimal, it will not be necessary to repeat this part of the argument.

I return to the case of the farmer and the cattle-raiser. The farmer would suffer increased damage to his crop as the size of the herd increased. Suppose that the size of the cattle-raiser's herd is 3 steers (and that this is the size of the herd that would be maintained if crop damage was not taken into account). Then the farmer would be willing to pay up to $3 if the cattle-raiser would reduce his herd to 2 steers, up to $5 if the herd were reduced to 1 steer and would pay up to $6 if cattle-raising was abandoned. The cattle-raiser would therefore receive $3 from the farmer if he kept 2 steers instead of 3. This $3 foregone is therefore part of the cost incurred in keeping the third steer. Whether the $3 is a payment which the cattle-raiser has to make if he adds the third steer to his herd (which it would be if the cattle-raiser was liable to the farmer for damage caused to the crop) or whether it is a sum of money which he would have received if he did not keep a third steer (which it would be if the r was not liable to the farmer for damage caused to the crop) does not ⸤ne final result. In both cases $3 is part of the cost of adding a third steer, to be included along with the other costs. If the increase in the value of production in cattle-raising through increasing the size of the herd from 2 to 3 is greater than the additional costs that have to be incurred (including the $3 damage to crops), the size of the herd will be increased. Otherwise, it will not. The size of the herd will be the same whether the cattle-raiser is liable for damage caused to the crop or not.

It may be argued that the assumed starting point—a herd of 3 steers—was arbitrary. And this is true. But the farmer would not wish to pay to avoid crop damage which the cattle-raiser would not be able to cause. For example, the maximum annual payment which the farmer could be induced to pay could not exceed $9, the annual cost of fencing. And the farmer would only be willing to pay this sum if it did not reduce his earnings to a level that would cause him to abandon cultivation of this particular tract of land. Furthermore, the farmer would only be willing to pay this amount if he believed that, in the absence of any payment by him, the size of the herd maintained by the cattle raiser would be 4 or more steers. Let us assume that this is the case. Then the farmer would be willing to pay up to $3 if the cattle raiser would reduce his herd to 3 steers, up to $6 if the herd were reduced to 2 steers, up to $8 if one steer only were kept and up to $9 if cattle-raising were abandoned. It will be noticed that the change in the starting point has not altered the amount which would accrue to the cattle-raiser if he reduced the size of his herd by any given amount. It is still true that the cattle-raiser could receive an additional $3 from the farmer if he agreed to reduce his herd from 3 steer to 2 and that the $3 represents the value of the crop that would be destroyed by adding the third steer to the herd. Although a different belief on the part of the farmer (whether justified or not) about the size of the herd that the cattle-raiser would maintain in the absence of payments from him may affect the total payment he can be induced to pay, it is not true that this different belief would have any effect on the size of the herd that the cattle-raiser will actually keep. This will be the same as it would be if the cattle-raiser had to pay for damage caused by his cattle, since a receipt foregone of a given amount is the equivalent of a payment of the same amount.

It might be thought that it would pay the cattle-raiser to increase his herd above the size that he would wish to maintain once a bargain had been made, in order to induce the farmer to make a larger total payment. And this may be true. It is similar in nature to the action of the farmer (when the cattle-raiser was liable for damage) in cultivating land on which, as a result of an agreement with the cattle-raiser, planting would subsequently be abandoned (including land which would not be cultivated at all in the absence of cattle-raising). But such manoeuvres are preliminaries to an agreement and do not affect the long-run equilibrium position, which is the same whether or not the cattle-raiser is held responsible for the crop damage brought about by his cattle.

It is necessary to know whether the damaging business is liable or not for damage caused since without the establishment of this initial delimitation of rights there can be no market transactions to transfer and recombine them. But the ultimate result (which maximises the value of production) is independent of the legal position if the pricing system is assumed to work without cost. . . .

The Cost of Market Transactions Taken into Account

The argument has proceeded up to this point on the assumption . . . that there were no costs involved in carrying out market transactions. This is, of course, a very unrealistic assumption. In order to carry out a market transaction it is necessary to discover who it is that one wishes to deal with, to inform people that one wishes to deal and on what terms, to conduct negotiations leading up to a bargain, to draw up the contract, to undertake the inspection needed to make sure that the terms of the contract are being observed, and so on. These operations are often extremely costly, sufficiently costly at any rate to prevent many transactions that would be carried out in a world in which the pricing system worked without cost.

In earlier sections, when dealing with the problem of the rearrangement of legal rights through the market, it was argued that such a rearrangement would be made through the market whenever this would lead to an increase in the value of production. But this assumed costless market transactions. Once the costs of carrying out market transactions are taken into account it is clear that such a rearrangement of rights will only be undertaken when the increase in the value of production consequent upon the rearrangement is greater than the costs which would be involved in bringing it about. When it is less, the granting of an injunction (or the knowledge that it would be granted) or the liability to pay damages may result in an activity being discontinued (or may prevent its being started) which would be undertaken if market transactions were costless. In these conditions the initial delimitation of legal rights does have an effect on the efficiency with which the economic system operates. One arrangement of rights may bring about a greater value of production than any other. But unless this is the arrangement of rights established by the legal system, the costs of reaching the same result by altering and combining rights through the market may be so great that this optimal arrangement of rights, and the greater value of production which it would bring, may never be achieved.

Notes and Questions

1. Why do Hardin's herders not enter into agreements to limit the number of animals that they place in the commons? What conditions would make such agreements relatively more likely? There is a vast literature about how the experience with common grazing grounds may not have been, or need not be, tragic. The major works include Robert C. Ellickson, *Order Without Law: How Neighbors Settle Disputes* (Cambridge, Mass.: Harvard University Press, 1991); Ellickson, "Property in Land," 102 *Yale Law Journal* 1315 (1993); Rose, "The Comedy of the Commons: Custom, Commerce, and Inherently Public Property," 53 *University of Chicago Law Review* 711 (1986); Glenn G. Stevenson, *Common Property Economics: A General Theory and Land Use Applications* (Cambridge: Cambridge University Press, 1991). This literature draws a sharp distinction between "open-access territories that anyone may enter and tracts that are accessible only to the members of a limited populace and their licensees." (Ellickson, supra, at 1381). What is the relevance of this distinction?

2. Are certain types of resources more susceptible to the tragedy of the commons than others? What characteristics make those resources particularly prone to degradation?

3. Potential solutions to the problem of the commons include the privatization of property (each herder gets a specific plot of land), taxes levied on each animal introduced into the commons, unitization (all herders become part of a unit, contributing their gains to the unit and receiving some distribution of the surplus according to an agreed-upon formula), government efforts to encourage cooperation among the herders (such as providing means of enforcing any agreements they might reach), and numerical constraints on the total number of animals that can be introduced into the commons. In the latter case, rights can be allocated by auction, lottery, queue (first-come, first-served), some notion of merit or need, or simply to the herders that had animals in the commons prior to the introduction of the constraints. Assess the relative desirability of these approaches.

4. Is the political process likely to solve the problem of the commons? Assume that two candidates run for a town council. One advocates leaving the status quo of an uncontrolled commons unchanged. The other advocates controls designed to prevent overgrazing. How should a rational herder vote? Is the vote dependent on the votes of other herders?

5. Is Hardin right that each person "is locked into a system that compels him to increase his herd without limit?" At what point would the rational herder decide not to add another animal? Is there overgrazing at this point?

6. Hardin's "tragedy of the commons" is conceptually equivalent to the structure of the prisoner's dilemma. Under a traditional prisoner's dilemma, two prisoners, Row and Column, are to be questioned by a prosecutor. If neither confesses, they each will get convicted of a relatively minor crime, carrying a three-year sentence; if both confess, there is sufficient evidence to convict each of a more serious crime, carrying a five-year sentence. If one of them confesses and the other does not, the defendant who confesses receives a one-year sentence (to reflect the value of his cooperation) whereas the other defendant gets a six-year sentence (to reflect her lack of cooperation). Table I.1 presents the various payoffs (the first number in each box is Row's sentence; the second number is Column's sentence).

Table I.1.

		Column	
		No confession	Confession
Row	No confession	3,3	6,1
	Confession	1,6	5,5

If the defendants act rationally, they will both confess, even though they would both have been better off if neither had confessed. Consider what happens if Row confesses. Column is better off confessing as well, so as to get a five-year sentence rather than a six-year sentence. If, instead, Row does not confess, Column is still better off confessing, so as to get a one-year sentence rather than a three-year sentence. Thus, regardless of what Row does, Column is best off confessing (the same, of course, is true for Row): in game theory parlance, confessing is a dominant strategy.

Consider the following portrayal of the tragedy of the commons as a prisoner's dilemma from Elinor Ostrom, *Governing the Commons: The Evolution of Institutions for Collective Action* (New York: Cambridge University Press, 1990), at 3–5. The greatest number of animals that can graze on a meadow without destroying it is L. The cooperative strategy is for two herders to each place $L/2$ animals in the commons. If they do so, each will get 10 units of profit, whereas if they do not, and each person places as many animals as she pleases, they will each get zero profit. If one limits his number to $L/2$ whereas the other grazes as many as he wants, the former's profits are -1 units whereas the latter's are 11 units. Table I.2 presents the various payoffs (the first number in each box is Row's profits; the second number is Column's profits).

Table I.2.

		Column	
		L/2	No limit
Row	*L*/2	10,10	−1,11
	No limit	11,−1	0,0

Explain in detail why neither herder will limit the number of animals that he places in the commons.

How might the strategy be different if the game is repeated (if, for example, the herders know that they will share meadows again in subsequent years), rather than played only once? For discussion, see Robert M. Axelrod, *The Evolution of Cooperation* (New York: Basic Books, 1984). How does the potential for repeated dealings affect the examples discussed by Hardin?

7. Ostrom also explains (supra pp. 5–7) that the tragedy of the commons is conceptually equivalent to the logic of collective action, a theory explaining why it is difficult for individuals to pursue their joint interest, which is generally traced to Mancur Olson, *The Logic of Collective Action* (Cambridge, Mass.: Harvard University Press, 1965). Olson's central argument is that an individual who cannot be excluded from obtaining the benefits of a collective good does not have an incentive to contribute to the provision of this good. Instead, he has an incentive to "free ride" on the efforts of others. How can the tragedy of the commons be explained in free-rider terms?

8. Restate Hardin's argument by reference to the concept of externalities.

9. A public good (or collective good) has two characteristics:
   (a) it cannot be supplied to a given individual without at the same time supplying it to a large number of other people; and
   (b) one cannot exclude individuals from its enjoyment.

What are some typical examples of public goods? Why does the market not supply an appropriate amount of public goods? How can the tragedy of the commons be explained in public goods terms?

10. A factory can produce its output by emitting between zero and five units of air pollution. Obviously, the lower the emissions, the higher the factory's pollution control costs. The emissions impose costs on a neighboring laundry, which must employ more costly production processes to achieve its desired standard of cleanliness. For the different possible levels of emissions, the factory's pollution control costs and resulting costs imposed on the laundry are set forth in the Table I.3. Assume that the parties can bargain costlessly.

Table I.3

Units of Emissions	Factory's Pollution Control Costs	Laundry's Costs from Pollution
0	25	0
1	16	4
2	9	8
3	4	12
4	1	16
5	0	20

What level of emissions maximizes social welfare? Assume, initially, that under the prevailing legal regime, the laundry is unable to enjoin the pollution. How much pollution will result if the parties can bargain without cost? What payment will the laundry make to the factory? What happens if one of the parties attempts to capture too much of the surplus from bargaining? Does it matter whether each party to the bargaining knows the costs faced by the other party?

Assume, conversely, that the legal regime gives the laundry the right to enjoin the pollution. Again, assuming costless bargaining, how much pollution will result? What payment will the factory make to the laundry? What happens if does not have sufficient assets to make the payment? Is the resulting problem, if any, cured by the existence of credit markets?

11. Assume now that, if the parties engage in bargaining, they each face transactions costs of $1.50. How much pollution will result if the factory has the right to pollute without constraints? How much pollution will result if the laundry has the right to enjoin the pollution? Redo the problem for transaction costs of $4.00 for each party. Redo the problem once again for transaction costs of $6.00 for each party. What is the highest level of transaction costs for which bargaining will lead to the socially desirable outcome?

12. Assume that a Pigouvian tax (named after A. C. Pigou, whose work in the 1920s and 1930s called for "internalizing" externalities, either by taxation or compensation) is levied on the factory and that there is no bargaining. Note that such a tax would be $4.00 per unit of pollution—the cost imposed on the laundry. How much pollution will result? What happens if the taxing authority incorrectly evaluates the laundry's damages and imposes a tax of $2.00 per unit of pollution? What happens if, instead, the tax is $6.00 per unit of pollution? What happens if, instead of the tax, the factory is ordered to pay damages under the common law of nuisance? For further discussion of liability, tax, and property remedies for externalities, see Calabresi and Melamed, "Property Rules, Liability Rules, and Inalienability: One View of the Cathedral," 85 *Harvard Law Review* 1089 (1972); Cooter, "The Cost of Coase," 11 *Journal of Legal Studies* 1 (1982); Demsetz, "When Does the Liability Rule Matter?," 1 *Journal of Legal Studies* 13 (1972).

13. What happens if, in the face of a Pigouvian tax of $4.00, the parties can engage in costless bargaining? The result illustrates that when bargaining is possible, Pigouvian taxes do not lead to the maximization of social welfare. This feature of the taxes led Coase to strongly criticize Pigou's prescription. In what circumstances is this problem likely to be worrisome?

14. Should one be indifferent to whether the factory or the laundry is making the payment? Is Coase indifferent? Should the order in which the firms started operating matter? Does it matter if the beneficiary of the pollution reduction is a group of breathers rather than the laundry? If it is a multimillionaire who highly values clean air around the tennis court on the grounds of his country home?

15. In general, our preferences are determined by our level of wealth. At different wealth levels, we choose to purchase different mixes of goods and services. How might the presence of such wealth effects change Coase's analysis? Consider separately, the invariance and efficiency prongs of Coase's argument. In connection with this question, think of the victim of pollution as an individual who suffers as a result of pollution emitted by a firm. For further discussion, see Cooter, supra, at 15.

16. Studies show an important disparity between the willingness to pay (WTP) for an environmental amenity that one does not yet have and the willingness to accept (WTA) the loss of that amenity. Typically, the latter valuation is considerably higher. The difference between the WTP and the WTA valuations is often referred to as an endowment effect. How does this phenomenon affect Coase's analysis? For further discussion, see Sunstein, "Endogenous Preferences, Environmental Law," 22 *Journal of Legal Studies* 217 (1993).

17. What are sources of transaction costs when there is a single polluter and a single pollutee? What are the sources of transaction costs when many pollutees must bargain with a single polluter?

18. How do free-rider problems affect the efficacy of common law solutions to environmental problems? What legal instruments ameliorate this problem? To what extent are these instruments effective? Are there some contexts in which common law solutions are likely to work better? More generally, what difficulties exist in relying on common law remedies to address environmental problems? These issues are also considered in Chapter VI.

19. Economists generally rank the relative desirability of different allocations of resources through one of two criteria. According to the criterion of Pareto superiority, a reallocation of resources is superior if at least one person is made better off and no person is made worse off. According to the criterion of Kaldor-Hicks superiority, a reallocation of resources is superior if the gainers from the reallocation could compensate the losers so that nobody is made worse off; there is no requirement, however, that such compensation actually take place. In which of these senses are Coasian solutions desirable? Which concept is more useful to evaluate environmental policy? What criticisms can be levied against the Kaldor-Hicks criterion?

II

Noneconomic Perspectives on Environmental Degradation

The readings in this chapter present noneconomic perspectives on the problem of environmental degradation. These perspectives all reject the normative goal of the economic perspective—the maximization of social welfare—but they otherwise differ in important respects. Most importantly, such writings can be classified as either human centered or nature centered. Human-centered works derive the appropriate conditions for the treatment of the environment by reference solely to the interests of human beings. Of course, human beings might value particular animals or plants a great deal, and might choose to pursue an environmental policy that is highly protective of natural objects. The fact remains, however, that under this approach nature is protected only because it is of value to human beings. The economic perspective is a special case of a human-centered theory; noneconomic versions of such theories have different social goals.

In contrast, nature-centered works maintain that the human species has independent obligations toward other forms of life. Some theorists in this tradition accord rights to animals, whereas others reject the view that animals (or plants) have rights but believe, nonetheless, that nature must be protected for its own sake, not merely because it is useful to human beings. The two excerpted readings in this chapter reflect the wide divergence among the noneconomic perspectives: Mark Sagoff's book is human-centered, whereas Paul Taylor's book is nature-centered.

Mark Sagoff's argument relies principally on two distinctions: between consumers and citizens and between pluralist and deliberative conceptions of the political process. He states that, as consumers, we concern ourselves with satisfying our personal preferences, but as citizens, we seek to promote the public interest rather than our personal interest. Second, consistent with the tradition of civic republicanism, he argues that the goal of the political process is to ascertain the public interest through a process of deliberation rather than simply to aggregate self-interested preferences.

Sagoff maintains that environmental regulation is not simply about correcting market failures by eliminating externalities. Instead, it embodies public values that our society chooses collectively; it is not the product of consumers satisfying their individual preferences but of citizens articulating a vision of a desirable society.

Sagoff considers three ways in which the costs of environmental regulation could affect regulatory standards. First, environmental laws could insist on ample margins of safety for the whole population even if strict adherence to such laws would impose ruinous costs on the economy. Second, consistent with the economic perspective, environmental laws could price the benefits of environmental quality and balance them against the resulting costs. Sagoff rejects both these polar solutions. He advocates a third, middle course, under which the costs of compliance with environmental standards can be taken into account so that standards are reasonable in light of the effort needed to achieve them.

Paul Taylor defines four beliefs that form the core of what he calls the biocentric outlook on nature: (1) that human beings are members of the earth's community of life on the same terms as other living things; (2) that all living species, both human and nonhuman, are interconnected; (3) that each organism is a unique individual pursuing its own good in its own way; and (4) that human beings are not inherently superior to other living things.

In light of this outlook, Taylor considers how to resolve conflicts between the interests of human beings and those of other species, and proposes five principles for resolving such conflicts. First, the principle of self-defense asserts that it is permissible for human beings to destroy dangerous or harmful organisms that threaten human life or health. The remaining principles rely on the distinction between basic and nonbasic interests. The second principle, the principle of proportionality, asserts that nonbasic interests cannot override basic interests, even if the nonbasic interests are those of human beings and the basic interests are not. Third, the principle of minimum wrong applies in cases in which nonbasic interests of human beings will harm the basic interest of nonhumans, but where the former are regarded as so important that harming nonhumans is favored even by individuals who have adopted a respect for nature. Fourth, the principle of distributive justice applies when the basic interests of humans and nonhumans are in conflict, for example because a good that could benefit either humans or nonhumans must be allocated to one or the other; in these circumstances, each party must get an equal share of the good. Fifth, the principle of restitutive justice calls for making up wrongs caused by the principles of minimum wrong and distributive justice.

The Economy of the Earth: Philosophy, Law, and the Environment

MARK SAGOFF

Let me begin by invoking the useful, if rough, distinction between economic and social regulation. By "economic regulation," I refer to federal programs that set prices, performance standards, entry requirements, schedules, and so on, in the railroad, trucking, securities, telecommunications, and other industries thought to be "affected with a public interest." Congress generally limits the jurisdiction of agencies that administer economic regulations—for example, the Federal Communications Commission—to specific industries. . . .

Agencies that administer social regulations—for example, the Environmental Protection Agency (EPA) and the Occupational Safety and Health Administration (OSHA)—in contrast, generally have jurisdiction over a large range of industries and are located in the executive branch of the government, where the president can exercise more control over what they do. Congress created these agencies not primarily to address the problems of competition in specific industries but to pursue broad ethical and social objectives, such as an unpolluted environment and a safer workplace, that cut across the entire economy. . . .

In this book, I shall argue against the use of the efficiency criterion in social regulation, and against the idea that work-place, consumer-product, and environmental problems exist largely because "commodities" like environmental pollution, workplace safety, and product safety are not traded in markets. I shall argue, in contrast, that these problems are primarily moral, aesthetic, cultural, and political and that they must be addressed in those terms. . . . The argument I shall give relies on . . . . . . distinctions, which I should like to identify here. These distinctions are drawn between: . . . the citizen and the consumer, [and] public and private interests. . . .

The Citizen and the Consumer

In *The Presentation of Self in Everyday Life*, Erving Goffman describes a variety of roles each individual plays and a variety of attitudes, values, beliefs, and expectations each person brings to those roles. This variety is familiar to us all: Each of us recognizes that he or she acts in different ways and expresses different thoughts in different roles and situations—with strangers or with close friends, with family members or with fellow professionals, and so on. . . .

In this book, I shall be concerned with two rather abstract social roles we all play, namely, the role of citizen and the role of consumer. As a *citizen*, I am concerned with the public interest, rather than my own interest; with the good of the community, rather than simply the well-being of my family. Thus, as a citizen, I might oppose a foreign adventure, like the Vietnam War, because I think it is tragic from the point of view of the nation as a whole. As a consumer or producer of goods and services, however, I might at the same time look at the war as a good thing for me if, for example, neither I nor my children must serve and I have a lot of investments in war-related industries.

In my role as a *consumer*, in other words, I concern myself with personal or self-regarding wants and interests; I pursue the goals I have as an individual. I put aside the community-regarding values I take seriously as a citizen, and I look out for Number One instead. I act upon those preferences on which my personal welfare depends; I may ignore the values that are mine only insofar as I consider myself a member of the community, that is, as *one of us*. . . .

I shall contend that social regulation should reflect the community-regarding values we express through the political process and not simply or primarily the self-regarding preferences we seek to satisfy in markets. I shall argue that the interests, goals, or preferences we entertain as citizens with respect to social regulation, moreover, differ *logically* from those we seek to satisfy as individuals. . . .

Public and Private Interests

In an essay on democracy in America, Cass Sunstein contrasts republican and pluralist conceptions of government. The republican vision, which Sunstein attributes to James Madison and the Federalists, emphasizes the distinction between the personal or private interests individuals pursue as individuals and the public or objective opinions they would defend as citizens. On this Madisonian view:

> Politics consisted of self-rule by the people; but it was not a scheme in which people impressed their private preferences on the government. It was instead a system in which the selection of preferences was the object of the government process. Preferences were not to be taken as exogenous, but to be developed and shaped through the political process.

"Distinct from the republican understanding of government," Sunstein writes, "is a competing conception that might be called pluralist." Under the pluralist view, "politics mediates the struggle among self-interested groups for scarce social resources. . . . Preferences are not shaped through governance, but enter into the process as exogenous variables."

The view Sunstein attributes to the framers of the American Constitution draws, of course, from Rousseau, who argued that the business of the political process is to pursue the *common* or *public* interest of the community, which is determined by vote after suitable public deliberation. This general interest is logically separate from the aggregate *private* interest of individuals, which we might define today in terms of a social calculus or an efficient market.

Since citizens cannot deliberate and vote on every policy issue, they delegate this responsibility, in part, to their political representatives, who enact statutes setting forth general goals, values, and policy decisions. These laws also authorize regulatory agencies to deliberate over and to decide specific questions in ways consistent with these larger values and purposes. Sunstein explains:

The underlying idea is that the administrator must attempt to identify and implement the public values that underlie the statute, and that, in the absence of statutory guidance, must be found through a process of deliberation. To say this is hardly to deny that the promotion of (particular) private interests is often a legitimate function of regulation; but it is to say that the administrator must deliberate about those interests, rather than responding mechanically to constituent pressures.

We are led, then, to a distinction between deliberation, on the one hand, and a mechanical or mathematical balancing of interests, on the other. . . .

What This Book Argues

This book defends a positive and a negative thesis. The negative thesis can be stated simply: Market failure is not the basis of social regulation. This thesis should not be surprising. The statutes that give authority to agencies like EPA and OSHA generally instruct them to achieve stated ethical, aesthetic, and cultural objectives such as a cleaner environment and a safer workplace. These laws do not, as a rule, instruct these agencies to improve, ensure, simulate, or attend to the efficiency of markets. Although we may construe some environmental, public health, and public safety problems to a limited extent in terms of market failures, to do so consistently requires a willing suspension of disbelief. Attempts to explain or justify popular social policies—for example, the protection of endangered species—as necessary to "correct" market failures are often so implausible that they must bring into disrepute either the policy or the explanation. . . .

Social regulation of safety in consumer products, the workplace, and the environment historically responds to a need to make markets more humane, not necessarily to make them more efficient. These laws—whether statutory or judge-made—strive primarily to prevent injury, grief, misery, and death, not to correct market failures or to compensate for unequal bargaining power. . . .

The positive thesis of this book is that social regulation expresses what we believe, what we are, what we stand for as a nation, not simply what we wish to buy as individuals. Social regulation reflects public values we choose collectively, and these may conflict with wants and interests we pursue individually. It is essential to the liberty we cherish, of course, that individuals are free to try to satisfy their personal preferences under open and equitable conditions. It is also part of our cherished conception of liberty that we are free to choose societal ideals together and free to accomplish these ideas in ways consistent with personal and political rights through the rule of law.

Social regulation most fundamentally has to do with the identity of a nation—a nation committed historically, for example, to appreciate and preserve a fabulous natural heritage and to pass it on reasonably undisturbed to future generations. This is not a question of what we *want*; it is not exactly a question of what we *believe in*; it is a question of what we *are*. There is no theoretical way to answer such a question; the answer has to do with our history, our destiny, and our self-perception as a people. And there is no methodology for making "hard decisions" and "trade-offs." We have to rely on the virtues of deliberation—open-mindedness, attention to detail, humor, and good sense. . . .

What We Want Versus What We Are

This book concerns the economic decisions we make about the environment. It also concerns our political decisions about the environment. Some people have suggested that, ideally, these should be the same—that every environmental problem should be understood as an economic one. William Baxter, for example, writes, "All our environmental problems are, in essence, specific instances of a problem of great familiarity: How can we arrange our society so as to make the most effective use of our resources?" He adds:

> To assert that there is a pollution problem or an environmental problem is to assert, at least implicitly, that one or more resources is not being used so as to maximize human satisfactions. In this respect at least environmental problems are economics problems, and better insight can be gained by the application of economic analysis.

On this view, there is really only one problem: the scarcity of resources. Environmental problems exist, then, only if environmental resources could be used more equitably or efficiently so that more people could have more of the things for which they are willing to pay. "To the economist," Arthur Okun writes, "efficiency means getting the most out of a given input." Okun explains: "This concept of efficiency implies that more is better, insofar as the 'more' consists of items people want to buy."

Environmental economists generally define "efficiency" as the "maximum consumption of goods and services given the available amount of resources." On this approach to environmental policy, it is the preferences of the consumer that are important. "The *benefit* of any good or service is simply its value to a consumer." The only values that count or that can be counted, on this view, are those that a market, actual or hypothetical, can price. "In principle, the ultimate measure of environmental quality," one text assures us, "is the value people place on these services . . . or their *willingness to pay*."

Willingness to pay. What is wrong with that? The rub is this. Not all of us think of ourselves primarily as consumers. Many of us regard ourselves as citizens as well. As consumers, we act to acquire what we want for ourselves individually; each of us follows his or her conception of *the good life*. As citizens, however, we may deliberate over and then seek to achieve together a conception of *the good society*.

In a liberal state, we are all free to pursue our personal ideas of the good life, for example, by buying the books we want to read. In a democracy, however, we are also free to pursue our ideal of ourselves as a good society, by trying to convince one another and our political representatives of a particular idea of our national goals and aspirations.

Americans, like citizens of other countries, have national goals and aspirations—a vision of what they stand for as a nation. They believe, for example, that each person must be secure in certain basic rights if he or she is to be able to form preferences that are not merely imposed but are autonomous and express personal values and uncoerced choice. Thus, the freedoms guaranteed by the Constitution are not to be construed as preferences for which they are willing to pay. They are, on the contrary, protections needed to form and express preferences that are truly one's own and that therefore may claim societal recognition and respect. . . .

Consumers who have to pay higher prices as a result, for example, nevertheless may favor safety regulations in the workplace, not as a matter of personal self-interest but as a matter of national pride and collective self-respect. I shall argue, likewise, that our environmental goals similarly derive not necessarily from self-interest—not from consumer willingness to pay in markets—but from a common recognition of national purposes and a memory even newcomers adopt of our long historical relationship to a magnificent natural environment.

Our environmental goals—cleaner air and water, the preservation of wilderness and wildlife, and the like—are not to be construed, then, simply as personal wants or preferences; they are not interests to be "priced" by markets or by cost-benefit analysis, but are views or beliefs that may find their way, as public values, into legislation. These goals stem from our character as a people, which is not something we choose, as we might choose a necktie or a cigarette, but something we recognize, something we are.

These goals presuppose the reality of public or shared values we can recognize together, values that are discussed and criticized on their merits and are not to be confused with preferences that are appropriately priced in markets. Our democratic political processes allow us to argue our beliefs on their merits—as distinct from pricing our interests at the margin. Our system of political representation and majority vote may be the best available device for deciding on these values, for "filtering the persuasive from the unpersuasive, the right from wrong, and the good from bad.". . .

The Allocation and Distribution of Resources

In a course I teach on environmental ethics, I ask students to read the opinion of the Supreme Court in *Sierra Club v. Morton*. This case involves an environmentalist challenge to a decision by the U.S. Forest Service to lease the Mineral King Valley, a quasi-wilderness area in the middle of Sequoia National Park, to Walt Disney Enterprises, to develop a ski resort. . . .

I asked how many of the students had visited Mineral King or thought they would visit it as long as it remained undeveloped. There were about six hands. Why so few, Too many mosquitoes, someone said. No movies, said another. Another offered to explain in scrupulous detail the difference between chilblain and trench foot. These young people came from Boston, New York, and Philadelphia. They were not eager to subsist, for any length of time, on pemmican and rye biscuits.

Then I asked how many students would like to visit the Mineral King Valley if it were developed in the way Disney planned. A lot more hands went up. Someone wanted to know if he had to ski if he went. No; I told him if he stayed indoors, he need miss nothing. He could get snow blindness from the sour cream. He could meet Ms. Right at the après-ski sauna and at encounter sessions. The class got really excited. Two students in back of the room stood on tiptoe, bent their wrists, and leaned forward, as if to ski. I hope I have left no doubt about where the consumer interests of these young people lay.

I brought the students to order by asking if they thought the government was right in giving Disney Enterprises a lease to develop Mineral King. I asked them, in other words, whether they thought that environmental policy, at least in this instance, should be based on the principle of satisfying consumer demand. Was there a connection between what the students as individuals wanted for themselves and what they thought we should do, collectively, as a nation?

The response was nearly unanimous. The students believed that the Disney plan was loathsome and despicable, that the Forest Service had violated a public trust by approving it, and that the values for which we stand as a nation compel us to preserve the little wilderness we have for its own sake and as a heritage for future generations. On these ethical and cultural grounds, and in spite of their consumer preferences, the students opposed the Disney plan to develop Mineral King.

Consumer and Citizen Preferences

The consumer interests or preferences of my students are typical of those of Americans in general. Most Americans like a warm bed better than a pile of wet leaves at night. They would rather have their meals prepared in a kitchen than cook them over a camp stove. Disney's market analysts knew all this. They found that the resort would attract more than fourteen thousand tourists a day, in summer and winter alike, which is a lot more people than now hike into Mineral King. The tourists would pay to use the valley, moreover, while the backpackers just walk in. . . .

You might think that the public would have enthusiastically supported the Disney plan. Yet the public's response to the Disney project was like that of my students —overwhelming opposition. Public opinion was so unfavorable, indeed, that Congress acted in 1978 to prohibit the project, by making the Mineral King Valley a part of Sequoia National Park.

Were the rights of the skiers and scenemakers to act freely within a market thwarted by the political action of the preservationists? Perhaps. But perhaps some of the swingers and skiers were themselves preservationists. Like my students, they may themselves condemn the likely consequences of their own consumer interests on cultural or ethical grounds.

I sympathize with my students. Like them and like members of the public generally, I, too, have divided preferences or conflicting "preference maps." Last year, I bribed a judge to fix a couple of traffic tickets, and I was glad to do so because I saved my license. Yet, at election time, I helped to vote the corrupt judge out of office. I speed on the highway; yet I want the police to enforce laws against speeding. I used to buy mixers in returnable bottles—but who can bother to return them? I buy only disposables now, but to soothe my conscience, I urge my state senator to outlaw one-way containers.

I love my car; I hate the bus. Yet I vote for candidates who promise to tax gasoline to pay for public transportation. I send my dues to the Sierra Club to protect areas in Alaska I shall never visit. And I support the work of the American League to Abolish Capital Punishment although, personally, I have nothing to gain one way or the other. (If I hang, I will hang myself.) And of course, I applaud the Endangered Species Act, although I have no earthly use for the Colorado squawfish or the Indiana bat. The political causes I support seem to have little or no basis in my interests as a consumer, because I take different points of view when I vote and when I shop. I have an "Ecology Now" sticker on a car that drips oil everywhere it's parked. . . .

Compromise and Community

. . . The students in my class found it fairly easy to resolve the tension between their consumer interests and their public values with respect to the example of Mineral King. They recognized that private ownership, individual freedom of choice, and the profit motive . . . would undoubtedly lead to the construction of the Disney paradise. They reasoned, nevertheless, that we should act on principle to preserve this wilderness, which has an enormous cultural meaning for us, since the resort, though profitable, would not serve important social ends. The students argued that because there are a lot of places for people to party, we do not need to make a ski resort of Sequoia National Park.

But what if the stakes were reversed? What if we should have to make enormous financial sacrifices to protect an environmentally insignificant landscape? Suppose industry would have to pay hundreds of millions of dollars to reduce air pollution by a small, perhaps an insignificant, amount? The students in my class, by and large, answered these questions the way they answered questions about Mineral King. Just as they rejected the dogma of the perfect market, they also rejected the dogma of the perfect environment.

The students recognized that compromise is essential if we are to act as a community to accomplish any goal, however pure or idealistic it may be. To improve air quality, for example, one needs not only a will but a way; one needs to express one's goals in parts per billion or, more generally, to deal with scientific uncertainties and technical constraints. The goal of environmental purity, like the goal of economic efficiency, can become a Holy Grail, in other words, suitable only as the object of an abstract religious quest. To make progress, we need to recognize that God dwells in the details—in parts per billion and in the minute particulars of testing, monitoring, and enforcement.

Although the students thought that social policy usually involves compromise, they kept faith with the ideals they held as citizens. They understood, moreover, that if we are to take these ideals seriously, we must evaluate them in the context of the means available to achieve them. To will the end, in other words, one must also will the means: One must set goals in relation to the obstacles—economic, political, legal, bureaucratic, scientific, technical, and institutional—that stand in the way of carrying them out. We do not become a functioning political community simply by sharing public goals and by celebrating a vision of harmony between nature and society, although ceremonies of this sort are a part of citizenship. To function as a community we must also reach the compromises necessary to move beyond incantation to political and economic achievement.

This is the reason that the Mineral King example—and the difference between citizen and consumer preferences it illustrates—may serve to introduce a course in environmental ethics, but it does not take us very far into the problems of environmental policy. The interesting problems arise when we move, in Winston Churchill's phrase, "from the wonderful cloudland of aspiration to the ugly scaffolding of attempt and achievement." Then we must chasten our goals by adjusting them to economic, legal, scientific, and political realities. How can we do this and still retain the ethical and aspirational nature of our objectives? How do we keep faith with the values of the citizen while recognizing the power of the consumer? . . .

Means and Ends in Social Regulation

. . . Twenty years ago, it was impossible to calibrate the goals of social regulation with the obstacles that stood in the way of achieving them. When Congress passed the Clean Air Act, it probably knew that no one could determine "safe" threshold levels for many dangerous pollutants. It required an "ample margin of safety" anyway, leaving it to others to make that language operational. Likewise, Congress wrote laws forcing the development of pollution control technology. It had little idea whether technology, when forced, could bring pollution within hoped-for "safe" thresholds. The mood then was to experiment and see how far the environmental "revolution" would extend.

Regulators soon confronted an array of problems in implementing the goals Congress had set. These difficulties include pervasive scientific uncertainty concerning the health effects of many pollutants, especially those acting synergistically; the prohibitive costs involved in monitoring; the limited ability of technology to control, measure, trace, and otherwise deal with emissions; and the expensive and time-consuming legal hurdles involved in enforcing the law. These difficulties and many others intrude between the legislative cup and the policy lip.

There appear to be three ways to bring the cup and the lip together, that is, three ways to adjust legislated ends to the means available to accomplish them. First, we may interpret the laws narrowly to restrict pollution to thresholds that provide an ample margin of safety even for the most sensitive groups. If this entails a no-risk society, so be it; industry will have to comply. To be sure, quick compliance may bring the economy to a halt; accordingly, the agencies may stretch out the timetable for eliminating effluents. Environmental groups, however, could control the pace of progress through litigation. They could contend, successfully, that statutes like the Clean Air Act insist on absolute purity, without regard to technical constraints or economic costs.

Although this approach might look good to environmental organizations at first, it would not serve their interests in the long run. For one thing, it would put them in an adversarial role with respect to EPA and other agencies, which must also respond to industry and political pressures. If environmental lobbies alienate the bureaucrats, they may achieve nothing except after long and expensive legal battles. EPA could retaliate, moreover, by "going by the book," that is, by insisting that all polluting activities stop forthwith. This would force Congress to rewrite the laws, possibly substituting economic goals for ethical and environmental ones.

Second, the agencies themselves could substitute economic goals, for example, allocatory efficiency, for ethical and environmental ones by attempting to "price" the benefits of environmental quality, health, and safety and "balance" these against the costs. If the agencies in this way made efficiency in the allocation of resources their target, however, they would come to a curious result. It is likely that an honest cost-benefit account—one that established a "value" for life, for example, in terms of unregulated markets—would show that we have far less pollution than is economically optimal; indeed, an "unfudged" analysis might establish that pre–1970 levels were already too strict to be cost-beneficial.

To interpret the Clean Air Act, therefore, as permitting a cost-benefit test is to suggest at least the possibility that the statute may "really" call for more rather than less air pollution. Since this is not even a possibility, we cannot admit this interpretation.

Third, we might permit an agency to take technological and economic factors into account, on a case-by-case basis, as long as it acts in good faith to make progress toward reducing and, we hope, eventually eliminating damage to the environment and risks to human safety and health. Although there is no methodology, of course, for deciding what is reasonable in this regard, this approach may encourage cooperation between the agencies, industry, and consumer and environmental groups in understanding the technical, scientific, and economic details that in fact determine the outcomes of regulatory decisions. The ends of regulation, then, would remain ethically motivated, but they would also conform to the means available to achieve them. . . .

Supererogation

Our efforts to achieve a cleaner, safer, more beautiful environment are constrained, of course, by economic costs; "ought" implies "can." What we "can" do—or can afford to do—is a relative matter, however, depending in part on the importance of the ethical duty or principle at stake. How important, morally speaking, are duties and principles involved in reducing risks and protecting the environment? How much must we do—how far do we have to go in controlling pollution—to remain consistent with those duties and principles? . . .

Now, the question arises whether the duties expressed in our environmental and social legislation involve perfect or imperfect obligations. Do they state goals we must achieve regardless of costs? Do they also involve "imperfect duties" and therefore allow us to consider some progress at some times as supererogatory?

Plainly, polluters have a perfect obligation not to kill; we are horrified to hear reports that a corporation negligently or even willfully vented toxic substances that killed identifiable individuals. When deaths due to vinyl chloride were discovered, for example, EPA moved swiftly to reduce exposure to that pollutant. Where identifiable deaths can be attributed to particular exposures, society must honor the right of innocent individuals not to be killed. The government has an obligation to prohibit this sort of serious incident through statutory and tort law.

With respect to background hazards and risks, however, it is different. No one has a right to a completely risk-free environment or to be protected from *de minimis* hazards even when they are caused by man. The highways, for example, can hardly be perfectly safe, and although each of us has a perfect obligation not to drive recklessly, we are not bound, therefore, to drive at ten miles per hour, even if that would reduce traffic fatalities by many thousands. No, there is a point at which a duty of obligation shades into a duty of virtue; at that point, providing safety may remain morally praiseworthy without being an ethical requirement.

A perfectly unpolluted environment, of course, is meritorious from a moral point of view, and a society acts virtuously in attempting to eliminate pollution, just as it acts virtuously in attempting to eliminate poverty. Yet a society that stops short of committing enormous resources to efforts of this kind does not necessarily violate moral obligations. A virtuous society, of course, makes it a policy to go the "extra mile" to eliminate the causes of poverty, pollution, and other evils. But we are permitted, at some point, to take economic costs and technical feasibility into account. . . .

The problem for social regulation today, as I have argued, is not to determine what is efficient from an economic point of view; it is to weigh ends and means together in order to set targets and standards that are reasonable in relation to the efforts necessary to achieve them. It is to make this appraisal, insofar as possible, to begin with, rather than to make compromises afterward, at the level of enforcement. We may do more, in this way, to achieve goals appropriate to a caring, compassionate nation that respects its natural environment.

Respect for Nature: A Theory of Environmental Ethics

PAUL W. TAYLOR

The Biocentric Outlook on Nature

The Biocentric Outlook and the Attitude of Respect for Nature

The attitude we think it appropriate to take toward living things depends on how we conceive of them and of our relationship to them. What moral significance the natural world has for us depends on the way we look at the whole system of nature and our role in it. With regard to the attitude of respect for nature, the belief-system that renders it intelligible and on which it depends for its justifiability is the biocentric outlook. . . .

The beliefs that form the core of the biocentric outlook are four in number:

a. The belief that humans are members of the Earth's Community of Life in the same sense and on the same terms in which other living things are members of that Community.
b. The belief that the human species, along with all other species, are integral elements in a system of interdependence such that the survival of each living thing, as well as its chances of faring well or poorly, is determined not only by the physical conditions of its environment but also by its relations to other living things.
c. The belief that all organisms are teleological centers of life in the sense that each is a unique individual pursuing its own good in its own way.
d. The belief that humans are not inherently superior to other living things. . . .

Humans as Members of the Earth's Community of Life

From the perspective of the biocentric outlook on nature we see human life as *an integral part of the natural order of the Earth's biosphere*. We thus conceive of the place of humans in the system of nature in the same way we conceive of the place of other species. There is a common relationship to the Earth that we share with wild animals and plants. Full awareness of this common relationship gives us a sense of true community with them. Let us see how this sense of community develops.

The first thing we do when we accept the biocentric outlook is to take the fact of our being members of a biological species to be a fundamental feature of our existence. We do not deny the differences between ourselves and other species, any more than we deny the differences among other species themselves. Rather, we put aside these differences and focus our attention upon our nature as biological creatures. As far as our relation to the Earth's ecosystems is concerned, we see ourselves as but one species-population among many. Thus we keep in the forefront of our consciousness the characteristics we share with all forms of life on Earth. Not only is our common origin in one evolutionary process fully acknowledged, but also the common environmental circumstances that surround us all. We view ourselves as one with them, not as set apart from them. We are then ready to affirm our fellowship with them as equal members of the whole Community of Life on Earth.

Recognition of our membership in that universal Community of Life is rooted in five realities: (a) We as well as they must face certain biological and physical requirements for our survival and well-being. (b) They as well as we have a good of their own, the realization of which depends on contingencies that are not always under either our or their control. (c) Although the concepts of free will, autonomy, and social freedom apply only to humans, there is a fourth sense of freedom that holds equally of them and of us, and this kind of freedom is of great importance in any living thing's struggle to realize its good, whether human or nonhuman. (d) As a species we humans are a recent arrival on our planet, a relative newcomer to an order of life that had been established for hundreds of millions of years before we came into existence. (e) Finally, there is the fact that, while we cannot do without them, they can do without us. . . .

The Natural World as a System of Interdependence

To accept the biocentric outlook and regard ourselves and the world from its perspective is to see the whole natural domain of living things and their environment as an order of interconnected objects and events. The interactions among species-populations and between those populations and the physical environment comprise a tightly woven web. A particular change occurring in the conditions of a living population's existence or in the environmental situation will cause adjustments to be made throughout the entire structure.

As an illustration of a system of interdependence *within* the natural world we might consider some aspects of the ecology of the Everglades of Florida. In that ecosystem alligators make large depressions in marshy areas when they feed and rest. These depressions become permanent pools of water, containing a variety of forms of marine life nourished by the alligators' droppings and by bits of their food. During dry spells these pools are the only place where the marine organisms can survive. When rains come, they spread out over the Everglades, contributing to the whole grassland ecology in essential ways.

Female alligators are integral to the Everglades ecosystem in a special way. They build their nests by forming large piles of sticks and mud. These become the core of small islands of dry soil after many years, and trees begin to grow on them. The trees in turn support nesting birds which live on the fish, insects, and amphibia of the area, maintaining the balance between animal and plant life in the grasslands. Although the alligators sometimes eat nesting birds, they also protect the birds from predators.

When the alligators are trapped and killed (to supply skins to the makers of expensive shoes and handbags), the whole Everglades ecosystem suffers. The pools dry up, the marine life disappears, and the balance of life in the watery grassland is destroyed. Certain species of fish die off and, during rainy seasons, other species intrude into the area in great numbers. They had formerly been kept in check by their natural predator, the alligators. Thus the entire area undergoes deep ecological changes which, if not reversed, will spell the end of the ecosystem itself.

When one accepts the biocentric outlook, the whole realm of life is understood to exemplify a vast complex of relationships of interdependence similar to that found in each ecosystem. All the different ecosystems that make up the Earth's biosphere fit together in such a way that if one is radically changed or totally destroyed, an adjustment takes place in others and the whole structure undergoes a certain shift. . . .

Individual Organisms as Teleological Centers of Life

So far the biocentric outlook has been presented as a belief-system that sets a framework for viewing ourselves in relation to other species and for understanding how we and they alike fit into the whole natural environment of our planet. The third component of that outlook, in contrast with the first two, focuses our attention on the lives of individual organisms. The biocentric outlook includes a certain way of conceiving of each entity that has a life of its own. To accept the outlook is to sharpen and deepen our awareness of what it means to be a particular living thing.

Our knowledge of individual organisms has expanded rapidly with advances in the biological and physical sciences in the past century. Organic chemistry and microbiology have brought us close to every cell and every molecule that make up the physical structure of the bodies of organisms. We have greatly increased our understanding of how living things function as physical and chemical systems. We are acquiring ever more accurate and complete explanations of why organisms behave as they do. As we thus come to know more about their life cycles, their interactions with other organisms and with the environment, we become increasingly aware of how each of them is carrying out its life functions according to the laws of its species-specific nature. But besides this, our increasing knowledge and understanding also enable us to grasp the uniqueness of each organism as an individual. Scientists who have made careful and detailed studies of particular plants and animals have often come to know their subjects as identifiable individuals. Close observation over extended periods, whether in the laboratory or in the field, has led them to an appreciation of the unique "personalities" of their subjects. Sometimes a scientist develops a special interest in a particular animal or plant, all the while remaining strictly objective in the gathering and recording of data. . . .

Understanding individual organisms as teleological centers of life does not mean that we are falsely anthropomorphizing. It does not involve "reading into" them human characteristics. We need not, for example, consider them to have consciousness. That a particular tree is a teleological center of life does not entail that it is intentionally aiming at preserving its existence, that it is exerting efforts to avoid death, or that it even cares whether it lives or dies. . . . All organisms, whether conscious or not, are teleological centers of life in the sense that each is a unified, coherently ordered system of goal-oriented activities that has a constant tendency to protect and maintain the organism's existence.

Under this conception of individual living things, each is seen to have a single, unique point of view. This point of view is determined by the organism's particular way of responding to its environment, interacting with other individual organisms, and undergoing the regular, lawlike transformations of the various stages of its species-specific life cycle. As it sustains its existence through time, it exemplifies all the functions and activities of its species in its own peculiar manner. When observed in detail, its way of existing is seen to be different from that of any other organism, including those of its species. To be aware of it not only as *a* center of life, but as *the* particular center of life that it is, is to be aware of its uniqueness and individuality. The organism is the individual it is precisely in virtue of its having its own idiosyncratic manner of carrying on its existence in the (not necessarily conscious) pursuit of its good.

This mode of understanding a particular individual is not possible with regard to inanimate objects. Although no two stones are exactly alike in their physical characteristics, stones do not have points of view. . . . The true reality of a stone's existence includes no point of view. This is not due to the fact that it lacks consciousness. . . . [P]lants and simple animal organisms also lack consciousness, but have points of view nonetheless. What makes our awareness of an individual stone fundamentally different from our awareness of a plant or animal is that the stone is not a teleological center of life, while the plant or animal is. The stone has no good of its own. We cannot benefit it by furthering its well-being or harm it by acting contrary to its well-being, since the concept of well-being simply does not apply to it. . . .

The Denial of Human Superiority

Of all the elements that make up the biocentric outlook, the fourth is the most important as far as taking the attitude of respect for nature is concerned. This fourth element consists in a total rejection of the idea that human beings are superior to other living things. Most people in our Western civilization are brought up within a belief-system according to which we humans possess a kind of value and dignity not present in "lower" forms of life. In virtue of our humanity we are held to be nobler beings than animals and plants. Our reason and free will, it is supposed, endow us with special worth because they enable us to live on a higher plane of existence than other living things are capable of. It is this belief, so deeply and pervasively ingrained in our cultural traditions, that is brought into question and finally denied outright in the fourth element of the biocentric outlook.

In what sense are human beings alleged to be superior to other animals? We are indeed different from them in having certain capacities that they lack. But why should these capacities be taken as signs of our superiority to them? From what point of view are they judged to be signs of superiority, and on what grounds? After all, many nonhuman species have capacities that humans lack. There is the flight of birds, the speed of a cheetah, the power of photosynthesis in the leaves of plants, the craftsmanship of spiders spinning their webs, the agility of a monkey in the tree tops. Why are not these to be taken as sings of their superiority over us?

One answer that comes immediately to mind is that these capacities of animals and plants are not as *valuable* as the human capacities that are claimed to make us superior. Such uniquely human characteristics as rationality, aesthetic creativity, individual autonomy, and free will, it might be held, are more valuable than any of the capacities of animals and plants. Yet we must ask: Valuable to whom and for what reason?

The human characteristics mentioned are all valuable to humans. They are of basic importance to the preservation and enrichment of human civilization. Clearly it is from the human standpoint that they are being judged as desirable and good. It is not difficult to recognize here a begging of the question. Humans are claiming superiority over nonhumans from a strictly human point of view, that is, a point of view in which the good of humans is taken as the standard of judgment. All we need to do is to look at the capacities of animals and plants from the standpoint of *their* good to find a contrary judgment of superiority. The speed of the cheetah, for example, is a sign of its superiority to humans when considered from the standpoint of a cheetah's good. If it were as slow a runner as a human it would not be able to catch its prey. And so for all the other abilities of animals and plants that further their good but are lacking in humans. In each case the judgment of human superiority would be rejected from a nonhuman standpoint. . . .

Now consider one of the most frequently repeated assertions concerning the superiority of humans over nonhumans. This is the claim that we humans are *morally* superior beings because we possess, while animals and plants lack, the capacities that give us the status of moral agents. Such capacities as free will, accountability, deliberation, and practical reason, it is said, endow us with the special nobility and dignity that belong only to morally responsible beings. Because human existence has this moral dimension it exemplifies a higher grade of being than is to be found in the amoral, irresponsible existence of animals and plants. In traditional terms, it is freedom of the will and the moral responsibility that goes with it that together raise human life above the level of the beasts.

There is a serious confusion of thought in this line of reasoning if the conclusion drawn is understood as asserting that humans are morally superior to nonhumans. One cannot validly argue that humans are morally superior beings on the ground that they possess, while others lack, the capacities of a moral agent. The reason is that, as far as moral standards are concerned, only beings that have the capacities of a moral agent can meaningfully be said to be *either* morally good *or* morally bad. Only moral agents can be judged to be morally better or worse than others, and the others in question must be moral agents themselves. Judgments of moral superiority are based on the comparative merits or deficiencies of the entities being judged, and these merits and deficiencies are all moral ones, that is, ones determined by moral standards. One entity is correctly judged morally superior to another if it is the case that, when valid moral standards are applied to both entities, the first fulfills them to a greater degree than the second. Both entities, therefore, must fall within the range of application of moral standards. This would not be the case, however, if humans were being judged superior to animals and plants, since the latter are not moral agents. . . .

Competing Claims and Priority Principles

The General Problem of Competing Claims

. . . I consider the moral dilemmas that arise when human rights and values conflict with the good of nonhumans. Such conflicts occur whenever actions and policies that further human interests or fulfill human rights are detrimental to the well-being of organisms, species-populations, and life communities in the Earth's natural ecosystems. To put it another way, such conflicts occur whenever preserving and protecting the good of wild living things involves some cost in terms of human benefit. Clear examples are given in the following situations:

> Cutting down a woodland to build a medical center.
> Destroying a fresh water ecosystem in establishing a resort by the shore of a lake.
> Replacing a stretch of cactus desert with a suburban housing development.
> Filling and dredging a tidal wetland to construct a marina and yacht club.
> Bulldozing a meadow full of wildflowers to make place for a shopping mall.
> Removing the side of a mountain in a strip-mining operation.
> Plowing up a prairie to plant fields of wheat and corn. . . .

We must . . . try to find priority principles for resolving conflicts between humans and nonhumans which do not assign greater inherent worth to humans, but consider all parties as having the same worth. The principles, in other words, must be consistent with the fundamental requirement of *species-impartiality*. For only then can there be genuine fairness in the resolution of such conflicts.

Five Priority Principles for the Fair Resolution of Conflicting Claims

I shall now consider in depth five such principles, to be designated as follows:

1. The principle of self-defense.
2. The principle of proportionality.
3. The principle of minimum wrong.
4. The principle of distributive justice.
5. The principle of restitutive justice.

1. The principle of self-defense. The principle of self-defense states that it is permissible for moral agents to protect themselves against dangerous or harmful organisms by destroying them. This holds, however, only when moral agents, using reasonable care, cannot avoid being exposed to such organisms and cannot prevent them from doing serious damage to the environmental conditions that make it possible for moral agents to exist and function as moral agents. Furthermore, the principle does not allow the use of just any means of self-protection, but only those means that will do the least possible harm to the organisms consistent with the purpose of preserving the existence and functioning of moral agents. There must be no available alternative that is known to be equally effective but to cause less harm to the "attacking" organisms.

The principle of self-defense permits actions that are absolutely required for maintaining the very existence of moral agents and for enabling them to exercise the capacities of moral agency. It does not permit actions that involve the destruction of organisms when those actions simply promote the interests or values which moral agents may have as persons. Self-defense is defense against *harmful* and *dangerous* organisms, and a harmful or dangerous organism in this context is understood to be one whose activities threaten the life or basic health of those entities which need normally functioning bodies to exist as moral agents. . . .

2. The principle of proportionality. Before considering in detail each of the four remaining priority principles, it is well to look at the way they are interrelated. First, all four principles apply to situations where the nonhuman organisms involved are *harmless.* If left alone their activities would not endanger or threaten human life and health. Thus all four principles apply to cases of conflict between humans and nonhumans that are not covered by the principle of self-defense.

Next we must make a distinction between basic and nonbasic interests. Using this distinction, the arrangement of the four principles can be set out as follows. The principles of proportionality and minimum wrong apply to cases in which there is a conflict between the *basic* interests of animals or plants and the *nonbasic* interests of humans. The principle of distributive justice, on the other hand, covers conflicts where the interests of all parties involved are *basic*. Finally, the principle of restitutive justice applies only where, in the past, either the principle of minimum wrong or that of distributive justice has been used. Each of those principles creates situations where some form of compensation or reparation must be made to nonhuman organisms, and thus the idea of restitution becomes applicable. . . .

I might note that with reference to humans, basic interests are what rational and factually enlightened people would value as an essential part of their very existence as *persons.* They are what people need if they are going to be able to pursue those goals and purposes that make life meaningful and worthwhile. Thus for human persons their basic interests are those interests which, when morally legitimate, they have a *right* to have fulfilled. . . . [W]e do not have a right to whatever will make us happy or contribute to the realization of our value system; we do have a right to the necessary conditions for the maintenance and development of our personhood. These conditions include subsistence and security ("the right to life"), autonomy, and liberty. A violation of people's moral rights is the worst thing that can happen to them, since it deprives them of what is essential to their being able to live a meaningful and worthwhile life. And since the fundamental, necessary conditions for such a life are the same for everyone, our human rights have to do with universal values or primary goods. They are the entitlement we all have as persons to what makes us persons and preserves our existence as persons.

In contrast with these universal values or primary goods that constitute our basic interests, our non-basic interests are the particular ends we consider worth seeking and the means we consider best for achieving them that make up our individual value systems. The nonbasic interests of humans thus vary from person to person, while their basic interests are common to all. . . .

The central idea of the principle of proportionality is that, in a conflict between human values and the good of (harmless) wild animals and plants, greater weight is to be given to basic than to nonbasic interests, no matter what species, human or other, the competing claims arise from. Within its proper range of application the principle prohibits us from allowing nonbasic interests to override basic ones, even if the nonbasic interests are those of humans and the basic are those of animals and plants.

The conditions of applicability of this principle are that the human interests concerned are nonbasic ones that are intrinsically incompatible with the attitude of respect for nature, that the competing claims arise from the basic interests of wild animals and/or plants, and that these animals and plants are harmless to humans (self-defense is not in question). . . . It should be noted that such practices as recreational fishing and hunting and buying luxury furs made from the pelts of wild creatures are actually accepted by millions of people as morally permissible. This fact merely shows the unquestioned, total anthropocentricity of their outlook on nature and their attitude toward wild creatures. It is clear, however, that from the standpoint of the life-centered system of environmental ethics defended in this book, such practices are to be condemned as being fundamentally exploitative of beings who have as much inherent worth as those who exploit them.

3. The principle of minimum wrong. The principle of minimum wrong applies to situations in which (a) the basic interests of animals and plants are unavoidably in competition with nonbasic interests of humans; (b) the human interests in question are *not* intrinsically incompatible with respect for nature; (c) actions needed to satisfy those interests, however, are detrimental to the basic interests of animals and plants; and (d) the human interests involved are so important that rational and factually informed people who have genuine respect for nature are not willing to relinquish the pursuit of those interests even when they take into account the undesirable consequences for wildlife.

Examples of such situations [include]: building a library or art museum where natural habitat must be destroyed; constructing an airport, railroad, harbor, or highway involving serious disturbance of a natural ecosystem; damming a river for a hydroelectric power project; replacing a wilderness forest with a timber plantation; landscaping a natural woodland to make a public park. The problem of priority in these situations is this: How can we tell when it is morally permissible for humans to pursue their nonbasic interests when doing so adversely affects the basic interests of wild animals and plants?

... The answer lies, first, in the role such interests play in the overall view of civilized life that rational and informed people tend to adopt autonomously as part of their total world outlook. Secondly, the special value given to these interests stems from the central place they occupy in people's rational conception of their own true good. The first point concerns the cultural or social aspect of the valued interests—more specifically, the importance of their contribution to human civilization seen from a broad historical perspective. The second concerns the relation of the valued interests to an individual's view of the kind of life which, given one's circumstances and capacities, is most worth living. ...

It is now time to make clear the content of the principle. The principle states that, when rational, informed, and autonomous persons *who have adopted the attitude of respect for nature* are nevertheless unwilling to forgo the two sorts of values mentioned above, even though they are aware that the consequences of pursuing those values will involve harm to wild animals and plants, it is permissible for them to pursue those values only so long as doing so involves fewer wrongs (violations of duties) than any alternative way of pursuing those values. ...

4. The principle of distributive justice. This fourth priority principle applies to competing claims between humans and nonhumans under two conditions. First, the nonhuman organisms are not harming us, so the principle of self-defense does not apply. Secondly, the interests that give rise to the competing claims are on the same level of comparative importance, all being *basic* interests, so the principles of proportionality and of minimum wrong do not apply. The range of application of the fourth principle covers cases that do not fall under the first three.

This principle is called the principle of distributive justice because it provides the criteria for a just distribution of interest-fulfillment among all parties to a conflict when the interests are all basic and hence of equal importance to those involved. Being of equal importance, they are counted as having the same moral weight. This equality of weight must be preserved in the conflict-resolving decision if it is to be fair to all. The principle of distributive justice requires that when the interests of the parties are all basic ones and there exists a natural source of good that can be used for the benefit of any of the parties, each party must be allotted an equal share. A fair share in those circumstances is an equal share.

When we try to put this principle of distributive justice into practice, however, we find that even the fairest methods of distribution cannot guarantee perfect equality of treatment to each individual organism. Consequently we are under the moral requirement to supplement all decisions grounded on distributive justice with a further duty imposed by the fifth priority principle, that of restitutive justice. Since we are not carrying out perfect fairness, we owe some measures of reparation or compensation to wild creatures as their due. As was true in the case of the principle of minimum wrong, recognition of wrongs being done to entities possessing inherent worth calls forth the additional obligation to do what we can to make up for these wrongs. In this way the idea of fairness will be preserved throughout the entire system of priority principles. ...

Sometimes, however, the clash between basic human interests and the equally basic interests of nonhumans cannot be avoided. Perhaps the most obvious case arises from the necessity of humans to consume nonhumans as food. Although it may be possible for most people to eat plants rather than animals, I shall point out in a moment that this is not true of all people. And why should eating plants be ethically more desirable than eating animals?

Let us first look at situations where, due to severe environmental conditions, humans must use wild animals as a source of food. In other words, they are situations where subsistence hunting and fishing are necessary for human survival. Consider, for example, the hunting of whales and seals in the Arctic, or the killing and eating of wild goats and sheep by those living at high altitudes in mountainous regions. In these cases it is impossible to raise enough domesticated animals to supply food for a culture's populace, and geographical conditions preclude dependence on plant life as a source of nutrition. The principle of distributive justice applies to circumstances of that kind. In such circumstances the principle entails that it is morally *permissible* for humans to kill wild animals for food. This follows from the equality of worth holding between humans and animals. For if humans refrained from eating animals in those circumstances they would in effect be sacrificing their lives for the sake of animals, and no requirement to do that is imposed by respect for nature. Animals are not of *greater* worth, so there is no obligation to further their interests at the cost of the basic interests of humans. . . .

5. The principle of restitutive justice. . . . As a priority principle in our present context, the principle of restitutive justice is applicable whenever the principles of minimum wrong and distributive justice have been followed. In both cases harm is done to animals and plants that are harmless, so some form of reparation or compensation is called for if our actions are to be fully consistent with the attitude of respect for nature. (In applying the minimum wrong and distributive justice principles, no harm is done to harmless *humans*, so there occurs an inequality of treatment between humans and nonhumans in these situations.) In its role as a priority principle for determining a fair way to resolve conflicts between humans and nonhumans, the principle of restitutive justice must therefore supplement those of minimum wrong and distributive justice.

What kinds of reparation or compensation are suitable? Two factors can guide us in this area. The first is the idea that the greater the harm done, the greater the compensation required. Any practice of promoting or protecting the good of animals and plants which is to serve to restore the balance of justice between humans and nonhumans must bring about an amount of good that is comparable (as far as can be reasonably estimated) to the amount of evil to be compensated for.

The second factor is to focus our concern on the soundness and health of whole ecosystems and their biotic communities, rather than on the good of particular individuals. As a practical measure this is the most effective means for furthering the good of the greatest number of organisms. Moreover, by setting aside certain natural habitats and by maintaining certain types of physical environments in their natural condition, compensation to wild creatures can be "paid" in an appropriate way.

The general practice of wilderness preservation can now be understood as a matter of fairness to wild animals and plants in two different respects. On the one hand it is a practice falling under the principle of distributive justice, and on the other it is a way of fulfilling the requirements of restitutive justice. In its first aspect the preservation of wilderness is simply a sharing of the bounties of nature with other creatures. . . .

In a second respect the fairness of wilderness preservation derives from its suitability as a way of compensating for the injustices perpetrated on wildlife by humans. To set aside habitat areas and protect environmental conditions in those areas so that wild communities of animals and plants can realize their good is the most appropriate way to restore the balance of justice with them, for it gives full expression to our respect for nature even when we have done harm to living things in order to benefit ourselves. We can, as it were, return the favor they do us by doing something for their sake. Thus we need not bear a burden of eternal guilt because we have used them—and will continue to use them—for our own ends. There is a way to make amends.

Notes and Questions

1. Assess the strength of Sagoff's distinction between economic and social regulation. Is it persuasive? How does one figure out whether a particular regulatory program is economic or social? What is the relevance of the following factors?

(a) the placement of certain agencies under more direct control of the president

(b) the multi-industry jurisdiction of certain agencies

Is Sagoff justified in finding these factors significant?

2. Sagoff argues that the environmental, and health and safety statutes "do not, as a rule, instruct . . . agencies to improve, ensure, simulate, or attend to the efficiency of markets." Are the goals of these statutes necessarily different from those that Sagoff categorizes as involving economic regulation?

3. What would Sagoff say if the political process, after having engaged in appropriate deliberation, picks a standard of environmental quality that is less stringent than the standard that would maximize social welfare? Is it likely that the political process would produce such an outcome? For further discussion, see the literature on public choice, which is presented in Chapter VIII.

4. Is it inconsistent with the economic perspective for Sagoff's students to oppose the Disney plan? They might value a pristine wilderness even if they never plan to go there. (The valuation of such nonuse (or existence) values is considered in Chapter IV.) How could Sagoff's discussion of the Disney plan be recast in economic terms? For a critique of Sagoff's sharp distinction between the political and the economic spheres, see Farber, "From Plastic Trees to Arrow's Theorem," 1986 *University of Illinois Law Review* 337; Rose, "Environmental Faust Succumbs to Temptations of Economic Mephistopheles, or, Value by Any Other Name Is Preference," 87 *Michigan Law Review* 1631 (1989). For further writings on the distinction between economic values and environmental values, see Tribe, "Ways Not To Think About Plastic Trees: New Foundations for Environmental Law," 83 *Yale Law Journal* 1315 (1974); Tribe, "Policy Science: Analysis or Ideology?" 2 *Philosophy and Public Affairs* 66 (1972).

5. Does Sagoff offer a substantive standard for measuring the adequacy of environmental regulations? Does his position provide any basis for attacking a regulation on substantive grounds?

6. Does Sagoff offer a procedural standard for measuring the adequacy of environmental regulations? If such a standard is embedded in Sagoff's work, how should it be applied to evaluate the decisions of the following decision makers?

    (a) a voter

    (b) a legislator

    (c) the Administrator of the Environmental Protection Agency (EPA)

7. Sagoff is opposed to making "enormous financial sacrifices to protect an environmentally insignificant landscape." On what basis would he determine that the landscape is insignificant? On what basis would the economic perspective make this determination?

8. Why does Sagoff believe that air pollution levels before the advent of regulation brought about by the Clean Air Act of 1970 might already have been too strict to be cost beneficial? Is this plausible? How would an advocate of the economic perspective respond to this statement?

9. To what extent would Sagoff take costs into account in setting environmental standards? To what extent would he take costs into account in deciding on timetables for compliance with environmental standards?

10. Sagoff appears to believe that when we are selfish we want less environmental quality, but when we are public minded we want more. Is this viewpoint compelling? Consider the case of a wealthy individual who favors stringent environmental quality, but who, through a well-functioning political process, is persuaded to support the views of his fellow citizens who prefer more jobs. How would Sagoff judge the actions of this citizen?

11. Are each of Taylor's four beliefs necessary to a biocentric outlook on nature? What kind of outlook could be constructed on the basis of the third and fourth beliefs alone?

12. Consider the principle of denial of human superiority. Is it consistent with Taylor's five priority principles for the fair resolution of conflicting claims? With which of these five principles is it arguably in tension?

13. Consider the principle of self-defense. Under this principle, is it justifiable for a human being who climbs into a lion cage at a zoo to shoot the lion when he is attacked? Can an individual shoot a lion that has escaped from its cage as a result of the negligence of the zookeeper? Should reasonable care play a role in Taylor's theory?

14. Are the principles of proportionality and of minimum wrong mutually consistent? Why is it acceptable to dam a river but not to fish recreationally? What is the basis for Taylor's principle of minimum wrong? Is it consistent with his biocentric outlook? Under the principle of minimum wrong, how does one determine whether the individuals who want to destroy a natural habitat have a genuine respect for nature?

15. What are the basic needs of humans? What are the basic needs of animals and plants? What types of everyday activities would be prohibited if the principle of proportionality were taken seriously?

16. Does Taylor's theory impose an obligation on humans to prevent animals from hurting one another? Does it impose an obligation on humans to prevent animals from hurting plants?

17. What useful prescriptions for environmental policy can be derived from Taylor's theory? How would Taylor assess the permissibility of constructing a power plant that would lower the price of electricity and bring jobs to an economically depressed region but cause acid rain, thus damaging fish and plant life?

18. The ecological perspective, generally traced to Aldo Leopold, *A Sand County Almanac* (New York: Oxford University Press, 1949), is a different nature-centered approach to environmental ethics. Leopold states (pp. 224–25): "A thing is right when it tends to preserve the integrity, stability, and beauty of the biotic community. It is wrong when it tends otherwise." More recent works within this tradition include William Ophuls, *Ecology and the Politics of Scarcity: Prologue to a Political Theory of the Steady State* (San Francisco: W. H. Freeman, 1977); Eric T. Freyfogle, *Justice and the Earth: Images for Our Planetary Survival* (New York: Free Press, 1993). Why is such equilibrium desirable? To what extent should the changing economic activities of human beings be viewed as part of the evolutionary process?

19. Taylor criticizes this perspective (p. 118):

> [S]uch a position is open to the objection that it gives no place to the good of individual organisms, other than how their pursuit of the their good contributes to the well-being of the system as a whole. This overlooks the fact that unless individuals have a good of their own that deserves the moral consideration of agents, no account of the organic system of nature-as-a-whole can explain why moral agents have a duty to preserve its good.

Is his criticism apt? In the last few decades, ecologists have largely rejected the equilibrium paradigm. See Tarlock, "The Nonequilibrium Paradigm in Ecology and the Partial Unraveling of Environmental Law," 27 *Loyola of Los Angeles Law Review* 1121 (1994). What are the implications of the nonequilibrium paradigm?

20. The deep ecology movement is influenced by Leopold's writings, believing, for example, that "[t]here is wisdom in the stability of natural processes unchanged by human intervention." (Devall, "The Deep Ecology Movement," 20 *Natural Resources Journal* 299, 311 [1980]) However, it makes more radical prescriptions. One of the central proponents of this view writes: "Man is an integral part of nature, not over or apart from nature. Man is a 'plain citizen' of the biosphere, not its conqueror or manager. . . . Man flows with the system of nature rather than attempting to control all of the rest of nature." (Id. at 310) The central tenets of the movement include the following:

> *Optimal human carrying capacity should be determined for the planet as a biosphere and for specific islands, valleys, and continents.* A drastic reduction of the rate of growth of population of *Homo sapiens* through humane birth control programs is required.

> *A new philosophical anthropology will draw on data of hunting/gathering societies for principles of healthy, ecologically viable societies.* Industrial society is not the end toward which all societies should aim or try to aim. Therefore the notion of "reinhabiting the land" with hunting-gathering, and gardening as a goal and standard for post-industrial society should be seriously considered. (Id. at 311–12).

The movement was inspired by Naess, "The Shallow and the Deep, Long-Range Ecology Movement: A Summary", 16 *Inquiry* 95 (1973). More recent writing include Bill Devall, *Simple in Means, Rich in Ends: Practicing Deep Ecology* (Salt Lake City, Utah: Peregrine Smith, 1988); Bill Devall and George Sessions, *Deep Ecology: Living as if Nature Mattered* (Salt Lake City, Utah: G.M. Smith, 1985). What arguments can be constructed in favor of deep ecology?

21. Consider the following description of ecofeminism:

> Ecofeminism argues that patriarchy, the domination of women by men, has been asso-
> ciated with the domination of nature. Men have justified their attempts to dominate
> nature by associating it with women, objectifying both women and nature by placing
> them in the category of "other." Patriarchy also involves a denial of human links with
> the natural world and of men's feminine side. Ecofeminists regard both the despolia-
> tion of the environment and violence and militarism as rooted in the culture of domi-
> nation; they argue that both have become major threats to the human race. Patriarchy,
> ecofeminists argue, must be replaced with an egalitarian form of social organization in
> which men and women have equal power, and by a social ecology in which the nat-
> ural environment is treated with respect and sustained rather than manipulated and
> destroyed. Ecofeminists also believe that capitalism is linked to domination and must
> be replaced by another social order; they envision small-scale economies and local,
> grass-roots democracy rather than the large-scale, state-directed societies and
> economies of existing socialist nations." (Epstein, "Ecofeminism and Grass-Roots
> Environmentalism in the United States," in Richard Hofrichter, ed., *Toxic Struggles:
> The Theory and Practice of Environmental Justice* [Philadelphia: New Society Pub-
> lishers, 1993])

The classic works in ecofeminism include Susan Griffin, *Woman and Nature: The Roaring Inside Her* (New York: Harper & Row, 1978); Mary Daly, *Gyn/Ecology: The Metaethics of Radical Feminism* (Boston: Beacon Press, 1978). What effects do you believe that greater gender equality would have on the environment? To what extent do the precepts of ecofeminism seem consistent with those of deep ecology? Might the tenets of deep ecology interfere with the move toward greater gender equality?

22. In a portion of the book that is not excerpted (pp. 295–96), Taylor makes the following case for vegetarianism:

> [A]nyone who has respect for nature will be on the side of vegetarianism, even
> though plants and animals are regarded as having the same inherent worth. The point
> that is crucial here is the amount of arable land needed for raising grain and other
> plants as food for those animals that are in turn to be eaten by humans when com-
> pared with the amount of land needed for raising grain and other plants for direct
> human consumption. . . . [O]ne acre of cereal grains to be used for human food can
> produce five times more protein than one acre used for meat production; one acre of
> legumes (peas, lentils, and beans) can produce ten times more; and one acre of leafy
> vegetables fifteen times more.

Contrast this view with that in Peter Singer, *Animal Liberation* (New York: Random House, 2d ed., 1990). Singer argues (p. 171):

> [T]he only legitimate boundary to our concern for the interests of other beings is the
> point at which it is no longer accurate to say that the other being has interests. To
> have interests, in a strict, nonmetaphorical sense, a being must be capable of suffer-
> ing or experiencing pleasure. If a being suffers, there can be no moral justification for
> disregarding that suffering, or for refusing to count it equally with the like suffering
> of any other being. But the converse of this is also true. If a being is not capable of
> suffering, or of enjoyment, there is nothing to take into account.

Singer suggests two indicators of the capacity of nonhumans to suffer: "the behavior of the being, whether it writhes, utters cries, attempts to escape from the source of pain, and so on; and the similarity of the nervous system of the being to our own." In the first edition of his book, Singer drew the line between a shrimp and a scallop, deeming it permissible to eat the latter but not the former. (In his second edition, he expresses concern that scallops and other mollusks might feel pain.)

Singer's central argument is against speciesism—a prejudice in favor of individuals of one's own species. Singer writes (p. 19):

> To avoid speciesism we must allow that beings who are similar in all relevant respects have a similar right to life—and mere membership in our own biological species cannot be a morally relevant criterion for this right. Within these limits we could still hold, for instance, that it is worse to kill a normal adult human, with a capacity for self-awareness and the ability to plan for the future and have meaningful relations with others, than it is to kill a mouse, which presumably does not share all these characteristics; or we might appeal to the close family and other personal ties that humans have but mice do not have to the same degree. . . .

> Whatever criteria we choose, however, we will have to admit that they do not follow precisely the boundary of our own species. We may legitimately hold that there are some features of certain beings that make their lives more valuable than those of other beings; but there will surely be some nonhuman animals whose lives, by any standards, are more valuable than the lives of some humans. A chimpanzee, dog, or pig, for instance, will have a higher degree of self-awareness and a greater capacity for meaningful relations with others than a severely retarded infant or someone in a state of advanced senility.

How compelling is Singer's theory? Recall Taylor's priority principles for the fair resolution of competing claims. What different principles might Singer offer? For further discussion of the moral relevance of consciousness, see chapter 8 of Robin Attfield, *The Ethics of Environmental Concern* (Athens: University of Georgia Press, 2d ed., 1991).

23. Taylor's respect for nature is based on respect for worth of all *living* objects. Evaluate the following argument for respecting inanimate objects as well:

> If we accept life as the intrinsic or inherent good, then a foundation is laid for claims on behalf of animate Things, but claims on behalf of inanimates are at best derivative. A river would not be morally considerate as such, but we would have duties to treat it in a certain way in order to fulfill our duties to the living things that depend on it. One might suppose that this distinction would be of little moment to an environmentalist. Life in one form or another is so ubiquitous that to protect life seems to be a handy and plausible way to protect the entire environment. If a lake were utterly lifeless, or unconnected to the support of any life, we would have no duty to it; but it is hard to imagine such a lake. Even the Dead Sea is not quite dead.

> Nonetheless, the use of life as a foundational good has drawbacks for an environmental movement. For one thing, it may be intuitively disingenuous, like pitching the case against cruelty to animals on human betterment. [Consider] the campaign to save Mono Lake from destruction as its freshwater feeder streams are increasingly diverted to meet Southern California's water demand. The lake is remote from popu-

lation centers and, aside from brine shrimp and gulls, only subtly life-supporting. If there were no brine shrimp and gulls, would morality have nothing to say?

More importantly, a life maximizing principle leads to odd choices. Suppose that two water plants were being weighed, each of which would supply the same amount of water at the same cost. In other words, imagine the present utility of the two projects to be identical. Further, imagine future generations to be indifferent as between which project we should choose. There would be this difference only: different amounts of life are to be affected, whether life be measured in terms of absolute bio-mass or diversity. The first would destroy a grand, unique geologic formation that took shape over thousands of years but supports very little life (it is all the more rare for that). The competing water delivery plan, by requiring reconstruction of some man-made water courses, would eliminate tons of life in the form of bacteria and algae. Does the fact that the second project would eliminate more *life* mean that the destruction of the unique formation is therefore preferred, and that all further moral conversation is cut off. (Christopher Stone, *Earth and Other Ethics: The Case for Moral Pluralism* [New York: Harper & Row, 1987], at 94–95)

Stone had earlier written an article, which was embraced in dissent by Justice Douglas in Sierra Club v. Morton, 405 U.S. 727 (1972), urging the Supreme Court to confer standing on natural objects. See Stone, "Should Trees Have Standing? Towards Legal Rights for Natural Objects," 45 *Southern California Law Review* 450 (1972). Does Stone offer a more com-pelling argument than Taylor?

24. One human-centered perspective to environmental ethics focuses on our duties to future generations. For a philosophical work in this vein, see John Arthur Passmore, *Man's Responsibility for Nature: Ecological Problems and Western Traditions* (New York, Scribner, 1974). This question is dealt with in Chapter XII in connection with the discussion of sustain-able development. How would Sagoff and Taylor deal with our obligation to future generations?

III

The Scientific Predicate for Environmental Regulation: Risk Assessment

This chapter analyzes the process of risk assessment, by which the administrative agencies responsible for regulating environmental harms determine a pollutant's risk to human health. Risk assessment is the first step in the regulatory process. Once a regulatory agency has made a judgment about the nature and extent of a risk, it must decide whether and how much the risk should be regulated—a process known as risk management, which is the focus of Chapter IV.

The article by William Ruckelshaus argues that a sharp distinction should be drawn between risk assessment and risk management. Ruckelshaus was the first administrator of the Environmental Protection Agency (EPA) in the early 1970s; in the mid-1980s, he was drafted to lead the agency again after the leadership of Anne Gorsuch had severely impaired its credibility. He advocates keeping political considerations out of the risk assessment process—an approach that many believe EPA had not followed under Gorsuch's leadership. According to Ruckelshaus, the appropriate place for political decisions is at the risk management stage.

Ruckelshaus recognizes that risk assessment is dependent on a variety of assumptions and that these assumptions will reflect the values of the individuals responsible for the choice, whether they are scientists, civil servants, or politicians. He argues that the discretion of individual risk assessors should be constrained through generic policies governing recurring issues. Such policies were adopted in the early 1980s by EPA and a number of other federal agencies. One prescribes a no-threshold model for carcinogens, for example, so that any exposure to a carcinogen is assumed to increase the probability of cancer. Ruckelshaus argues that such policies make the process of risk assessment more uniform and less likely to be influenced by political considerations.

   The article by Alon Rosenthal, George Gray, and John Graham explains the process of quantitative risk assessment, which, as its name implies, quantifies the human health risks attributable to environmental hazards. For carcinogens, this process takes place in four stages: hazard identification, dose-response evaluation, exposure assessment, and risk characterization.

   Hazard identification—determining whether a substance is hazardous to human health—is generally conducted through human epidemiological studies or long-term animal bioassays. Epidemiological studies monitor the health of human populations that have been exposed to the substance (such as a community surrounding a factory from which the compound was accidentally released). In long-term animal bioassays, large doses of the substance are administered to controlled groups of animals, generally rodents, and the resulting reaction is monitored.

   Dose-response evaluation involves determining the relationship between the dose of the substance to which human beings are exposed and the probability of adverse health effects. This determination is performed by extrapolating from the effects of the large doses to which rodents are exposed in animal bioassays to the far lower doses that a plausible regulatory regime is likely to regard as permissible, as well as extrapolating from the effects on rodents to the likely effects on human beings.

   Exposure assessment consists of determining the extent to which human populations are exposed to hazardous substances. For example, how much groundwater contaminated by a hazardous waste site do individuals in a surrounding community drink?

   Finally, *risk characterization* involves providing a numerical estimate of the health risk. In the case of carcinogens, this number is often expressed as the increased probability of cancer from a lifetime exposure to the harmful substance.

   The excerpt from the book by John Graham, Laura Green, and Marc Roberts uses the risk assessment of formaldehyde as a case study to highlight the policy choices that must be made in dose-response evaluation. (In parts that are not excerpted, the authors show that similar choices underlie other important components of the risk assessment process, such as the relationship between the dose administered in animal bioassays and the effective dose or dose actually reaching the target tissues.)

   Two important conclusions emerge from the excerpt. First, even the generic policies advocated by Ruckelshaus leave enormous room for policy judgments. Far from eliminating the uncertainty, they shift it from more visible decisions, such as what model to use, to less visible ones. Second, the generic policies often lead to results inconsistent with the best scientific judgments.

   The excerpt from the book by Justice Stephen Breyer, which was published shortly before his appointment to the Supreme Court, focuses on the vast disparity between how the public assesses risk and how experts do. This problem was made salient by the publication of two reports by EPA, in 1987 and 1990, which showed that problems that experts regarded as most serious, such as radon contamination in homes, were regarded as relatively unimportant by the public, whereas problems that experts ranked as far less serious, such as contamination from hazardous waste sites, were perceived by the public as most significant. Other studies confirm that lay and expert perceptions of risk are significantly different.

   Justice Breyer explores the reasons why the public might have different reactions to risk than experts. In particular, human beings use heuristic devices as shortcuts to characterize risk, give greater prominence to unusual events than to everyday risks, have greater feelings of moral obligation toward those who are close to them, distrust experts, are reluctant to change their minds, and have difficulty understanding the mathematical probabilities involved in assessing risk. While some commentators have argued that the solution lies in more effective communication about risk, Justice Breyer is skeptical about whether such educative approaches are likely to significantly change the cognitive processes that lead to the gap between scientific and lay perceptions of risk.

Risk, Science, and Democracy

WILLIAM D. RUCKELSHAUS

"Risk" is the key concept here. It was hardly mentioned in the early years of EPA, and it does not have an important place in the Clean Air or Clean Water Acts passed in that period. Of the events that contributed to this change, the most important were the focus of public attention on PCBs and asbestos (two substances that are ubiquitous in the American environment and that are capable of causing cancer) and the realization that exposure to a very large number of unfamiliar and largely untested chemicals is universal. The discovery by cancer epidemiologists that cancer rates vary with environment suggested that pollution might play a role in causing this disease. And finally, the cancer risk was pushed to the forefront by the emergence of abandoned dumps of toxic chemicals as a consuming public issue. As a direct result of this shift in attention, the relation of EPA to its science base was altered; the problem of uncertainty was moved from the periphery to the center.

This shift occurred because the risks of effects from typical environmental exposures to toxic substances—unlike the touchable, visible, and malodorous pollution that stimulated the initial environmental revolution—are largely constructs or projections based on scientific findings. We would know nothing at all about chronic risk attributable to most toxic substances if scientists had not detected and evaluated them. Our response to such risks, therefore, must be based on a set of scientific findings. Science, however, is hardly ever unambiguous or unanimous, especially when the data on which definitive science must be founded scarcely exist. The toxic effects on health of many of the chemicals EPA considers for regulation fall into this class.

"Risk assessment" is the device that government agencies such as EPA have adopted to deal with this quandary. It is the attempt to quantify the degree of hazard that might result from human activities—for example, the risks to human health and the environment from industrial chemicals. Essentially, it is a kind of pretense; to avoid the paralysis of protective action that would result from waiting for "definitive" data, we assume that we have greater knowledge than scientists actually possess and make decisions based on those assumptions.

Of course, not all risk assessment is on the controversial outer edge of science. We have been looking at the phenomenon of toxic risk from environmental levels of chemicals for a number of years, and as evidence has accumulated for certain chemicals, controversy has diminished and consensus among scientists has become easier to obtain. For other substances—and these are the ones that naturally figure most prominently in public debate—the data remain ambiguous.

In such cases, risk assessment is something of an intellectual orphan. Scientists are uncomfortable with it when the method must use scientific information in a way that is outside the normal constraints of science. They are encroaching on political judgments and they know it. As Alvin Weinberg has written:

> Attempts to deal with social problems through the procedures of science hang on the answers to questions that can be asked of science and yet which cannot be answered by science. I propose the term *trans-scientific* for these questions. . . . Scientists have no monopoly on wisdom where this kind of trans-science is involved; they shall have to accommodate the will of the public and its representatives.

However, the representatives of the public, in this instance policy officials in protective agencies, have their problems with risk assessment as well. The very act of quantifying risk tends to reify dreaded outcomes in the public mind and may make it more difficult to gain public acceptance for policy decision or push those decisions in unwise directions. It is hard to describe, say, one cancer case in 70 years among a population of a million as an "acceptable risk" when such a description may too easily summon up for any individual the image of some close relative on his deathbed. Also, the use of risk assessment as a policy basis inevitably provokes endless arguments about the validity of the estimates, which can seriously disrupt the regulatory timetables such officials must live by.

Despite this uneasiness. there appears to be no substitute for risk assessment, in that some sort of risk finding is what tells us that there is any basis for regulatory action in the first place. The alternative to not performing risk assessment is to adopt a policy of either reducing all *potentially* toxic emissions to the greatest degree technology allows (of which more later) or banning all substances for which there is any evidence of harmful effect, a policy that no technological society could long survive. Beyond that, risk assessment is an irreplaceable tool for setting priorities among the tens of thousands of substances that could be subjects of control actions—substances that vary enormously in their apparent potential for causing disease. In my view, therefore, we must use and improve risk assessment with full recognition of its current shortcomings.

This accommodation would be much easier from a public policy viewpoint were it possible to establish for all pollutants the environmental levels that present zero risk. This is prevented, however, by an important limitation of the current technique; the difficulty of establishing definitive no-effect levels for exposure to most carcinogens. Consequently, whenever there is any exposure to such substances, there is a calculable risk of disease. The environmentalist ethos, which is reflected in many of our environmental laws, and which requires that zero-risk levels of pollutant exposure be established, is thus shown to be an impossible goal for an industrial society, as long as we retain the no-threshold model for carcinogenesis. . . .

This situation has given rise to two conflicting viewpoints on protection. The first, usually proffered by the regulated community, argues that regulation ought not to be based on a set of unprovable assumptions but only on connections between pollutants and health effects that can be demonstrated under the canons of science in the strict sense. It points out that for the vast majority of chemical species, we have no evidence at all that suggests effects on human health from exposures at environmental levels. Because many important risk assessments are based on assumptions that are scientifically untestable, the method is too susceptible to manipulation for political ends and, the regulated community contends, it has been so manipulated by environmentalists.

The second viewpoint, which has been adopted by some environmentalists, counters that waiting for firm evidence of human health effects amounts to using the nation's people as guinea pigs, and that is morally unacceptable. It proposes that far from overestimating the risks from toxic substances, conventional risk assessments underestimate them, for there may be effects from chemicals in combination that are greater than would be expected from the sum effects of all chemicals acting independently. While approving of risk assessment as a priority-setting tool, this viewpoint rejects the idea that we can use risk assessment to distinguish between "significant" and "insignificant" risks. Any identifiable risk ought to be eliminated up to the capacity of available technology to do so.

It is impossible to evaluate the merits of these positions without first drawing a distinction between the assessment of risk and the process of deciding what to do about it, which is "risk management." The arguments in the form sketched here are really directed at both these processes, a common confusion that has long stood in the way of sensible policymaking.

Risk assessment is an exercise that combines available data on a substance's potency in causing adverse health effects with information about likely human exposure, and through the use of plausible assumptions, it generates an estimate of human health risk. Risk management is the process by which a protective agency decides what action to take in the face of such estimates. Ideally the action is based on such factors as the goals of public health and environmental protection, relevant legislation, legal precedent, and application of social, economic, and political values. *Risk Assessment in the Federal Government*, a National Research Council (NRC) document, recommends that regulatory agencies establish a strict distinction between the two processes, to allay any confusion between them. In my view Congress should do the same in all statutes seeking to deal with risk.

Returning now to the opposing viewpoints we see that both reflect the fear that risk assessment may be imbued with values repugnant to one or more of the parties involved. That is, some people in the regulated community believe that the structure of risk assessment inherently exaggerates risk, while many environmentalists believe that it will not capture all the risk that may actually exist. As we have seen, this disagreement is not resolvable in the short run through recourse to science. Risk assessment is necessarily dependent on choices made among a host of assumptions, and these choices will inevitably be affected by the values of the choosers, whether they be scientists, civil servants, or politicians.

The NRC report suggests that this problem can be substantially alleviated by the establishment of formal public rules guiding the necessary inferences and assumptions. These rules should be based on the best available information concerning the underlying scientific mechanisms. Adoption of such guidelines reduces the possibility that an EPA administrator may manipulate the findings of some risk assessment so as to avoid making the difficult, and perhaps politically unpopular, choices involved in a risk-management decision. Both industry and environmentalists fear this manipulation—from different brands of administrator, needless to say. Although we cannot remove values from risk assessment, we can and should keep those values from shifting arbitrarily with the political winds.

The explicit and open codification suggested by the NRC will also ensure that the assumptions used in risk assessment will at least be uniform among all agencies that adopt them, will be plausible scientifically, and will reflect a predictable and relatively constant policy amid this complex and chaotic hybrid discipline. It also offers the possibility that one day all the protective agencies of government will speak with one voice when they address risks, so that estimates of risk will be comparable among agencies and the public at last will be able to make a fair comparison of the individual risk-management decisions of separate agencies.

The remaining points of both positions are really about risk management and on this issue both are flawed. At its extreme, the first position—that regulation should be based solely on scientifically provable connections between pollutants and health effects—would allow the release of unlimited quantities of substances that cause cancer in animals, on the assumption that there will be no analogous effect on people and that there must be thresholds for carcinogenesis. I expect that most Americans would reject that assumption as imprudent, given our current knowledge about carcinogenesis (for example, the similarity of cancer causing genes across species). At some level we have to regard the possibility that we are controlling somewhat in excess of the true risk as a kind of insurance, with the cost of control as its premium. The effort to reduce apprehension, even so-called unreasonable apprehension, about the future results of current practices is a valid social function. Risk-management agencies such as EPA could be chartered to do precisely that. If so, we had better make clear what we are doing, and establish rules for doing it.

The weakness of the second viewpoint, that any identifiable risk ought to be eliminated up to the capacity of available technology to do so, lies in the concept of a best available technology that must invariably be applied where risk is discovered. "Best" and "available" are terms as infinitely debatable as the assumptions of risk assessment. There is always a technology conceivable that is an improvement on a previous one, and as the last increments of pollution are removed, the cost of each successive fix goes up very steeply. Because, according to the no-threshold assumption, even minute quantities of carcinogens can be projected out to cause cases of disease, arguments about technology reduce in the end to arguments about risk and cost: technology A allows a residual risk of 10^{-5} and costs \$1 million; technology B allows a residual risk of 10^{-6} and costs \$10 million, and so on ad infinitum. It is specious to pretend that costs do not matter, because it is always possible to show that at a certain level of removal, costs in fact do matter: technology Z allows residual risks of 10^{-15} and costs \$1 trillion.

Once this is admitted, as it almost always is when we come down to debating actual regulations, the position is reduced to arguments about affordability. This too is treacherous ground. Firms vary in their ability to pay, and what is affordable for one may bankrupt another. If requirements are adjusted so as not to cripple the poorest firms, the policy amounts to an environmental subsidy to the less efficient players in our economy. . . .

My point is that in confronting any risk there is no way to escape the question "Is controlling it worth it?" We must ask this question not only in terms of the relationship of the risk reduced and the cost to the economy but also as it applies to the resources of the agency involved. Policy attention is the most precious commodity in government, and a regulation that marginally protects only 20 people may take up as much attention as a regulation that surely protects a million.

"Is it worth it?" That this question must be asked and asked carefully is a token of how the main force of the environmental idea has been modified by the recent focus on toxic risk to human health. In truth this question should always have been asked, but because the early goals of environmentalism were so obviously good, the requirement to ask, "Is it worth it?" was not firmly built into all our environmental laws. Who would dare to question the worth of saving Lake Erie? Environmentalism at its inception was a grand vision, one that nearly all Americans willingly shared. Somehow that vision of the essential unity of nature and of the need for bringing industrial society into harmony with it has been lost among the parts per billion, and with it we have lost the capacity to reach social consensus on environmental policy.

Legislating Acceptable Cancer Risk from Exposure to Toxic Chemicals

ALON ROSENTHAL, GEORGE M. GRAY, AND JOHN D. GRAHAM

EPA uses risk assessment to predict the probability of developing cancer as a result of exposure to a particular agent. As currently practiced, risk assessment of a carcinogen takes place in four steps: hazard identification, dose-response evaluation, exposure assessment, and risk characterization.

The first step, hazard identification, is the process of determining whether an "agent" (for example, an industrial chemical, a natural product in the environment, or a particular lifestyle) increases a person's risk of developing cancer. The second step, dose-response evaluation, reveals how the likelihood of cancer changes with the level of exposure. A risk assessor might estimate, for example, how the probability of lung cancer changes with the number of cigarettes smoked. The third step, exposure assessment, quantifies the amount, or dose, of the carcinogen to which people may be exposed. This may be the amount of a chemical in the air near a factory, the concentration of radon in the basement of a home, or the amounts of various foods and beverages which an individual consumes each day.

After these quantitative inputs to a risk assessment have been determined, the numbers are combined to yield an overall estimate of risk, the basic component of the final step, risk characterization. A risk characterization is usually expressed numerically as the incremental lifetime risk of cancer due to a particular agent at a particular level of exposure (also referred to as an incremental risk). This is the number that a risk manager might compare to a legislated bright line. Good risk characterizations contain not only a final risk number but also a discussion of the uncertainties in and the assumptions behind the assessment, but unfortunately this step is rarely taken. . . .

Reprinted by permission of the authors and the University of California Press Journals Department from *Ecology Law Quarterly*, Vol. 19, No. 2, pp. 269–362, Copyright © 1992 by *Ecology Law Quarterly*.

Hazard Identification

The most definitive way to determine whether a compound can cause human cancer is through the science of epidemiology. Cancer epidemiology attempts to establish associations between human exposure to a suspected cancer causing agent and the frequency of cancer in the human population. The major drawback of epidemiological studies is that they cannot measure risks before those who are exposed develop cancer, but merely identify effects which have already occurred. Risk managers want to identify human carcinogens before cancer develops, before they can be discovered by epidemiology.

Furthermore, cancer epidemiology is fraught with interpretive difficulty. Cancer is a disease with a long latency period that arises from many causes, only some of which are known. Human exposures to potential carcinogens are often complex, uncertain, and poorly documented. If exposures are mismeasured, the epidemiologist will have a difficult time detecting any association between exposure and disease, even if one exists. Moreover, epidemiological studies are often plagued by confounding factors, such as smoking, by a lack of suitable control groups, and by alternative interpretations of data. Due to practical limitations on the size of studies and the large background risk of cancer, epidemiologists usually cannot detect modest cancer risks that would still be of concern to risk managers. While some epidemiological studies of animal carcinogens have been "negative," this may simply reflect the inadequate sample sizes in these studies. When epidemiologists do detect human cancer risks, they usually do so in occupational settings where historical levels of exposure have been quite high. If findings from the workplace are to be extrapolated to environmental settings, epidemiologists must resolve uncertainties about how to extrapolate tumors observed at relatively high doses to the tumors that might occur at low levels of environmental exposure.

Credible epidemiological studies, especially several showing the same positive result, are considered adequate evidence of human carcinogenicity. Such results are difficult to obtain except when studying very potent carcinogens or carcinogens which cause an unusual type of tumor. For example, epidemiological studies identified vinyl chloride as a human carcinogen because it causes liver angiosarcoma, an extremely rare type of tumor. By contrast, there is little consensus within the scientific community on how much weight to give negative epidemiological reports, or on how to resolve controversies when there are both positive and negative epidemiological studies of a compound. As a result, fewer than sixty chemicals and mixtures have been identified as known human carcinogens.

In light of the limits of epidemiology and the need to identify hazards before they cause serious harm, scientists have resorted to animal experiments in an effort to identify agents that are potential human carcinogens. The key laboratory test used in hazard identification is the long-term rodent bioassay, which is conducted on the assumption that a rodent carcinogen may also be a human carcinogen. In addition, laboratory tests of the biological properties of chemicals provide information which can help scientists assess a chemical's potential for human carcinogenicity.

The National Toxicology Program (the NTP) of the U.S. Department of Health and Human Services has established rigorous guidelines for the conduct of rodent carcinogen bioassays. Under the NTP's guidelines, a researcher must expose fifty animals of each sex of two species (usually rats and mice) to several dose levels of the suspected carcinogen for virtually their entire lives. The dose levels selected are the maximum tolerated dose (the MTD) and fractions thereof, usually MTD/2 or MTD/4. The MTD is the highest dose that the animals can tolerate without becoming so sick that the test will not be useful in detecting tumors. High dose levels are chosen to compensate for the small number of rodents, which are expensive to house and feed. Since most bioassays are performed with only fifty animals at each dose level, the animals must be given the highest dose that they can tolerate if the researchers are to maximize their chances of seeing a statistically significant response. However, the small number of animals used greatly limits the sensitivity of the assay. For example, if a dose of a carcinogen causes an increased cancer risk of one in 100 in a rodent's lifetime, it is unlikely to be detected in a cohort of fifty rodents.

Tumors observed at the MTD are considered relevant on the theory that cancer is a disease that can be caused by a single molecule of a carcinogen interacting with the DNA in a single cell, and therefore, the response of a carcinogen at the MTD can be extrapolated to the much lower levels of exposure that humans experience. However, there is controversy within the scientific community about whether results from rodent bioassays performed at or near the MTD are applicable to the much lower level of exposure typically faced by humans.

Several hundred compounds have been shown to cause cancer in animal tests. The usefulness of these studies in predicting human carcinogenicity depends on the accuracy of certain assumptions. These include the assumption that humans respond in a similar manner to rodents; the assumption that results of exposure to high doses over the relatively short lifetimes of animals are functionally equivalent to the results of exposure to low doses over human lifetimes; and the assumption that cross-species scaling methods accurately extrapolate doses given to small test animals to reflect comparable human doses. These assumptions are hotly contested within the scientific and regulatory communities, but a frequently stated rationale is that, while they may not be accurate, they are conservative—reliance upon them will minimize the chance that a carcinogen will be falsely exonerated. On the other hand, carcinogens are unlikely to be classified as carcinogens until enough high-quality, large-sample testing has been done in a variety of rodent strains and species to reveal their carcinogenic activity. . . .

Dose-Response Evaluation

Once a carcinogenic hazard has been identified, the second step in assessing cancer risks is the determination of the relationship between the dose of the agent and the probability of developing cancer. We will discuss dose response analysis of both carcinogens and noncarcinogens, since some scientists believe that, contrary to current agency practice, a similar method should be used to assess both types of toxic responses.

Toxicologists have for many years engaged in efforts to determine what dose of a chemical is safe and what is harmful. The data they have discovered describing these dose-response relationships have been used in occupational health, environmental protection, and medicine to protect people from the toxic effects of chemicals. Central to these efforts to determine a safe level of exposure is the concept of a response threshold.

The threshold is the dose of the toxicant below which no adverse effects will occur. Above the threshold, adverse effects do occur. There are two types of thresholds: population and individual. A population threshold is the dose of a compound below which absolutely no one in the population will show a response. An individual threshold is the dose below which an individual will not have a response. Individual thresholds vary from person to person and from toxin to toxin. The population threshold can be thought of as the threshold for the most sensitive individual in the population.

The dose-response relationship for a chemical is usually determined by tests on rodents, exposing them to a variety of doses of the compound and observing any toxic responses. The lowest dose producing an adverse effect on the animals is called the lowest observable adverse effect level (LOAEL), and the next tested dose below the LOAEL is called the no observable adverse effect level (NOAEL). The threshold dose in the experiments, then, is assumed to be somewhere between the LOAEL and the NOAEL, although its actual value is unknown.

When the rodent dose-response relationship is used to establish safe human doses, the NOAEL is divided by a safety factor. This safety factor accounts for potential differences in human and rodent response, protects potentially sensitive segments of the human population, and accounts for lack of knowledge of human response when there is little or no human data. The safety factor is usually 100 or 1000, which means that toxicologists set the safe level of exposure for humans at 1/100 or 1/1000 of the animal NOAEL.

Dose-response evaluation for carcinogens differs from that used in traditional toxicology. With suspected carcinogens, the threshold concept is essentially discarded—the threshold dose below which no risk may be seen is assumed to be zero. The no-threshold model, which is prominently used in cancer risk assessment, postulates that cancer can arise from a single change to the DNA of a single cell. In other words, theoretically, a single molecule of a carcinogen has some nonzero probability of causing cancer. For this reason, assessors of cancer risk assume that any dose of a carcinogen, however small, increases the probability of tumor formation.

Further complications arise in collecting and interpreting data from rodent tests of carcinogenicity. Chemicals may exhibit carcinogenic activity in some rodent species but not in others. The same chemical may even test positive in one strain of rats while testing negative in another strain of rats. Pathologists may disagree about the classification of tumors, especially when hyperplasia (a pretumor condition), benign tumors, and malignant tumors must be distinguished. Chemicals may cause tumors in one or more sites in the rodent's body which have no obvious human counterpart.

Scientists must make judgment calls to complete a dose-response evaluation of any particular animal carcinogen. The important judgments include (a) which set of animal data (e.g. which animal species response from which bioassay) to use in the modeling process; (b) which tumor types (e.g., benign and/or malignant) and tumor sites (e.g., liver and/or Zymbal gland) in the animal to count; (c) how to extrapolate the high-dose findings from animal bioassays or occupational epidemiology to the low doses humans encounter in daily life; and (d) how to scale the doses between species, adjust for different routes of exposure (e.g., ingestion in animals versus inhalation in humans), and account for variable durations or patterns of exposure. None of these judgments can currently be resolved solely on the basis of science. In the face of this uncertainty, agency scientists make quasi-policy judgments that reflect values about how protective or conservative they should be.

Perhaps the most contentious judgment in carcinogen risk assessment is how to extend the dose-response curve from the high doses to which animals are exposed in the laboratory to the lower doses to which humans are exposed in the environment. There are several well-known statistical models for fitting the animal data and extrapolating the dose-response curve to low doses. Often each model will fit the experimental animal data quite well and have at least some plausible basis in biology. The models nonetheless may yield low-dose risk estimates for the same chemical or even from the same data set, that vary enormously, by factors of hundreds or even of thousands.

As a default position based primarily on policy considerations, EPA requires use of the linearized multistage (LMS) model in all risk assessments. Agency risk assessors can choose another model only if there is persuasive evidence to support their choice; EPA guidelines do not indicate what sort of evidence would be persuasive. EPA favors the LMS model because it is generally considered to be a conservative method of estimating low-dose risks. Among biologically plausible models, few produce higher estimates of risks than does the LMS model. Scientists derive the critical low-dose potency parameter, the so-called q_1^*, by applying LMS to the tumor incidence data in rodents. The q_1^* is the upper ninety-five percent confidence limit on the linear term of the dose-response function. This linear term is produced by the LMS model's linearization of the data: the model assumes that the dose-response relationship is linear at low doses, regardless of the shape of the dose-response curve within the range of tested doses. The q_1^*, which EPA calls the "cancer potency factor" (the CPF), is an estimate of the carcinogenic strength of a compound based on the LMS model. The cancer potency factor reflects the fact that not all carcinogenic agents are equal; CPF's differ by factors of as much as a million. . . .

Exposure Assessment

Exposure assessment is the phase of a risk assessment that determines just how much exposure to a carcinogen people actually confront. Exposure can occur through a variety of routes, including inhalation, dermal absorption, and ingestion of contaminated food or water. While some sources of pollution cause human exposure through more than one such pathway, EPA risk assessments do not always consider this possibility. More recent risk assessments, however, indicate a trend to account for as many sources and routes of exposure as possible.

Exposure assessment permits evaluation of two risk parameters: population risk (incidence) and maximum individual risk (MIR). Population risk is the traditional public health measure that reports the number of cases of disease in the population attributable to a specific source or contaminant. The person at maximum individual risk is the individual who suffers the largest incremental risk due to a particular source or contaminant. In theory, the MIR should reflect scientific information about variability in human exposure and sensitivity to chemical carcinogens.

Since little is known about which people are most sensitive to chemical carcinogens, EPA usually assumes that the person at MIR is the maximally exposed individual (the MEI). The MEI is the (usually hypothetical) person expected to receive the greatest lifetime exposure from a particular source. The MEI may be the resident living closest to a factory that emits the suspected carcinogen, or the resident who draws his or her drinking water from the well closest to a Superfund site that is leaking a suspected carcinogen.

EPA generally uses predictive models, rather than direct measurements, to calculate the exposure of the MEI. In the case of a resident at a factory fenceline, a mathematical dispersion model might estimate the air concentration of the carcinogen 200 meters from the source (EPA typically assumes in such scenarios that the fenceline, and the residence of the MEI, are 200 meters from the source). In addition, the models often assume that the MEI is outdoors breathing air at this predicted concentration twenty-four hours a day for seventy years. Although no one spends his or her entire life outdoors at the fenceline of the factory, and although few factories produce the same products, or even exist, for seventy years, the MEI calculation is designed to be conservative. By overstating probable actual exposure, it provides a safety margin, giving an upper bound on the true lifetime exposure.

Use of the hypothetical MEI to set standards is extremely controversial. Critics of MEI-based standards argue that it is unsound to regulate, often at very great cost, on the basis of an inflated exposure scenario that never occurs. Supporters argue that highly exposed people, even if they are few in number, have a right to protection, and that the conservatism in MEI scenarios may be appropriate given the other uncertainties in risk assessment. . . .

Risk Characterization

When a risk assessor has the three important pieces of information—an identified hazard, an estimate of the dose-response relationship q_1^*, and estimates of exposure (or dose)—he or she can make a numerical estimate of risk. Essentially all the assessor does is multiply the q_1^* the cancer potency factor derived from the LMS procedure, by the measured or predicted exposure. The q_1^* is usually expressed in units of increased lifetime probability of cancer per milligram of carcinogen per kilogram of body weight per day of exposure, and the exposure is expressed in units of milligrams of carcinogen per kilogram of body weight per day. The calculation therefore leads to an estimate of the increase in the lifetime probability of cancer from the particular level of exposure. For a properly performed risk-characterization, this number is only the beginning.

The meaning of EPA's risk estimates cannot be accurately conveyed except in light of the numerous assumptions that have been made. As two commentators have stated, risk estimates from analyses done according to EPA procedures "do not give certainty in the scientific sense, nor can they be used to establish precise numbers of persons who will be stricken with some disease." However, the number that comes from the risk characterization step is often reported and used without qualification. Advocates of risk assessment constantly call for analysts to quantify and report the full range of uncertainty in a risk assessment. In fact, because of the numerous conservative assumptions built into the EPA risk assessment process (so-called "compounded conservatism"), EPA has stated that a risk estimate produced in accord with its procedures should be regarded as a plausible upper bound on risk. That is, the actual risk will almost certainly lie somewhere between the EPA risk estimate and zero. The actual risk is very unlikely to be greater than the EPA risk estimate, is probably lower than the EPA estimate, and may even be zero.

Therefore, EPA states that, in addition to the risk number, a risk characterization should contain: (a) a discussion of the "weight of the evidence" for human carcinogenicity (e.g., the EPA carcinogen classification); (b) a summary of the various sources of uncertainty in the risk estimate, including those arising from hazard identification, dose-response evaluation, and exposure assessment; and (c) a report of the range of risk using EPA's risk estimate as the upper limit and zero as the lower limit.

In Search of Safety: Chemicals and Cancer Risk
JOHN D. GRAHAM, LAURA C. GREEN, AND MARC J. ROBERTS

Quantitative risk assessments have not proven to be the panacea that they were intended to be. Agency risk estimates convey both too much and too little confidence in science. The estimates convey overconfidence in the sense that the true extent of biological uncertainty and scientific conflict is not reflected in the published numbers. At the same time, the procedures used by agency analysts often do not incorporate all of the relevant biological information and the final numbers often fly in the face of technical intuition and judgment. To better understand the nature of these difficulties, it is useful to explore the major analytical decisions in risk assessment and the procedures currently employed by federal agencies.

The Essential Problem

The difficulties of doing quantitative risk assessment arise from the lack of information about human dose-response curves at the doses that are important to regulatory policy. For a variety of reasons, the low-dose region of the human dose-response curve for cancer is generally unobservable. Epidemiologists often must work with inadequate data. The human data that are available typically involve historic exposures that are both poorly documented and generally much higher than those of current regulatory concern. Similarly, we have little data from laboratory animal bioassays at the dose levels of regulatory interest. Questions also arise about how the animal data should be interpreted and the extent to which humans resemble the test species.

Whether the risk analyst is trying to extrapolate from high to low doses, from mouse to man, or both, the challenge is to use the known to predict the unknown (or at least the observed to forecast the unobserved). If all chemicals caused cancer through the same biological mechanisms, we might be able to accumulate enough knowledge of the process to model a particular case with relatively little data. For example, if the physiologic and metabolic differences between animals and humans were the same for all substances, then we could apply the same interspecies "correction factor" to each chemical.

The world is not this simple. Our cases suggest that the processes by which chemical exposures produce cancer vary widely. Benzene does not act like formaldehyde; leukemia is not like squamous cell nasal carcinoma. There are often significant differences in the way particular compounds affect different species. Rats don't contract leukemia from breathing benzene; humans do. Rats get nasal carcinomas from formaldehyde, but people may not. Hence, no single interspecies extrapolation rule will be accurate (or even equally conservative) in all cases.

Since cancer risks are always assessed in the face of such great uncertainty, disagreements arise about how to proceed. Indeed, it is likely that no responsible scientist would even try to make (or take much interest in) such numerical estimates were it not for the pressure to do so for policy reasons. As a result, the boundary between science and policy runs through many of the methodological disputes about cancer risk assessment. In the heat of particular controversies, the disputants often fail (even after much effort) to distinguish among the grounds for their arguments. Methods are sometimes defended on the grounds that they are "scientific" and at other times on the grounds that they are "conservative." Confusion is rampant, as we shall see.

As a context for what follows, it is worth noting that there is a conceptual framework that purports to tell us how such problems should be handled. That framework is a branch of applied mathematics called "decision theory." It offers two main principles. First, risk assessors should make a conscious and self-critical attempt to distinguish between estimates of the extent of cancer risk and values concerning how much effort society should devote to reducing various levels of risk. Having made this distinction, the risk assessor should then try to quantify explicitly the magnitude of uncertainty about his or her estimates of cancer risks. A single, or "point," estimate is not sufficiently informative. The full extent of uncertainty—biological and statistical—must be characterized quantitatively. Making specific estimates of the probabilities of various outcomes can help the regulator think about the extent of potential gains from acquiring better information about the chemical. In addition, the regulator should try to quantify explicitly both how "risk averse" (or conservative) he or she wants to be by quantifying just how unacceptable various rates of cancer would be relative to each other and to other kinds of policy outcomes. Thus, the decision analyst urges regulators to quantify *separately* the likelihood of various health outcomes and their degree of happiness or unhappiness if the various outcomes were to occur.

One striking feature of the review we are about to provide is how thoroughly such decision-theoretical ideas are ignored by cancer risk assessors. In part the problem is with decision theory, which does not yet offer a clear program for obtaining all of the necessary quantification. But part of the problem lies with risk assessors in federal agencies, where there are institutional incentives to conceal uncertainty and neglect scientific evidence and judgments that run counter to prevailing guidelines. It is therefore useful to describe how quantitative cancer risk assessment is currently practiced by federal agencies.

Extrapolating from High to Low Doses

The task of extrapolating from high to low doses arises with both animal and human data. Standard practice is to "fit" some sort of mathematical function to the observed dose-response data and use that function to project responses into the low-dose region, where actual responses have not been observed. This requires two main choices: what mathematical model to fit and what statistical method to use when fitting it.

Low-Dose Extrapolation of Animal Data

At least half a dozen tractable mathematical models can be used with experimental bioassay data. Generally they all fit such data fairly well and have some bases in cancer biology. It is apparent, though, that the models often predict quite different carcinogenic responses at doses below the experimental range. . . .

Although there is considerable uncertainty about which model best reflects the biology of any particular chemical carcinogen, the multi-stage model is by far the most widely used in the federal government. All three major federal agencies—CPSC, EPA, and OSHA[1]—used it in their formaldehyde deliberations. . . . The multistage model is an exponential model of the following form:

$$P = 1 - \exp[-(q_0 + q_1 d + q_2 d^2 + \cdots + q_n d^n)]$$

where P is the probability of developing cancer after lifetime exposure to dose d of a chemical and the qs are coefficients estimated from the dose-response data. The underlying biological theory is that a developing tumor progresses through several different stages before it becomes clinically detectable, and each stage of the process can be influenced by exposure to carcinogens.

It is a flexible model in that it can take different forms—in response to different data sets—by assuming a different number of dose related stages. Each power of

---

[1][Consumer Product Safety Commission, Environmental Protection Agency, and Occupational Safety and Health Administration—ED.]

dose in the above equation corresponds to a stage. When the model was originally designed, the number of stages was assumed to be no greater than one less than the number of dose levels in the experiment that generates the dose-response data.

Once a mathematical form is selected for low-dose extrapolation, a statistical technique must be selected to estimate the model's coefficients. Least squares regression, a common statistical method, is generally not helpful because of the nonlinear forms of the model and the small number of experimental data points. The standard practice among risk assessors is to use the "maximum likelihood" technique, which determines—for a given model and assumed random error component—what coefficient estimates make it most likely that the observed experimental data would be generated.

To provide some measure of statistical uncertainty, risk assessors often generate confidence intervals around the maximum likelihood estimates, or MLEs. A 90 to 95 percent upper confidence limit (UCL) is commonly reported, although there is no particular reason—other than that they are "round numbers"—for choosing these particular limits. The confidence limits, also called upper bounds, capture only the random errors, given the model. They do not convey any uncertainty about whether the model was properly specified or whether the model selected was the correct one. They therefore understate the overall degree of uncertainty. . . .

Suppose, for the sake of argument, that we are willing to accept multistage as the appropriate model and maximum likelihood as the appropriate estimation technique. What risk estimates for formaldehyde are generated based on CIIT[2] rat data? Table III.1 provides two sets of estimates based on subtly different analytical choices, which lead to some significant differences in risk estimates. For example, OSHAs MLEs are consistently one to two orders of magnitude smaller than those reported by Clement Associates, a private consulting firm. The UCLs, though, vary by a factor of only about two.

The major reason for the disparate estimates is the number of stages assumed in the two analyses. Clement used a version of the multistage equation in which only d^3 was multiplied by a nonzero coefficient (q_3). OSHA used dose to the fourth and fifth powers, which yields a fitted dose-response function that is more nonlinear. Since the fitted curve can now come closer to the small observed response at 5.6 ppm, it also tends to predict smaller risks at low doses.

Table III.1   Formaldehyde Risk Estimates Based on the Multistage Model

| Exposure Level (ppm) | Excess Lifetime Cancers per 100,000 Exposed Workers | | | |
| | OSHA | | Clement | |
	MLE	UCL	MLE	UCL
3	71	834	620	930
1	0.6	264	23	130
0.5	0.03	132	2.8	58
0.1	0.001	26	< 0.1	11

[2][Chemical Industry Institute of Toxicology—ED.]

Proponents of a fifth-degree polynomial argue that the fourth and fifth powers of dose are necessary to capture the extreme nonlinearity in the data points from the CIIT rat study. In short, five stages fit the experimental data better than do three stages. Advocates of three stages respond that the multistage model was not designed to allow more stages than the number of experimental dose levels. (What matters statistically is the number of terms relative to the number of data points— that is, the degrees of freedom available—and not the size of the exponent on the largest term.) They argue also that the fit is not "much" better and that there is no biological reason to prefer five stages to three. . . .

It should be emphasized, however, that these sensitivities to model selection in the case of formaldehyde are "small" relative to those for extrapolations of data on many other chemicals. This is because occupational exposures to formaldehyde are much closer to the bioassay doses than is often the case. The greater the difference between actual doses and bioassay doses, the more opportunity there is for the various models to diverge in their low-dose risk estimates.

Federal agencies have developed a strong commitment to the multistage model, but the reasons for this preference are not entirely clear. It is also apparent that EPA and CPSC (but not OSHA) have decided to rely on the 95 percent upper confidence limits instead of the MLEs. Here the mixing of policy and science comes to the forefront. Agencies want to be sure not to underestimate cancer risk. So rather than relying on MLEs or some overall "best guess" of risk, they intentionally aim high in order to be conservative. And instead of always calling the upper limit an upper limit, agencies sometimes call it an "estimate" from the linearized multistage model. This terminology is based on the observation that the 95 percent upper limit of risk from the multistage model tends to be a linear function of dose at sufficiently low doses, even though the MLE may behave in a nonlinear fashion over all doses. EPA frequently emphasizes that the linearized multistage model is conservative and provides a "plausible upper limit" on actual risk.

When criticized for mixing science and policy, EPA responds that low-dose linearity is scientifically the most plausible hypothesis. Since humans are exposed to a large background of carcinogenic exposure, low doses of any particular carcinogen may add to the overall load. This may produce a linear segment of the dose-response curve for each carcinogen. One advocate of this view, statistician Kenneth Crump, has stated, concerning formaldehyde, that "I hold to the view that the true response is probably linear at low doses . . . we can't rule out the possibility that the risks are about as large as those predicted by the linearized multistage model."

An alternative view, held by statistician Robert Sielken, is that the linearized multistage model produces upper bounds on risk, not estimates of risk. He argues further that the computer programs used by federal agencies, such as GLOBAL 79, do not produce normal upper confidence limits (numbers based solely on variability in the fitting technique). Instead the procedures used to generate upper confidence limits are designed to "force" the maximum amount of linearity into the multistage model without creating a very poor fit to the data. In the case of formaldehyde, he argues that the dose-response function is very likely to be nonlinear at low doses— thus arguing against giving credence to the 95 percent upper limit. . . .

Interspecies Extrapolation

Given the length of latency periods and the limited power of epidemiology to detect human risks, regulators generally cannot wait until they have positive evidence of health effects in people. That would entail too great a risk of failing to regulate some truly harmful substances. The question thus becomes how to utilize alternative data, namely data from laboratory animal bioassays, to predict human cancer risk.

Our cases have called attention to the following general issues in interspecies extrapolation:

> If bioassay data are available on more than one strain of a nonhuman species or on more than one nonhuman species, which data should be used for human risk estimation?
>
> How should data on different types of tumors, especially benign versus malignant tumors, be incorporated into human risk estimates?
>
> When responses in a test species are observed at multiple sites or target organs, which responses should be used for estimating human risk?
>
> How should doses administered to animals be converted to equivalent doses for humans?
>
> How should suspected variations in susceptibility to cancer risk among humans and test animals be incorporated into human risk estimates?

Answering these questions involves a mixture of technical and policy considerations in the face of great uncertainty, and no definitive answers have emerged. Agencies have tried to devise and rely on generic (and rather arbitrary) guidelines to help them decide what to do in specific cases. Yet estimates based on these guidelines are often ignored when they lead to results that are "implausible.". . .

In the case of formaldehyde, the tumor site and species used for extrapolation to human risk was dictated largely by the data. Nasal carcinomas are the only confirmed carcinogenic response to formaldehyde in test animals. They occur in both rats and mice, but the results for mice occur only at the highest exposure level. The CIIT investigators believe that rats are more sensitive than mice primarily because mice alter their breathing patterns in response to formaldehyde. The implication is that rats and mice do have comparable responses, if one calculates effective, as opposed to administered, doses to nasal tissues in the two species.

But what does this suggest about potential human risk? Do humans also alter their breathing patterns when exposed to formaldehyde or do they behave more like rats? OSHA's stated view is that "an employee performing a job would generally be unable to lower his or her breathing rate." The rat data are thus "preferred over the results for mice" because "they may represent a situation more typical of human exposure." At very low doses, when formaldehyde cannot be detected by smell or felt by irritation, this assumption seems particularly reasonable.

What about the broader issue of human susceptibility to formaldehyde? Some analysts contend that humans are not obligate nose breathers, which may mean that the effective dose to nasal tissues is lower. Others point out that rats "have approximately twice the relative [nasal] surface area [of] man for filtering inspired air," indicating that "man would receive a smaller target dose than rodents if both were exposed to a similar concentration." Hence, some speculate that humans are less sensitive than rats in this respect.

An alternative view is that humans are presumably more heterogeneous in susceptibility to cancer than any specific strain of rat and are also exposed to many more carcinogenic insults on a daily basis. These considerations lead some observers to speculate that "proportionately more humans (than rats) should develop cancer [from formaldehyde] at lower doses." On this question the Consensus Workshop on Formaldehyde said only: "Although there are differences in formaldehyde carcinogenicity among different species, at present there is no reason to assume that humans would be more or less susceptible than the rat." OSHA, EPA, and CPSC made the same assumption.

An even more controversial issue is the treatment of the benign tumors (polypoid adenomas) in the CIIT rat study. Federal regulatory guidelines generally call for including benign tumors in quantitative risk assessments. But many scientists argue that formaldehyde should be an exception. . . . Although polypoid adenomas in general progress to malignant adenocarcinomas, the rate at which they do so in both rats and humans appears to be "very low." And papillomas, the benign counterparts of squamous cell carcinomas, were not observed in the CIIT rat study, although they did appear in [another] study.

Others argue that the polypoid adenomas should be included. One analyst believes that a malignant neoplasm found in one rat in the CIIT's high-dose group exhibits "similar morphologic features to many of the polypoid adenomas," indicating that "it may represent the malignant counterpart of the polypoid adenoma." Another observer counted two adenocarcinomas in the high-dose group. . . .

In defense of this approach, OSHA argues that "changes occurring in the nasal passages, leading to squamous cell carcinomas, could have masked the appearance of adenomas at higher doses." EPA speculates that the "cell type needed for these tumors [polypoid adenomas] to occur is lost sooner and to a greater extent with increasing dose." Hence they argue for excluding higher dose data. . . .

The line here between scientific and policy judgment is quite blurred. It is conservative to combine the two types of tumors, but this approach may grossly inflate the actual human cancer risk. To ignore the benign tumor data is to run the risk of underestimating actual human risk. If risk estimates are reported separately for the two tumor types, the regulator is left in a quandry about what to make of the disparate estimates. It may make little sense to perform low-dose extrapolation at all for such an apparently odd-shaped dose-response curve—whose shape may be odd because the data are "noisy" and have little biological meaning. Laboratory scientists presented with the polypoid adenoma data for formaldehyde might conclude that the experiment should be repeated. The policy maker may feel that he does not have time for that luxury. . . .

Conclusion

. . . Our view is that risk assessment has become too formalized and mechanical in light of the limited data. Little is gained from the sophisticated massaging of weak data. The key challenge is to improve the quality of the data used in risk assessment. But how can this be done if direct measurement of the relevant risks is impractical or impossible? . . . The answer lies partly in developing a better understanding of the mechanisms that produce cancer and partly in eliciting careful scientific judgments. Better information on mechanisms might allow extrapolation to be performed more intelligently on a chemical-by-chemical basis. Giving the judgments of working scientists a greater role in risk assessment would allow for revision of estimates that go against scientific intuition.

Society must face a large and irreducible element of uncertainty in the estimation of cancer risk. No amount of data on the mechanisms of carcinogenesis or scientific intuition can substitute for direct observation of low-dose effects in humans. The uncertainties—both those that are reducible and those that are inevitable—need to be acknowledged and (when possible) quantified by risk assessors. Otherwise, policy makers and the public will be misled. . . .

Breaking the Vicious Circle: Toward Effective Risk Regulation
STEPHEN BREYER

Study after study shows that the public's evaluation of risk problems differs radically from any consensus of experts in the field. Risks associated with toxic waste dumps and nuclear power appear near the bottom of most expert lists; they appear near the top of the public's list of concerns, which more directly influences regulatory agendas (see Table III.2 and Table III.3). To some extent, these differences may reflect that the public fears certain risks more than others with the same probability of harm. As previously pointed out, of two equal risks, one could rationally dislike or fear more the risk that is involuntarily suffered, new, unobservable, uncontrollable, catastrophic, delayed, a threat to future generations, or likely accompanied by pain or dread.

Still, these differences in the source, quality, or nature of a risk may not account for the different ranking by the public and the experts. A typical member of the public would like to minimize risks of death to himself, to his family, to his neighbors; he would normally prefer that regulation buy more safety for a given expenditure or the same amount of safety for less. Not many of us would like to shift resources to increase overall risks of death significantly in order to increase the likelihood that death will occur on a bicycle or in a fire, rather than through disease. There is a far simpler explanation for the public's aversion to toxic waste

Table III.2   How the Public and EPA Rate Health Risks Associated with Environmental Problems

Public	EPA Experts
1. Hazardous waste sites	Medium-to-low
2. Exposure to worksite chemicals	High
3. Industrial pollution of waterways	Low
4. Nuclear accident radiation	Not ranked
5. Radioactive waste	Not ranked
6. Chemical leaks from underground storage tanks	Medium-to-low
7. Pesticides	High
8. Pollution from industrial accidents	Medium-to-low
9. Water pollution from farm runoff	Medium
10. Tap water contamination	High
11. Industrial air pollution	High
12. Ozone layer destruction	High
13. Coastal water contamination	Low
14. Sewage-plant water pollution	Medium-to-low
15. Vehicle exhaust	High
16. Oil spills	Medium-to-low
17. Acid rain	High
18. Water pollution from urban runoff	Medium
19. Damaged wetlands	Low
20. Genetic alteration	Low
21. Non-hazardous waste sites	Medium-to-low
22. Greenhouse effect	Low
23. Indoor air pollution	High
24. X-ray radiation	Not ranked
25. Indoor radon	High
26. Microwave oven radiation	Not ranked

dumps than an enormous desire for supersafety, or a strong aversion to the tiniest risk of harm—namely, the public does not *believe* that the risks are tiny. The public's "nonexpert" reactions reflect not different values but different understandings about the underlying risk-related facts.

My assumption that the public assigns "rational" values to risks, however, does not entail rational public reactions to risk. Psychologists have found several examples of thinking that impede rational understanding, but may have helped us survive as we lived throughout much of prehistory, in small groups of hunter-gatherers, depending upon grain, honey, and animals for sustenance. The following, rather well-documented aspects of risk perception are probably familiar.

1. Rules of thumb. In daily life most of us do not weigh all the pros and cons of feasible alternatives. We use rules of thumb, more formally called "heuristic devices." We simplify radically; we reason with the help of a few readily understandable examples; we categorize (events and other people) in simple ways that tend to create binary choices—yes/no, friend/foe, eat/abstain, safe/dangerous, act/don't act—and may reflect deeply rooted aversions, such as fear of poisons. The resulting categorizations do not always accurately describe another person or circumstance, but they help us make quick decisions, most of which prove helpful.

Table III.3   Ordering of Perceived Risk for Thirty Activities and Technologies

Activity or Technology	League of Women Voters	College Students	Active Club Members	Experts
Nuclear power	1	1	8	20
Motor vehicles	2	5	3	1
Handguns	3	2	1	4
Smoking	4	3	4	2
Motorcycles	5	6	2	6
Alcoholic beverages	6	7	5	3
General (private) aviation	7	15	11	12
Police work	8	8	7	17
Pesticides	9	4	15	8
Surgery	10	11	9	5
Firefighting	11	10	6	18
Large construction	12	14	13	13
Hunting	13	18	10	23
Spray cans	14	13	23	26
Mountain climbing	15	22	12	29
Bicycles	16	24	14	15
Commercial aviation	17	16	18	16
Electric power (non-nuclear)	18	19	19	9
Swimming	19	30	17	10
Contraceptives	20	9	22	11
Skiing	21	25	16	30
X-rays	22	17	24	7
High school and college football	23	26	21	27
Railroads	24	23	29	19
Food preservatives	25	12	28	14
Food coloring	26	20	30	21
Power mowers	27	28	25	28
Prescription antibiotics	28	21	26	24
Home appliances	29	27	27	22
Vaccinations	30	29	29	25

This kind of quick decision-making may help cut a swath through the modern information jungle, but it oversimplifies dramatically and thereby inhibits an understanding of risks, particularly small risks.

2. Prominence. People react more strongly, and give greater importance, to events that stand out from the background. Unusual events are striking. We more likely notice the (low-risk) nuclear waste disposal truck driving past the school than the (much higher-risk) gasoline delivery trucks on their way to local service stations. Journalists, whose job is to write interesting stories, know this psychological fact well. The American Medical Association examined how the press treated two similar stories, one finding increased leukemia rates among nuclear workers, the other finding no increased cancer rates among those living near nuclear plants. More than half of the newspapers in the study mentioned the first story but not the second; and more than half of those that mentioned both emphasized the first.

3. Ethics. The strength of our feelings of ethical obligation seems to diminish with distance. That is to say, feelings of obligation are stronger (or we have different, more time-consuming obligations) toward family, neighbors, friends, community, and those with whom we have direct contact, those whom we see, than toward those who live in distant places, whom we do not see but only read or hear about.

4. Trust in experts. People cannot easily judge between experts when those experts disagree with each other. The public, since the mid-1960s, has shown increasing distrust of experts and the institutions, private, academic, or governmental, that employ them.

5. Fixed decisions. A person who has made up his or her mind about something is very reluctant to change it.

6. Mathematics. Most people have considerable difficulty understanding the mathematical probabilities involved in assessing risk (see Table III.4). People consistently overestimate small probabilities. What is the likelihood of death by botulism? (One in two million.) They underestimate large ones. What is the likelihood of death by diabetes? (One in fifty thousand.) People cannot detect inconsistencies in their own risk-related choices. Ask a friend two questions: (1) "Which escape route, Path A or Path B, should the general take with his six hundred soldiers? Path A means two hundred soldiers are saved for certain; Path B means a two-thirds chance that all die, a one-third chance that all are saved." Most people prefer A. (2) "Which escape route, Path A or Path B, should the general take with his six hundred soldiers? Path A means four hundred soldiers die for certain; Path B means a one-third chance that all are saved and a two-thirds chance that all will die." Most people prefer B. They do not realize that the two questions are the same and that their answers are inconsistent. The words "saved" and "die" make the difference. The way one phrases a question can determine people's preferences. Moreover, people do not understand the counterintuitive consequence of certain important statistical propositions. Consider the statistical fact that after any unusually high or low result, the next instance

Table III.4.   The Frequency of Dramatic and Sensational Lethal Events Tends to Be Overestimated; the Frequency of Unspectacular Events Tends to Be Underestimated

Most Overestimated	Most Underestimated
All accidents	Smallpox vaccination
Motor vehicle accidents	Diabetes
Pregnancy, childbirth, and abortion	Stomach cancer
Tornadoes	Lightning
Flood	Stroke
Botulism	Tuberculosis
All cancer	Asthma
Fire and flames	Emphysema
Venomous bite or sting	
Homicide	

will tend to "deviate toward the mean," which is why parents who are particularly intelligent, or particularly stupid, tend to have more normal children. Now imagine a teacher who rewards a student's unusually good performance; the next time the student will likely do worse (just because of deviation toward the mean). The same teacher punishes an unusually bad performance; the next time the student will likely do better (just because of the law of averages). The teacher, seeing these results, thinks there is something wrong with the teaching theory of "positive reinforcement." There isn't. Instead, the statistical deviation toward the mean is positively reinforcing the teacher's negative reinforcement, and negatively reinforcing the positive reinforcement. Now, do you understand why students sometimes complain about the harshness of law school professors?

These few, near-commonsense propositions, with strong statistical support in the technical literature, verify Oliver Wendell Holmes's own observation that "most people think dramatically, not quantitatively." They also have important consequences. Consider the public reaction to toxic waste dumps. Start with the mathematical facts about the probability of various occurrences: In 1985 a New Jersey woman won the state lottery twice. What are the odds against this, billions to one? Given the vast number of lotteries in the world, the odds come close to favoring someone somewhere winning a lottery twice. Given the population of the world, and the number of dreams each night, the odds favor someone somewhere dreaming he marries a girl who looks very much like the girl he meets the next day and marries. Given the number of toxic waste dumps in the United States (26,000) and the number of places with above-average cancer rates (half of all places), obviously many cities, towns, and rural areas near toxic waste dumps must also have seriously elevated cancer rates ("mathematics").

Add what sells newspapers—interesting stories—and you can be fairly certain the press will write about the double lottery-winner, perhaps the dreamer, and, if the mathematical evidence is somewhat less crude than my example, the toxic waste dump ("prominence"). Will it be easy to convince the cancer victim that the waste dump (water that is "pure" or "not pure") had nothing to do with the disease ("rules of thumb")? And how will the public react to the image of the angry family member on nightly television ("ethics"), particularly if experts disagree ("trust in experts")—as they might, for the relation between the disease and the toxic site may not be strictly chance (the lottery, too, might be fixed). If further study exonerates the dump, will the viewing public change its mind ("fixed decision")?

When we think about nuclear power controversies, we should take account of the fact that hearing about an accident is what psychologists tell us is an heuristic "tip-off" of danger, whether or not anyone is hurt. We have "seen" Chernobyl and Three Mile Island, and we may therefore doubt nuclear power's safety, whether or not experts tell us that the reactor at Chernobyl was not properly designed, that the accident at Three Mile Island hurt no one, that military weapons, not electric power generators, are responsible for 99 percent of all nuclear waste, that nuclear power's risks are minuscule compared to the risks of coal-generated power. Add a few disagreements among experts and the fact that most members of the public made up their minds long ago, and one can understand nuclear power's position on the public perception risk charts.

These few propositions suggest that better "risk communications," such as efforts to explain risks to the public at open meetings, may not suffice to alleviate risk regulation problems. It is not surprising that, after the EPA Administrator William Ruckelshaus spent days at such meetings in Tacoma, Washington, explaining why an ASARCO chemical plant that was leaking small amounts of arsenic could remain open, he was misunderstood, criticized, and accused of trying to drive a wedge between environmentalists and blue collar workers. The plant eventually closed, although perhaps for other reasons. Nor is it surprising that after special public discussions of nuclear power plants were held in Sweden, surveys of the eighty thousand Swedes who participated showed no consensus, but increased confusion.

There is little reason to hope for better risk communication over time. To the contrary, as science improves, scientists may more easily detect and identify ever tinier risks—the risks associated, for example, with the migration of a single molecule of plastic from a container into a soft drink; they may more easily identify geographical areas near toxic waste dumps with higher than average cancer rates. As international communications improve, the press will have an ever larger pool of unusual, and therefore more interesting, accident stories to write about. Why should we not expect an outcry from a public that reads about Love Canal, Times Beach, Alar, Chilean grapes laced with cyanide, and the leaflet of Villejuif, whether or not such examples reflect meaningful danger? (At the same time, how can one expect public reaction to potentially greater but more mundane problems, of which it is unaware?)

It is hard to make the normal human mind grapple with this inhuman type of problem. To change public reaction, one would either have to institute widespread public education in risk analysis or generate greater public trust in some particular group of experts or the institutions that employ them. The first alternative seems unlikely. The second, over the past thirty years, has not occurred.

Notes and Questions

1. As Ruckelshaus makes clear, for many substances, there is substantial scientific uncertainty about the relationship between exposure to the substance and the onset of adverse health effects. In the face of such uncertainty, there are two polar choices:
   (a) not regulating until there is scientific certainty concerning this relationship
   (b) regulating whenever there is at least some evidence linking the exposure to adverse health effects

Doing the former means that, if the link is ultimately established, many individuals will have been exposed, perhaps for a period of several decades, to an unnecessary risk. In contrast, following the latter course means that, if the substance is ultimately found to be benign, pollution control costs will have been incurred needlessly. What criteria should be used to decide between the two courses of action? Who should make the decision about whether regulation is appropriate?

2. To what extent are risk management judgments about how to regulate under uncertainty intertwined with the risk assessment decision. For example, as is generally the case, risk assessment might label a substance a carcinogen despite less-than-conclusive evidence of carcinogenicity. Are risk management considerations (for example, the cost of reducing human exposure to the substance) relevant to the determination of whether the substance

should be labeled a carcinogen? Otherwise, how should this decision be made? Should a substance be labeled carcinogenic when the probability that it would, in fact, have carcinogenic effects on humans is 5 percent? 25 percent? 51 percent? 75 percent? How should a risk assessor proceed if, under the applicable statutory scheme, the consequence of calling the substance carcinogenic is that stringent controls will have to be imposed? For an argument that, in the face of scientific uncertainties, social policy considerations should inform the risk assessment process, see Latin, "Good Science, Bad Regulation, and Toxic Risk Assessment," 5 *Yale Journal on Regulation* 89 (1988).

3. What interactions should risk managers have with risk assessors? Should the risk manager inquire about how confident the risk assessor was in labeling the substance carcinogenic? Should the risk assessor inquire about the costs of the regulation required if risk assessment results in a particular finding? Would Ruckelshaus find such interactions desirable? Is it likely that such interactions take place in federal regulatory agencies?

4. Consider two possible reasons for not regulating a particular substance:
   (a) the risks associated with the substance are insignificant
   (b) the risks are significant but regulation would impose unacceptably high costs on industry.
Which reason will a decision maker want to give the public? Is it inevitable that the decision maker will seek to influence the risk assessment process so as to be able to give the first reason?

5. Consider the following issues regarding the risk assessment of potential carcinogens:
   (a) Should negative epidemiological evidence of carcinogenicity be disregarded in light of positive evidence from long-term animal bioassays?
   (b) Should benign tumors be aggregated with malignant tumors to determine the dose-response relationship?
   (c) What models should be used to extrapolate from high doses to low doses?
   (d) What models should be used to extrapolate from rodents to human beings?
   (e) How should inconsistent results in different species of rodents be treated?
There is no scientific consensus on these issues. How should a regulator reach a decision? For each issue, should a regulator always make the most conservative assumption?

6. In the case of noncarcinogens, should the NOAEL be 1/100 or 1/1000 of the LOAEL, or some other amount? What academic discipline has the most to contribute in answering this question? On the basis of what criteria should the choice be made?

7. Which measure of exposure assessment is more relevant for risk assessment: population risk or maximum individual risk? Which measure should be used for risk management decisions? What are the possible distributional effects of ignoring maximum individual risk. The environmental justice critique of risk assessment is discussed in Chapter V.

8. Rosenthal, Gray, and Graham's article states that plausible models used to extrapolate from high doses to low doses may lead to estimates that vary by factors of hundreds or even thousands. In a portion of his book that is not excerpted (p. 45), Justice Breyer reports that "[t]wo scientifically plausible models for the risk associated with aflatoxin in peanuts or grain may show risk levels differing by a factor of 40,000." He further reports (p. 47) that, according to the Office of Management and Budget (OMB), agencies "'often' overstate risks by factors of a thousand or even a million or more" because they pile conservative assumption upon conservative assumption. In light of these disparities, is risk assessment at all useful? Can the problem be corrected by taking seriously the plea by Rosenthal, Gray, and Graham that "[g]ood risk characterizations [should] contain not only a final risk number but also a discussion of the uncertainties in and the assumptions behind the assessment"?

9. Is it plausible that a regulatory system would ever resolve all uncertainties by making conservative assumptions at every turn? Consider the discussion in the Graham, Green, and Roberts excerpt concerning the choice between maximum likelihood estimates (MLEs) or of upper confidence limits (UCLs). If a regulatory agency chooses to use UCLs, should the confidence interval be 90 or 95 percent? Why should it not be 99 percent? Is there any logical stopping point?

10. In light of the uncertainties surrounding risk assessment, what do you think of Ruckelshaus' plea for insulating risk assessment from political pressures? Who should decide how to resolve such uncertainties: Congress, congressional staffers, the EPA Administrator, a scientific panel within EPA, a consensus of independent scientists?

11. Ruckelshaus advocates the use of standardized methods to resolve uncertainties in order to reduce the politicization of risk assessments. Graham, Green, and Roberts argue that such policies might diminish the scientific validity of the exercise. Who has the better argument?

12. Justice Breyer's theoretical discussion about the discrepancy between expert and lay perceptions of risk is derived from a large academic literature on the subject. Representative works include Fischhoff, "Managing Risk Perceptions," *Issues in Science and Technology,* Fall 1985, at 83; Noll and Krier, "Some Implications of Cognitive Psychology for Risk Regulation," 19 *Journal of Legal Studies* 747 (1990); Slovic, "Perception of Risk," 236 *Science* 280 (1987); Slovic, Fischhoff, and Lichtenstein, "Rating the Risks," 21 *Environment* 14 (1979).

Slovic and colleagues define eight characteristics that affect lay perceptions of risks:

(a) *voluntariness of risk*: involuntary risks are perceived as more serious;
(b) *immediacy of effect*: risks with delayed effects (such as cancer) are perceived as more serious;
(c) *knowledge about risk*: unknown risks are perceived as more serious;
(d) *control over risk*: risks that individuals cannot reduce through their own actions are perceived as more serious;
(e) *newness*: new risks are perceived as more serious;
(f) *chronic/catastrophic*: catastrophic risks—risks that kill large numbers of people at once—are perceived as more serious;
(g) *common/dread*: dread risks—risks that people have not yet learned to deal with—are perceived as more serious;
(h) *severity of consequence*: fatal risks are perceived as more serious.

Given these factors, for what types of environmental risks would you expect that the lay perceptions would be greater than the experts' assessments?

13. Justice Breyer's factual views about the discrepancy between expert and lay evaluations of risk are informed to a large extent by two EPA reports: *Unfinished Business: A Comparative Assessment of Environmental Problems*, published in 1987, and *Reducing Risk: Setting Priorities and Strategies for Environmental Protection*, published in 1990. *Unfinished Business*, from which Table II.2 in Justice Breyer's excerpt is derived, is the report of a special task force of senior career managers and technical experts at EPA, commissioned by the administrator to compare the risks of different environmental problems. The group chose to focus on thirty-one problems and, for each problem, considered four types of risks: cancer risks, noncancer health risks, ecological effects, and welfare effects (such as visibility impairment and materials damage). The study reached two important conclusions (p. xix–xx):

The [group's] rankings by risk . . . do not correspond well with EPA's current program priorities. Areas of relatively high risk but low EPA effort include: indoor radon; indoor air pollution; stratospheric ozone depletion; global warming; non-point sources; discharges to estuaries, coastal waters, and oceans; other pesticide risks; accidental releases of toxics; consumer products; and worker exposures. Areas of high EPA effort but relatively low or medium risks include: RCRA sites; Superfund; underground storage tanks; and municipal non-hazardous waste sites.

Second, the group concluded:

Overall, EPA's priorities appear more closely aligned with public opinion than with our estimated risks. Recent national polling data rank areas of public concern about environmental issues as follows:

High: chemical waste disposal, water pollution, chemical plant accidents, and air pollution;
Medium: oil spills, worker exposure, pesticides, and drinking water;
Low: indoor air pollution, consumer products, genetic radiation (except nuclear power), and global warming.

What prescriptions for regulation should one draw from these conclusions? What prescriptions does Justice Breyer draw?

14. *Unfinished Business* is more agnostic than Justice Breyer about whether regulatory attention should be more closely aligned with expert perceptions of risk. The report states (p. xix):

This divergence between what we found in terms of relative risks and EPA's priorities is not necessarily inappropriate. Some problems appear to pose relatively low risks precisely because of the high levels of program effort that have been devoted to controlling them. And these high levels of attention may remain necessary in order to hold risks to current levels.

Air pollution and water pollution control, the first two large-scale environmental programs, probably fall in this category.

Moreover, the report expressly declined to consider the economic or technological controllability of the risks. Also, in some of the areas studied by the report—particularly consumer products and worker exposure to toxic chemicals—EPA shared jurisdiction with other government agencies, and for other areas—particularly stratospheric ozone depletion and global warming—international coordination is necessary for a successful response (this issue is considered in Chapter XII).

Is it necessarily inconsistent with the economic perspective for regulation to focus on certain less serious risks rather than on other more serious ones? For further discussion, see Viscusi, "Equivalent Frames of Reference for Judging Risk Regulation Policies," 3 *New York University Environmental Law Journal* 431 (1995). What additional information would you need to determine whether EPA's priorities are appropriate?

15. The second EPA study, *Reducing Risk*, was undertaken by EPA's Science Advisory Board (SAB). The SAB was asked to review the findings in *Unfinished Business* and to develop recommendations for prioritizing EPA's efforts. Most importantly, the SAB called on EPA (p. 6) to "target its environmental protection efforts on the basis of opportunities for the greatest risk reduction." Why is this objective desirable? Should the relative costs of achieving different risk reductions be taken into account? Should it matter whether the risks are vol-

untary or involuntary? Should it matter whether the environmental problem is fully internalized (such as indoor radon in owner-occupied houses)?

Elsewhere in the report, the SAB formulates this recommendation in two other ways (App. C, p. 28). First, it states: "We should set priorities for environmental protection based on an explicit comparison of the relative risk posed by different environmental problems and, more specifically, the opportunities for cost-effective risk reduction." Second, it states: "[T]o the extent that EPA has discretion to emphasize one environmental protection program over another, it should emphasize the program that reduces the most environmental risk at the lowest overall cost to society." Are these three formulations consistent? Under the third formulation, is it necessarily the case that the most serious risks would be addressed first?

The process of comparing risks so as to address the most serious ones first has come to be known as comparative risk assessment or comparative risk analysis (CRA). For academic criticism of CRA, see Hattis and Goble, "Current Priority-Setting Methodology: Too Little Rationality or Too Much?," in Adam M. Finkel and Dominic Golding, eds., *Worst Things First?: The Debate Over Risk-Based National Environmental Priorities* (Washington, D.C.: Resources for the Future, 1994), at 107; Hornstein, "Reclaiming Environmental Law: A Normative Critique of Comparative Risk Analysis," 92 *Columbia Law Review* 562 (1992); Silbergeld, "The Risks of Comparing Risks," 3 *New York University Environmental Law Journal* 405 (1995).

16. If it cannot do both, should environmental regulation address risks that experts think are serious or risks the public thinks are serious? What is the economic perspective's view on this question? What would Mark Sagoff say? The economic and noneconomic perspectives on environmental degradation are discussed in Chapters I and II, respectively. If people are being compensated for exposure to risk, to what extent should public perceptions affect the amount of compensation?

17. Is the divergence between lay and expert perceptions of risk a case of public misperception that should be solved by changing the public's perceptions? Or should the regulatory system accept the divergence and seek to adjust the decision-making process to accommodate the differing perceptions? In particular, should lay perceptions of risk count in risk assessments? For discussion, see Freeman and Godsil, "The Question of Risk: Incorporating Community Perceptions Into Environmental Risk Assessments," 21 *Fordham Urban Law Journal* 547 (1994).

18. In some areas of the law, the courts have tried to address the divergence between lay and expert perceptions of risk by, for example, prohibiting awards of damages in nuisance cases for loss of property value due to scientifically unfounded fear of contamination. Should the response of the legislative and administrative processes be different from the response of the courts?

19. Do you agree with Justice Breyer that risk communication is unlikely to significantly narrow the difference between expert and lay perceptions of risk? Have your views about risk changed as a result of the readings in this chapter?

20. Justice Breyer sees the problem of misperception of risk as one of three elements in a "vicious circle"—therein the title to his book (the other two elements are the congressional reaction to this misperception and uncertainties in the risk assessment process). As a solution to these problems, he advocates the creation of a small, elite career path staffed by individuals who would rotate among positions in the agencies with direct responsibility for environmental protection, OMB, and Congress. He urges that this group enjoy the following attributes:

    (a) the mission of building a coherent risk-regulating system

    (b) interagency jurisdiction so that it can transfer resources to where they are likely to do the most good

    (c) political insulation from both the public and Congress guaranteed, at a minimum, through civil service protection

    (d) prestige necessary to attract a capable staff

    (e) authority, both formal and informal, to accomplish its objectives

To what extent do you think that Justice Breyer's proposal would reduce the problem of differing lay and expert perceptions of risk? Justice Breyer suggests (p. 63) that insofar as his proposal "produces technically better results, the decision will become somewhat more legitimate, and thereby earn the regulator a small additional amount of prestige, which may mean an added small amount of public confidence." Do you agree with him?

21. Three principal themes have been explored in these notes and questions:

    (a) the feasibility and desirability of insulating risk assessment from politics

    (b) the uncertainties embedded in risk assessment

    (c) the implications of the divergence between expert and lay rankings of risk

Are there solutions to these problems other than those proposed by Ruckelshaus, Graham and his co-authors, and Justice Breyer?

IV

The Objectives of Environmental Regulation: Risk Management

This chapter examines the process of risk management, by which regulators make decisions about which risks are worth addressing and about the extent to which these risks should be controlled. As discussed in Chapter III, risk management is generally seen as a second step in the regulatory process, following the completion of a risk assessment. At the risk management stage, regulators have an estimate of the magnitude of the environmental risks and must decide about the extent, if any, to which these risks should be controlled.

The excerpt from the book by Baruch Fischhoff, Sarah Lichtenstein, Paul Slovic, Stephen Derby, and Ralph Keeney examines the process by which particular risks are deemed acceptable. The authors argue that simplistic solutions are likely to be undesirable. For example, a no-risk command might be unattractive if there is a competing alternative that poses only slightly greater risk but is far less costly. Similarly, setting a standard so that the probability of some important adverse consequence is lower than a given threshold (say one in a million) might lead to an undesirable outcome if an alternative standard, with a probability only slightly above this threshold, would be far less costly. Also problematic, according to the authors, is a fixed value trade-off between cost and risk. For example, one might be willing to pay more to reduce a unit of risk when risk is high than when it is low.

The excerpt from the book by Lester Lave sets forth and analyzes nine competing frameworks for making regulatory decisions:

1. *market regulation*, under which individuals choose, through market transactions, the risk that they wish to be exposed to;
2. *no-risk* under which activities posing any risk at all are banned;

3. *technology-based standards*, under which regulation is set by reference to what can be achieved through the use of the best available control technology;
4. *risk-risk (direct)*, under which the risks to consumers of a product eliminated by regulation are balanced against the risks to these consumers that would increase as a result of this regulation;
5. *risk-risk (indirect)*, under which the risk to individuals other than the consumer of the product would also be taken into account;
6. *risk-benefit*, under which both the benefits and burdens of a regulation would be taken into account;
7. *cost-effectiveness*, under which regulation would seek to achieve a given goal at least cost;
8. *regulatory budget*, under which regulation could not impose on affected parties costs that exceeded a predetermined figure; and
9. *benefit-cost* (sometimes termed cost-benefit), under which the benefits and costs of regulation would be quantified and compared.

Lave offers four criteria by which to choose among the competing frameworks: whether all relevant issues are taken into account; whether there is an adequate intellectual foundation for the use of the framework; whether the resources necessary to implement the framework are excessive; and whether the framework presents the public policy issues in a manner that is comprehesible.

The article by Steven Kelman presents a philosophical critique of cost-benefit analysis. Kelman opposes the monetization of environmental benefits. He argues that the methods used by economists to perform these valuations are all subject to criticism on technical grounds. Moreover, the valuation of benefits is usually performed by ascertaining an individual's willingness to pay (WTP) for an environmental amenity; Kelman notes that this measure is likely to understate the true value of the resource because studies show that one's willingness to pay to obtain something that one does not have is generally lower than one's willingness to accept (WTA) compensation to give up something that one has. Kelman also argues that the WTP values imputed by economists are ascertained by reference to public transactions but then are used to provide guidance for public decisions; consistent with the position that Mark Sagoff espoused in Chapter II, Kelman maintains that there is a difference between the attitudes that people express in the public and private spheres. Finally, he argues that the mere act of pricing certain non-market commodities reduces their value.

As further background to this chapter, it is useful to present an overview of the risk management frameworks in the federal environmental statues, trace the shift—which began in the early 1980s—toward the application of cost-benefit analysis, and present a brief overview of the major techniques for the valuation of environmental benefits.

The federal environmental statutes employ a variety of risk management frameworks, often combining several of the categories described above. For example, the National Ambient Air Quality Standards (NAAQS) under the Clean Air Act, which apply to criteria pollutants (as opposed to hazardous pollutants), must be set at the levels that, "allowing an adequate margin of safety, are requisite to protect the public health" (42 U.S.C. §7409). The Environmental Protection Agency (EPA) has employed a "critical populations, critical effects" approach, seeking to protect the most sensitive members of the population against every adverse health effect. Moreover, the courts have held that the costs of pollution reduction cannot be taken into account in setting these standards. Thus, at least in theory (the practical problems with this approach are discussed in Chapter IX) the NAAQS are set by reference to a no-risk framework. Technology-based standards are also employed, but primarily as a mechanism to allocate the pollution control burden required by the NAAQS.

The Clean Air Act takes a different approach with respect to hazardous air pollutants. The pollutants are currently controlled by means of technology-based standards. Over the course of the next decade, however, EPA must promulgate more stringent standards if the technology-based standards "do not reduce lifetime excess cancer risks to the individual most exposed to emissions . . . to less than one in one million" (see 42 U.S.C. §7412[f][2]). These more stringent standards correspond to a negligible risk framework (a variant of a no-risk framework). The resulting standards are therefore the *more* stringent of those that would result from the technology-based and negligible risk frameworks, respectively.

Under the Safe Drinking Water Act, a different two-part procedure is used. EPA first sets maximum contaminant level goals "at the level at which no known or anticipated adverse effects on the health of persons occur and which allows an adequate margin of safety." Next, EPA promulgates the national primary drinking water regulations, which are the enforceable standard, "as close to the maximum contaminant level goal as is feasible." The statute defines "feasible" as "feasible with the use of the best technology . . . (taking cost into consideration)" (see 42 U.S.C. §300g–1[b][4], [5]). Thus, these standards are the *less* stringent of those that would result from the technology-based and risk threshold frameworks, respectively.

The Federal Insecticide, Fungicide, and Rodenticide Act (FIFRA) provides that pesticides can be marketed only if they "will not generally cause unreasonable adverse effects on the environment" (7 U.S.C. §136a[5]). Such effects, in turn, are defined as "unreasonable risk to man or the environment, taking into account the economic, social, and environmental costs and benefits" (id. §136[bb]). This approach has elements of both the risk-risk framework, in that it takes into account both environmental costs and benefits of the pesticide and of the risk-benefit framework, in that it considers both the benefits and burdens of regulation.

Since the early 1980s, the risk management frameworks of the various federal environmental statutes have been significantly affected by Executive Order 12,291, which was promulgated by President Reagan and remained in effect during the Bush administration. Procedurally, it established a centralized process for the prepromulgation review of any major rule (defined as a rule with an annual effect on the economy of $100 million or more) by the Office of Management and Budget (OMB). Substantively, it required that "to the extent permitted by law," "[r]egulatory action . . . not be undertaken unless the potential benefits to society of the regulation outweigh the potential costs to society," "[r]egulatory objectives . . . be chosen to maximize the net benefits to society," and "[a]mong alternative approaches to any given regulatory objective, the alternative involving the least net cost to society . . . be chosen."

President Clinton replaced this order with Executive Order 12,866, which retains the centralized review by OMB, and provides, consistent with the approach of its predecessor, that "in choosing among alternative regulatory approaches, agencies should select those approaches that maximize net benefits . . . unless a statute requires another regulatory approach." Thus, these orders push EPA in the direction of cost-benefit analysis, except in those instances, such as promulgation of NAAQS under the Clean Air Act, where the statute prohibits the application of this framework.

The move toward cost-benefit analysis as the preeminent risk-management technique in environmental law was a plank in the "Contract with America," which led the Republican party, in the 1994 elections, to capture the House of Representatives for the first time in several decades. Early in its new session, the House passed H.R. 1022, entitled the Risk Assessment and Cost-Benefit Act of 1995. It applies to the promulgation of any rule with an annual effect on the economy of $25 million or more. Under the provisions of the bill, no such rule can be promulgated unless "the incremental risk reduction or other benefits . . . will be likely to justify, and be reasonably related to, the incremental costs incurred by State, local, and tribal governments, the Federal Government, and other public and private entities." The bill further provides that "costs and benefits shall be quantified to the extent feasible and appropriate and may otherwise be qualitatively described." Signficantly, unlike the Executive Orders, this bill provides that this cost-benefit framework "shall supplement and, to the extent there is a conflict, supersede the decision criteria for rulemaking otherwise applicable under the statute pursuant to which the rule is promulgated."

During the last few decades, economists have developed different techniques for the valuation of different types of environmental benefits. These benefits can be categorized broadly as benefits to human life and health on the one hand and benefits to the environment on the other. The latter, in turn, are generally subdivided into three categories: use values, option values, and existence values. Use values arise with respect to environmental resources, for example, pristine lakes that one values because one plans to actually use them. Option values arise with respect to resources that one does not currently use but that one may want to use in the future. Existence values, sometimes called non–use values, arise with respect to resources that one does not currently use or expect to use in the future but the existence of which nonetheless gives rise to utility, for example, one can derive utility from knowing that a wilderness area that one never plans to visit will remain undeveloped.

For the purposes of cost-benefit analysis, the standard technique for the valuation of lives is the "willingness to pay" approach. Typically, this approach first identifies jobs that are similar except for different exposures to risk. Second, it determines the wage premiums obtained by workers at the riskier jobs. Third, it estimates the additional probability of death that results from the riskier job. Finally, it extrapolates, from the wage premium obtained for this additional probability of death, to the value of a life. In turn, use values for environmental resources such as parks are often estimated through the travel cost method. This method estimates the expenditures, in terms of time and out-of-pocket disbursements, that individuals are willing to incur in order to visit the site. Finally, option and existence values are generally ascertained by means of the contingent valuation methodology (CVM). Under CVM, individual valuations are determined by means of surveys, and the resulting figures are multiplied by the number of affected individuals.

Acceptable Risk

BARUCH FISCHHOFF, SARAH LICHTENSTEIN, PAUL SLOVIC, STEPHEN L. DERBY, AND RALPH L. KEENEY

Acceptable-risk problems are decision problems; that is, they require a choice among alternative courses of action. What distinguishes an acceptable-risk problem from other decision problems is that at least one alternative option includes a threat to life or health among its consequences. We shall define *risk* as the existence of such threats.

Whether formal or informal, examination of the options in a decision problem involves the following five interdependent steps:

1. specifying the objectives by which to measure the desirability of consequences
2. defining the possible options, which may include "do nothing"
3. identifying the possible consequences of each option and their likelihood of occurring should that option be adopted, including, but not restricted to, risky consequences
4. specifying the desirability of the various consequences
5. analyzing the options and selecting the best one

This final step prescribes the option that should be selected, given the logic of the process; thus, it identifies the *most acceptable* option. If its recommendation is followed, then that seemingly best alternative will be adopted or accepted. Of course, one need not do so unless one feels that the decision-making process was adequately comprehensive and defensible.

The act of adopting an option does not in and of itself mean that its attendant risks are acceptable in any absolute sense. Strictly speaking, one does not accept risks. One accepts options that entail some level of risk among their consequences. Whenever the decision-making process has taken into account benefits or other (nonrisk) costs, the most acceptable option need not be the one with the least risk. Indeed, one might choose (or accept) the option with the highest risk if it had enough compensating benefits. The attractiveness of an option depends upon its full set of relevant positive and negative consequences.

Deciding which of a set of options is most attractive is inherently situation specific. That is, *there are no universally acceptable options* (or risks, costs, or benefits). The choice of an option (and its associated risks, costs, and benefits) depends on the set of options, consequences, values, and facts examined in the decision-making process. In different situations, different options, values, and information may be relevant. Over time, any of a number of changes could lead to a change in the relative attractiveness of any given option: Errors in the analysis may be discovered, new safety devices may be invented, values may change, additional information may come to light, and so forth. Even in the same situation and at a single time, different people with different values, beliefs, objectives, or decision methods might disagree on which option is best. In short, the search for absolute acceptability is misguided.

Thus, the acceptance of a risk is contingent on many different things, including the other features of the option with which it is associated, the other options considered, and so forth. One could use the term *acceptable risk* to refer to the risk associated with the most acceptable option in a particular decision problem. However, it may be quite difficult to bear in mind the context upon which such a choice is based in its full richness. Fearing that the term "acceptable risk" will tend to connote absolute rather than contingent acceptability, we have chosen to use it only as an adjective, describing a kind of decision problem, and not as a noun describing one feature of the option chosen in such a problem. Hence, we shall refer to "acceptable-risk problems" and not to "acceptable risks."

Illustrations

A decision-making perspective offers a common language for treating some recurrent issues in acceptable-risk problems, as shown in Figures IV.1 to IV.4. Assume that a single individual is empowered to make each decision, that all risks and costs can be identified, characterized, and assessed with certainty, and that the benefits of all the options are identical. The options differ only in their cost and level of risk; 0 is the best level for each of these dimensions. As concrete examples, consider an individual choosing among automobiles or among surgical procedures that differ only in cost and risk.

Figure IV.1 shows how the set of options considered affects the choice of the most acceptable option. If K and L are the only options available, then the choice is between high cost with low risk (K) and low cost with high risk (L). The level of risk accepted would then be that level associated with either K or L, depending on which was chosen. If another option having lower cost and lower risk (M) became available, then it should be preferred to either K or L. The risk accepted would then be the level associated with the new option.

Figure IV.2 illustrates how determination of the most acceptable option depends upon decision makers objectives. If the goal is minimizing risk, then option K would be chosen. Minimizing cost, on the other hand, entails the choice of option L and its higher level of risk.

Figure IV.3 relaxes the assumption of perfect knowledge. New information drastically alters the decision maker's appraisal of the costs and risks of M. Had M already been selected, then the accepted level of risk would prove to be much higher than that originally anticipated. If the decision had yet to be made, then the choice would revert to K or L, with their associated risk levels.

The decision rules used in Figure IV.2, minimize cost and minimize risk, were rather simplistic. The two broken-line indifference curves in Figure IV.4 present more believable preferences. Each point on such a curve would be equally attractive to an individual whose preferences it represents. Case 1 reflects a willingness to incur large costs in return for small reductions in risk. By this criterion, option K is preferred to L; the cost saving of L is achieved at the price of too great an increase in risk. Indeed, this individual would prefer K even if L's cost was zero. Case 2 reflects less willingness to increase costs in exchange for reduced risk; option L is now the best choice.

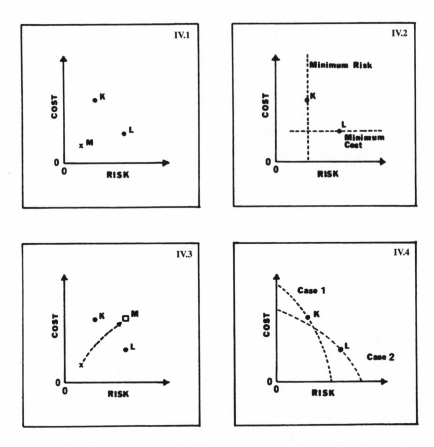

Figures IV.1–IV.4. Exemplary choices among alternative risky options. Figure IV.1 shows the effect of the options considered on the choice made; Figure IV.2 shows the effect of the decision makers' values; Figure IV.3 shows the effect of changing information; Figure IV.4 shows the effects of more complicated preferences.

Apparently Easy Solutions

Viewing acceptable-risk problems as decision problems also helps illuminate the flaws in some simplistic solutions. For example, it may be tempting to claim that no risk should be tolerated. However, the decision-making perspective forces one to ask, What is the cost of absolute safety? Applied strictly, total abhorrence of risk could lead to rather dubious decisions, such as preferring option A to option B in Figure IV.5, thereby incurring great cost for a minor reduction in risk.

Rather than paying for safety, one might propose doing without the substance, activity, or technology in question. A decision-making perspective requires one to ask what option is to be chosen in its stead. When that option has risks of its own,

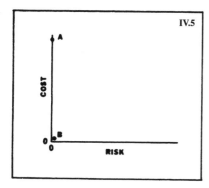

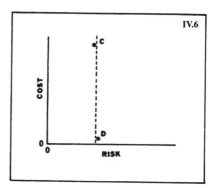

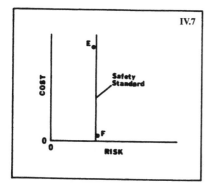

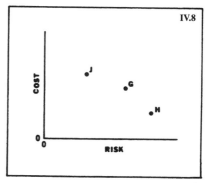

Figures IV.5–IV.8. Exemplary choices among risky options, clarifying the pitfalls of some seemingly easy solutions. Figure IV.5 shows an implication of wanting no risk; Figure IV.6 shows an implication of deciding that the option adopted should be as safe as possible; Figure IV.7 considers the adoption of an absolute standard for maximum allowable risk; Figure IV.8 shows an implication of specifying fixed risk-benefit trade-

the gain in safety may prove illusory. For example, if diabetics have a need for sweeteners, banning saccharin may eliminate one possible cancer risk in return for increased risk from the consumption of sugar.

A variant of the desire for absolute safety is the unqualified insistence that the chosen option be as safe as possible. Option C in Figure IV.6 provides less risk than option D, but at a large incremental cost. Most people would tolerate some small increase in risk for a large reduction in cost (at least if those bearing the risk also received the cost savings).

Another simplification calls for expressing the answer to, How safe is safe enough? by a small number (like 10^{-7}) representing the maximum allowable probability of some important adverse consequence. Figure IV.7 illustrates one situation in which this solution would appear inappropriate. Suppose that options E and F lie just on opposite sides of the designated standard, and that E costs substantially more than F. In practice, most people might prefer F to E despite the fact that F is above the safety standard.

A more sophisticated solution is to specify fixed value tradeoffs between cost and risk. For example, one could adopt any safety measure costing less than $1 million per expected life saved. But Figure IV.8 suggests that that, too, could be an oversimplification. When risk is very high, one might be willing to incur great cost to reduce it. Thus, one might prefer G to H even though the shift to G doubles the cost in order to reduce the risk by only one-fourth. At the same time, one might be more reluctant to pay for added safety when risks are low. Thus one might not prefer J to G even though that shift buys more safety for less cost than the change from H to G. Such preferences are consistent if one feels that different value tradeoffs are appropriate at different levels of risk.

The Strategy of Social Regulation: Decision Frameworks for Policy

LESTER B. LAVE

Six frameworks for making regulatory decisions are currently being used and two have been proposed. The frameworks range, roughly, from those requiring the least theory, data, and analysis and offering the least flexibility to those at the opposite pole; they include market regulation, no-risk, technology-based standards, risk-risk (proposed), risk-benefit, cost-effectiveness, regulatory budget (proposed), and benefit-cost.

Market Regulation

Economic theory has formalized the 200-year-old insight of Adam Smith that competitive markets are efficient. In particular (under a set of stringent assumptions including complete information, no transaction costs, rational consumers and producers, no economies of scale in production, and no externalities), a competitive market produces an efficient (or Pareto optimal) equilibrium in the sense that no one can be made better off without making at least one person worse off. This efficiency principle also holds for situations involving risk, such as hazardous products or jobs, although still more stringent assumptions are needed.

Reprinted by permission of the author and publisher from Lester B. Lave, "The Strategy of Social Regulation: Decision Frameworks for Policy," Copyright © 1981, Brookings Institution.

Each person in such an economy presumably would decide what is best for him by looking at the array of available products and jobs. Since risk is an undesirable attribute, all risky products and jobs having no compensating attributes would be eliminated, and individuals would scrutinize those risky products and jobs that offered higher pay or some other advantage to determine which should be taken. Under the restrictive assumptions, government regulation would be unnecessary.

Clearly the U.S. economy does not satisfy the host of restrictive assumptions; both buyers and sellers can often influence price, many effects are transmitted outside the marketplace, and often buyers and sellers are woefully ignorant of the health and safety implications of a product. Market equilibrium is inefficient and a case can be made for government intervention. Some economists caution Americans to eschew perfection, arguing that they would be better off in the long run by tolerating these relatively minor evils instead of erecting a huge, self-defeating regulatory structure. Regulation requires resources, but more important, it is virtually impossible to regulate so that incentives are not distorted, and this often leads to even greater inefficiency than in the unregulated market—for example, transportation regulation, particularly of airlines and trains. Many economists argue that regulation is justified only when serious violations of the assumptions occur, and then only if the regulation can be relatively efficient. . . .

In summary, the decision to use the market to regulate risk puts faith in consumer information and judgments. It sees the costs of bureaucracy constraining private decisions as larger than costs arising from market imperfections and advises accepting current imperfections rather than creating a regulatory morass.

No-Risk

The philosophy behind the Delaney Clause of the Food, Drug, and Cosmetic Act, and food additive amendments generally, is that the public should be exposed to no additional or unnecessary risk. Carcinogens cannot be added to foods or remain as residuals in meat since this might increase the risk of cancer; according to the Clean Air Act Amendments of 1970, air pollution levels must be sufficiently low to protect the population from adverse effects, presumably even the most sensitive members.

This approach has great appeal as rhetoric. To argue that carcinogens ought to be permitted in the food supply is to argue that society must allow higher than necessary risks of cancer. Why should any unnecessary exposure be tolerated, even if the risk appears to be small?

The no-risk framework has the advantage of requiring little data and analysis and precludes agonizing about the decision to be made. According to the Delaney Clause, the only question is whether a food additive has been shown to be a carcinogen in humans or animals. Thus data (on the quality, variety, and price of food) concerning the consequences of banning may not be considered. The Delaney Clause has a simple, straightforward answer to a complicated question: ban a substance if there is evidence of carcinogenicity. Frameworks other than market regulation require answers to a set of complicated questions: What level of risks are acceptable? What benefits would serve to offset the risks? Can animal bioassays be relied on to demonstrate human carcinogenicity? Can the potency of a substance for humans be demonstrated under current exposure levels? If one requires simple answers to these questions or distrusts the complicated answers given by experts, no-risk offers an appealing solution.

Unfortunately the answers are too simple. Virtually all "natural" foods contain trace elements of carcinogens, including biological contaminants and pesticides. The Food and Drug Administration treats natural foods differently than food additives; apparently it is less troublesome to die from a cancer induced by a natural food than from one induced by a food additive. Does anyone seriously propose to ban all foods with trace levels of carcinogens? Does it make sense to treat those with trace amounts in the same way as those with large amounts of potent carcinogens? . . .

The three principal objections to this framework are the current misallocation of resources, the closing of the door to future solutions, and the inconsistency in government policy. In addition this framework cannot distinguish between a toxin that is extremely weak and to which few people are exposed and a potent carcinogen to which nearly the entire population is exposed. Insofar as there are many carcinogens and it is costly to ban at least some of them, this framework does not help to develop priorities—which substance should be treated first?—or guidelines—what level of safety ought to be sought where banning is infeasible? Instead, it sends regulators scurrying off to devote much of their attention to relatively benign substances by giving all toxins equal priority. Thus the framework is a pernicious guide to regulators confronted with complicated problems.

Although Congress has written the no-risk framework into legislation, it is a straw man unworthy of serious consideration. Even the attempt to maintain the facade is increasingly recognized by the regulatory agencies to be impossible. For example, the Food and Drug Administration has attempted to define a "negligible" risk level; any risk below a level of one in one million lifetimes would be considered to be zero for regulatory purposes.

Technology-Based Standards

Recognizing the difficulty of attempting to estimate the health and safety effects of a proposed standard (much less the problem of quantifying these effects), a number of agencies have placed their reliance on engineering judgments. The best available control technology has been required extensively by the Environmental Protection Agency in regulating air and water pollution. This framework has the simplicity of requiring the estimation of neither benefits nor costs. The data and analysis required are for identifying a hazard and then for making the engineering judgment as to the best available control technology. This framework requires a second set of information for determining the best available control technology in addition to the carcinogenicity data required for the no-risk framework.

In practice, however, there is never a best technology but only successively more expensive and stringent technologies. For example, the effectiveness of an electrostatic precipitator in removing suspended particles from air is proportional to the collector plate area; effectiveness can be increased by increasing the area. In practice, engineering judgment defines best available control technology as a finite collector plate area, even though further increases in plate area would improve (minutely) the effectiveness of collection. At some point additional abatement is unwarranted because social costs exceed social benefits; but even then technology is available that would abate emissions further. In practice, best available control technology embodies implicit assumptions about the benefits and costs of further abatement.

The crucial issue in implementing this framework at present is the financial burden each industry can bear. As long as an industry is not in danger of bankruptcy, a technology that lowered emissions would be considered acceptable. Sufficient uncertainty exists about what cost level would endanger an industry that regulators rarely impose standards that come close to doing so.

In summary, the primary advantage of technology-based standards is that they require no formal evidence on costs or benefits; the only data required are those necessary for good engineering judgments. The resulting standard, however, will depend on regulators' perceptions of industry profitability. If an area is populated by an industry teetering on the brink of bankruptcy, best available control technology will be weak and few emissions will be abated. If the industry is profitable, it will require large expenditures. There is more than a theoretical possibility that the first regulation in an industry would press it to the limit of its ability to afford regulation, leaving no financial resources to handle later regulations that might be far more important. Rather than being a framework for lowering risk or even for using engineering judgments, technology-based standards is a framework for regulating economic activity through imposing costs arbitrarily among industries until all are at the same minimal level of profit.

Risk-Risk: Direct

Even if maximum protection were desired, the Delaney Clause would be a poor framework because it requires banning carcinogens. Some toxic substances, such as food additives and fungicides, prevent contamination of food, and thus it is desirable to weigh one risk against the other, as recognized by the Food and Drug Administration and the Department of Agriculture in the proposed risk-risk analysis. Balancing the toxicity of a substance against the enhanced protection it brings can be done from either of two perspectives. The narrow perspective is that of balancing the risk to the consumer of the additive against the direct health benefits. Sodium nitrite may be a carcinogen, but it protects against botulism; the risk of cancer must be balanced against that of botulism. The broad perspective takes account of both producers and consumers as shown below.

Since the risk-risk framework allows beneficial health effects to be considered along with adverse health effects, it is more flexible than no-risk. It and the remaining frameworks are qualitatively different from no-risk in that they require quantification of risk and at least partial estimation of benefits. If quantification were impossible, this framework could not be implemented because there would be no method for balancing unmatched risks (for example, chronic respiratory disease versus broken legs). Quantification is particularly difficult for the effects of toxic substances; thus this and the remaining frameworks are subject to the caution of those who contend that potency cannot be estimated from animal bioassays, or at least that potency for humans at low doses cannot be inferred reliably. . . .

While the risk-risk framework provides somewhat greater flexibility, it still precludes consideration of nonhealth effects. Conceptually it is a small step since it merely includes both the health risks and health benefits of a proposal. In practice it appears to be a major improvement over the no-risk framework—where it is applicable. Cases such as sodium nitrite where the risk-risk framework is invaluable, are the exception. Few substances offer a direct health benefit to the consumer other than drugs, products for which the Food and Drug Administration already uses this framework. The framework is of limited interest because it is of such limited applicability.

Risk-Risk: Indirect

The advantage of the risk-risk framework over the no-risk framework is that it permits wider analysis of risks. One way of stating the objective is that society desires to minimize the adverse health effects associated with a given food such as bacon. Thus society would permit nitrite in bacon if the improvement in the health of consumers from botulism protection exceeded the decrement in health from the risk of cancer. Yet it is evident that the direct risk-risk framework takes only the first step of considering the health of the person consuming the food. People are also associated with the production and distribution of food; society desires to minimize the adverse health effects associated with producing as well as consuming bacon (for a fixed level of production). Workers would not countenance a regulation that offered consumers a small amount of protection at the cost of a large increase in risk to workers.

Since every human activity is risky, a regulation that requires more man-hours to produce a unit of food would increase the exposure and presumably the occupational risk of workers. The indirect risk-risk framework includes occupational risks associated with each additive or contaminant. . . .

Risk-Benefit

Unlike the risk-benefit framework, the three previous ones do not allow consideration of nonhealth effects. The folly of refusing to consider these effects is illustrated by examining one's own choices. For example, most people are willing to risk the minute chance of biological contamination rather than to be bothered with boiling drinking water. They are willing to undertake additional risks in order to get rewards such as additional income and recreational stimulation. For example, there is a risk premium in the pay of workers in hazardous occupations to attract them in the face of the higher risks. These premiums can be extremely high, as for test pilots, steeplejacks, and divers working deep in the ocean. If the effect of a regulation is to lower risk minutely at the cost of a vast increase in price, a lessening of choice or convenience, harm to the environment, or a sacrifice in social goals generally, society should not be satisfied. The frameworks previously mentioned suffer from their lack of recognition of other social goals such as the ecosystem, endangered species, and individual freedoms.

Under the risk-benefit framework, regulators would be enjoined to balance the general benefits of a proposed regulation against its general risks. This framework is intended to be somewhat vague, with all effects being enumerated, but with full quantification and valuation being left to the general wisdom of the regulators. The framework may account for cost, convenience, and even preferences in an attempt to balance benefits against risks. A vast array of frameworks can come under the risk-benefit heading, from balancing health risks against health benefits (like the risk-risk indirect framework) to consideration of all risks, costs, and benefits. The framework has an immediate appeal to congressmen and regulators since it is a general instruction to consider all social factors in arriving at a decision. While no one can oppose considering all relevant factors, no one has specified precisely how this is to be done.

The intellectual difficulty with this framework is its lack of precise definition. Are only health risks to be considered, or are risks to the present and future environment (air, water, louseworts, snail darters, and tundra) relevant? If they are not, the framework is no more complete than the previous one, and if they are, how can the risks to louseworts be added to those to the health of our great grandchildren and of current workers? Similarly, there is no guidance about how to quantify benefits: what is the value of an increase in the supply of food or electricity? . . .

Cost-Effectiveness

Many organizations, private and public, find themselves attempting to increase output even though their current budget is fixed. The intellectual contributions in defining this problem and developing rules to solve it have come from the Department of Defense. Although cost-effectiveness is often thought erroneously to refer to getting some specific project done at lowest cost, the concept is much broader, referring to accomplishing some general objective at lowest cost. President Eisenhower's secretary of defense, Charles Wilson, described the goal succinctly as an attempt to "get the most bang for the buck."

How can a goal be achieved within a fixed budget? For example, the goal of the National Cancer Institute is to lower the cancer death rate. It might achieve this goal by devoting resources to basic research, clinical trials testing new treatment techniques, public education, prevention by lowering the amounts of carcinogens in the environment, early detection of cancer, or the provision of more treatment. How should the fixed budget be allocated among these competing programs to lower both the incidence of cancer and the occurrence of death and lesser effects?

Mathematically, this is a problem of maximization under constraints; the solution is to equate the effectiveness of the last dollar spent on each activity. For this example, the National Cancer Institute ought to allocate funds among the programs (taking care that the most effective projects are done first within each program) by testing the effectiveness of each dollar. The first increment of funds should be given to the program where it would be estimated to save the most lives. The second increment of funds should be allocated by the same criterion, perhaps going to the same program. As each successive increment of funds is allocated, the number of lives it saves should fall (since the best projects were done first). When all funds have been allocated, it should be true that the last increment of funds to each program would be expected to save approximately the same number of lives. If not, then funds should be reallocated by recalling them from the program where they are least effective and giving them to the program where they are most effective. Mathematically, the ratio of lives saved to dollars expended (for the last increment of funds) should be equal across programs when all funds have been allocated. As long as the ratios are not equal, additional lives could be saved for the same budget by removing funds from the program with the lowest ratio and adding them to the program with the highest ratio. . . .

Cost-effectiveness offers a major advantage over benefit-cost analysis in that it does not require an explicit value for the social cost of premature death (or other untraced goods). Assumptions about these values are built into the goal and budget (for example, maximize lives saved for a fixed budget) but need not be stated explicitly. The flip side of this advantage, however, is that errors in stating the goal or in determining the budget can lead to bad decisions, and there is no internal mechanism for showing the errors in these decisions and the changes in goals or budget that are necessary.

Regulatory Budget

Cost-effectiveness is a good framework if the relevant costs are being measured in the analysis. Unfortunately when the only costs considered are those of the regulatory agency, the framework will misallocate resources because only one subset of the total costs of the regulation to the entire economy is being considered. The agencies have little or no reason to consider the costs that their regulations impose on others unless the costs are so high that industry bankruptcy is a relevant possibility. The agencies are instructed to protect the environment, consumers, or workers without any apparent limits on their ability to impose costs on others. That the resulting regulations are not universally perceived as desirable can be judged from the comments of the affected companies and the fact that the federal government has often exempted itself from the regulations or has been slow in implementing them.

An idea originating in the Council of Economic Advisers under Charles Schultze was to give each regulatory agency an implementation budget in the form of a limit on the total annual costs that its regulations could impose. For example, the Environmental Protection Agency might be given an implementation budget of $10 billion a year, which would mean that the costs of implementing its air, water, solid waste, radiation, and pesticide regulations could not exceed $10 billion in that year. Each agency would develop an implementation budget request, just as it currently develops its operating budget request. The administration would coordinate and impose priorities on the agencies, and then Congress would react to these requests, modifying them as necessary.

The regulatory budget is one method of implementing cost-effectiveness analysis. The goals needed for the framework are stated in the legislation for each agency, supplemented by whatever informal instructions arise from hearings, appropriations, Office of Management and Budget directives, or presidential intervention. The internal and implementation budgets would be considered and approved by Congress, based on each agency's data on effectiveness. A major advantage of the framework is that it would elicit from the agencies a clearer indication of their priorities and would enable Congress to make more intelligent decisions regarding social values.

The principal difficulties with the framework are in estimating the costs and effects of each regulation. Where a control device must be added to a smokestack, there is debate about the cost of the device and about its expected lifetime, maintenance, and reliability. For a new piece of technology, these difficulties might perhaps introduce a factor-of-two difference in estimated costs. When the regulation will require a change in process or result in banning a substance, the costs become much more uncertain. If there is a factor-of-five-or-ten difference between reasonable high and low estimates of implementation costs, the regulatory budget cannot provide a helpful constraint. . . .

Benefit-Cost

This framework is similar to the general balancing of risks against benefits; the principal difference is that it is more quantitative and formal. In addition to enumerating the various benefits of the regulation and then subjectively balancing benefits against costs, this framework would require quantification of the extent to which the benefits and costs vary with the level of regulation, and then would require each of these effects to be translated into dollars.

There are many controversial aspects to its application, including putting an explicit value on prolonging a life, quantifying other benefits, deciding the rate at which effects in the future are discounted to make them equivalent to current effects, and redistributing income. Valuing benefits, or even deciding what is a benefit, runs into the diversity of cultural backgrounds, personal goals, fears, and time horizons. . . .

Benefit-cost analysis is a sufficiently broad framework to be adapted to consider virtually any aspect of a regulation or public decision. The implications for those who gain or lose can be folded into the analysis. None of the objections to the framework have the effect of showing an inherent bias or blind spot in the analysis.

In practice, however, the picture is quite different. Benefit-cost analysis is often viewed, correctly, as a tool for defending the status quo. It is rarely used to consider who benefits or pays, and it focuses on the present, giving short shrift to even the near-term future with no importance for events more than a few decades in the future. Adjustment costs are often estimated to be higher than would be observed, reflecting a prejudice that the current situation must be the best one (when adjustment costs are not considered, the analysis is biased toward change). Finally, a number of simplifying assumptions are made that bias the analysis against change. . . .

A Comparison of Frameworks

. . . Four criteria might be used to compare frameworks:

The first is comprehensiveness. Are all the relevant issues encompassed within the framework? No-risk considers only carcinogenesis (or other health attributes);

risk-risk considers all health consequences either to the consumer (direct) or more generally (indirect). Cost-effectiveness and the regulatory budget require examination of costs as well as health, but they can be considered only within the goals of the agency. Benefit-cost and risk-benefit are the most encompassing, although even they are not used in practice to address equity questions.

The second criterion is the intellectual foundation required of each framework. One can be most certain about the foundation for the simple frameworks, but drawing in additional considerations requires more knowledge, assumptions, and value judgments. The wider coverage comes at a price. In some cases there is insufficient knowledge to be able to quantify or even explore these other considerations; if so, there is no alternative to a simple framework or an ad hoc decision.

The third criterion is the resources required to implement the frame work. The more complicated frameworks require exploration of further aspects of the problem, which in turn requires more data collection and analysis. Generally the resources available to analyze alternative regulations constitute a small proportion of those available for drafting and defending the regulations, and a minuscule proportion of the cost of carrying out the regulation. If additional analysis can result in even a tiny improvement in the quality of the regulation, the reduction in implementation and other costs should more than pay for the effort.

The fourth criterion is felicitousness. The world is complicated; it changes so rapidly that an agency rarely gets to second-order priority issues. The most important issues must be treated first, and they must be raised in easily comprehended fashion. If the issues are posed in a confused or obscure manner, the decision is likely to be made on an ad hoc basis. The felicitousness of the framework is more important than its comprehensiveness.

None of these frameworks is sufficiently complete and sound to serve as an automatic way of making decisions. The current Delaney Clause framework would appear to be the most concrete; even it, however, becomes mired in controversy over proving carcinogenicity. . . .

The other frameworks have the more difficult task of quantifying risk and of attempting to quantify other aspects of the issue (for example the value of greater choice). In all cases judgment is required to examine the suitability of the quantification, the factors that could not be quantified, and the valuation of the aspects that were quantified. These issues are far too complicated for a mechanical decision-making framework to be appropriate—for example one of pursuing a project if and only if estimated benefits exceed costs.

The real question is the extent to which each of these frameworks can prove helpful in informing the decisionmaker. Must all effects be quantified accurately and all valuations be agreed upon before benefit-cost analysis is helpful? If complete quantification is not possible or if there are difficulties in estimating risk, is it better to slip back to a less demanding framework, possibly back to the no-risk framework? The answer depends on both the amount of uncertainty and the extent to which the general nature of the uncertainty is known. No analysis of health and safety regulations has managed to quantify all aspects of the issue, and it is evident that no future analysis can be expected to be complete. If this lack of completeness is deemed fatal, there is no point in considering benefit-cost analysis further.

Cost-Benefit Analysis: An Ethical Critique
STEVEN KELMAN

At the broadest and vaguest level, cost-benefit analysis may be regarded simply as systematic thinking about decision-making. Who can oppose, economists some-times ask, efforts to think in a systematic way about the consequences of different courses of action? The alternative, it would appear, is unexamined decision-making. But defining cost-benefit analysis so simply leaves it with few implications for actual regulatory decision-making. Presumably, therefore, those who urge regulators to make greater use of the technique have a more extensive prescription in mind. I assume here that their prescription includes the following views:

1. There exists a strong presumption that an act should not be undertaken unless its benefits outweigh its costs.
2. In order to determine whether benefits outweigh costs, it is desirable to attempt to express all benefits and costs in a common scale or denominator, so that they can be compared with each other, even when some benefits and costs are not traded on markets and hence have no established dollar values.
3. Getting decision-makers to make more use of cost-benefit techniques is impor-tant enough to warrant both the expense required to gather the data for improved cost-benefit estimation and the political efforts needed to give the activity higher priority compared to other activities, also valuable in and of themselves.

My focus is on cost-benefit analysis as applied to environmental, safety, and health regulation. In that context, I examine each of the above propositions from the perspective of formal ethical theory, that is, the study of what actions it is morally right to undertake. My conclusions are:

1. In areas of environmental, safety, and health regulation, there may be many instances where a certain decision might be right even though its benefits do not outweigh its costs.
2. There are good reasons to oppose efforts to put dollar values on non-marketed benefits and costs.
3. Given the relative frequency of occasions in the areas of environmental, safety, and health regulation where one would not wish to use a benefits-outweigh-costs test as a decision rule, and given the reasons to oppose the monetizing of non-marketed benefits or costs that is a prerequisite for cost-benefit analysis, it is not justifiable to devote major resources to the generation of data for cost-benefit calculations or to undertake efforts to "spread the gospel" of cost-benefit analysis further. . . .

In order for cost-benefit calculations to be performed the way they are supposed to be, all costs and benefits must be expressed in a common measure, typically dollars, including things not normally bought and sold on markets, and to which dollar prices are therefore not attached. The most dramatic example of such things is human life itself; but many of the other benefits achieved or preserved by

Reprinted by permission of the author and publisher from *Regulation*, January/February 1981, at 33.

environmental policy—such as peace and quiet, fresh-smelling air, swimmable rivers, spectacular vistas—are not traded on markets either.

Economists who do cost-benefit analysis regard the quest after dollar values for non-market things as a difficult challenge—but one to be met with relish. They have tried to develop methods for imputing a person's "willingness to pay" for such things, their approach generally involving a search for bundled goods that *are* traded on markets and that vary as to whether they include a feature that is, *by itself*, not marketed. Thus, fresh air is not marketed, but houses in different parts of Los Angeles that are similar except for the degree of smog are. Peace and quiet is not marketed, but similar houses inside and outside airport flight paths are. The risk of death is not marketed, but similar jobs that have different levels of risk are. Economists have produced many often ingenious efforts to impute dollar prices to non-marketed things by observing the premiums accorded homes in clean air areas over similar homes in dirty areas or the premiums paid for risky jobs over similar nonrisky jobs.

These ingenious efforts are subject to criticism on a number of technical grounds. It may be difficult to control for all the dimensions of quality other than the presence or absence of the non-marketed thing. More important, in a world where people have different preferences and are subject to different constraints as they make their choices, the dollar value imputed to the non-market things that most people would wish to avoid will be lower than otherwise, because people with unusually weak aversion to those things or unusually strong constraints on their choices will be willing to take the bundled good in question at less of a discount than the average person. Thus, to use the property value discount of homes near airports as a measure of people's willingness to pay for quiet means to accept as a proxy for the rest of us the behavior of those least sensitive to noise, of airport employees (who value the convenience of a near-airport location) or of others who are susceptible to an agent's assurances that "it's not so bad," To use the wage premiums accorded hazardous work as a measure of the value of life means to accept as proxies for the rest of us the choices of people who do not have many choices or who are exceptional risk-seekers.

A second problem is that the attempts of economists to measure people's willingness to pay for non-marketed things assume that there is no difference between the price a person would require for *giving up* something to which he has a preexisting right and the price he would pay to *gain* something to which he enjoys no right. Thus, the analysis assumes no difference between how much a homeowner would need to be paid in order to give up an unobstructed mountain view that he already enjoys and how much he would be willing to pay to get an obstruction moved once it is already in place. Available evidence suggests that most people would insist on being paid far more to assent to a worsening of their situation than they would be willing to pay to improve their situation. The difference arises from such factors as being accustomed to and psychologically attached to that which one believes one enjoys by right. But this creates a circularity problem for any attempt to use cost-benefit analysis to determine *whether* to assign to, say, the homeowner the right to an unobstructed mountain view. For willingness to pay will be different depending on whether the right is assigned initially or not. The value judgment about whether to assign the right must thus be made first. (In order to set an upper bound on the value of the benefit, one might hypothetically assign the right to the person and determine how much he would need to be paid to give it up.)

Third, the efforts of economists to impute willingness to pay invariably involve bundled goods exchanged in *private* transactions. Those who use figures garnered from such analysis to provide guidance for *public* decisions assume no difference between how people value certain things in private individual transactions and how they would wish those same things to be valued in public collective decisions. In making such assumptions, economists insidiously slip into their analysis an important and controversial value judgment, growing naturally out of the highly individualistic microeconomic tradition—namely, the view that there should be no difference between private behavior and the behavior we display in public social life. An alternative view—one that enjoys, I would suggest, wide resonance among citizens—would be that public, social decisions provide an opportunity to give certain things a higher valuation than we choose, for one reason or another, to give them in our private activities.

Thus, opponents of stricter regulation of health risks often argue that we show by our daily risk-taking behavior that we do not value life infinitely, and therefore our public decisions should not reflect the high value of life that proponents of strict regulation propose. However, an alternative view is equally plausible. Precisely because we fail, for whatever reasons, to give life-saving the value in everyday personal decisions that we in some general terms believe we should give it, we may wish our social decisions to provide us the occasion to display the reverence for life that we espouse but do not always show. By this view, people do not have fixed unambiguous "preferences" to which they give expression through private activities and which therefore should be given expression in public decisions. Rather, they may have what they themselves regard as "higher" and "lower" preferences. The latter may come to the fore in private decisions, but people may want the former to come to the fore in public decisions. They may sometimes display racial prejudice, but support antidiscrimination laws. They may buy a certain product after seeing a seductive ad, but be skeptical enough of advertising to want the government to keep a close eye on it. In such cases, the use of private behavior to impute the values that should be entered for public decisions, as is done by using willingness to pay in private transactions, commits grievous offense against a view of the behavior of the citizen that is deeply engrained in our democratic tradition. It is a view that denudes politics of any independent role in society, reducing it to a mechanistic, mimicking recalculation based on private behavior.

Finally, one may oppose the effort to place prices on a non-market thing and hence in effect incorporate it into the market system out of a fear that the very act of doing so will reduce the thing's perceived value. To place a price on the benefit may, in other words, reduce the value of that benefit. Cost-benefit analysis thus may be like the thermometer that, when placed in a liquid to be measured, itself changes the liquid's temperature. . . .

The first reason that pricing something decreases its perceived value is that, in many circumstances, non-market exchange is associated with the production of certain values not associated with market exchange. These may include spontaneity and various other feelings that come from personal relationships. If a good becomes less associated with the production of positively valued feelings because of market exchange, the perceived value of the good declines to the extent that those feelings are valued. This can be seen clearly in instances where a thing may be transferred both by market and by non-market mechanisms. . . .

Furthermore, if one values in a general sense the existence of a non-market sector because of its connection with the production of certain valued feelings, then one ascribes added value to any non-marketed good simply as a repository of values represented by the non-market sector one wishes to preserve. This seems certainly to be the case for things in nature, such as pristine streams or undisturbed forests: for many people who value them, part of their value comes from their position as repositories of values the non-market sector represents.

The second way in which placing a market price on a thing decreases its perceived-value is by removing the possibility of proclaiming that the thing is "not for sale," since things on the market by definition are for sale. The very statement that something is not for sale affirms, enhances, and protects a thing's value in a number of ways. To begin with, the statement is a way of showing that a thing is valued for its own sake, whereas selling a thing for money demonstrates that it was valued only instrumentally. Furthermore, to say that something cannot be transferred in that way places it in the exceptional category—which requires the person interested in obtaining that thing to be able to offer something else that is exceptional, rather than allowing him the easier alternative of obtaining the thing for money that could have been obtained in an infinity of ways. This enhances its value. If I am willing to say "You're a really kind person" to whoever pays me to do so, my praise loses the value that attaches to it from being exchangeable only for an act of kindness.

In addition, if we have already decided we value something highly, one way of stamping it with a cachet affirming its high value is to announce that it is "not for sale." Such an announcement does more, however, than just reflect a preexisting high valuation. It signals a thing's distinctive value to others and helps us persuade them to value the thing more highly than they otherwise might. It also expresses our resolution to safeguard that distinctive value. To state that something is not for sale is thus also a source of value for that thing, since if a thing's value is easy to affirm or protect, it will be worth more than an otherwise similar thing without such attributes.

If we proclaim that something is not for sale, we make a once-and-for-all judgment of its special value. When something is priced, the issue of its perceived value is constantly coming up, as a standing invitation to reconsider that original judgment. Were people constantly faced with questions such as "how much money could get you to give up your freedom of speech?" or "how much would you sell your vote for if you could?", the perceived value of the freedom to speak or the right to vote would soon become devastated as, in moments of weakness, people started saying "maybe it's not worth *so much* after all." Better not to be faced with the constant questioning in the first place. Something similar did in fact occur when the slogan "better red than dead" was launched by some pacifists during the Cold War. Critics pointed out that the very posing of this stark choice—in effect, "would you *really* be willing to give up your life in exchange for not living under communism?"—reduced the value people attached to freedom and thus diminished resistance to attacks on freedom.

Finally, of some things valued very highly it is stated that they are "priceless" or that they have "infinite value." Such expressions are reserved for a subset of things not for sale, such as life or health. Economists tend to scoff at talk of pricelessness. For them, saying that something is priceless is to state a willingness to trade off an infinite quantity of all other goods for one unit of the priceless good, a situation that empirically appears highly unlikely. For most people, however, the word priceless is pregnant with meaning. Its value-affirming and value-protecting functions cannot be bestowed on expressions that merely denote a determinate, albeit high, valuation. John Kennedy in his inaugural address proclaimed that the nation was ready to "pay any price [and] bear any burden . . . to assure the survival and the success of liberty." Had he said instead that we were willing to "pay a high price" or "bear a large burden" for liberty, the statement would have rung hollow. . . .

An objection that advocates of cost-benefit analysis might well make to the preceding argument should be considered. I noted earlier that, in cases where various non-utility-based duties or rights conflict with the maximization of utility, it is necessary to make a deliberative judgment about what act is finally right. I also argued earlier that the search for commensurability might not always be a desirable one, that the attempt to go beyond expressing benefits in terms of (say) lives saved and costs in terms of dollars is not something devoutly to be wished.

In situations involving things that are not expressed in a common measure, advocates of cost-benefit analysis argue that people making judgments "in effect" perform cost-benefit calculations anyway. If government regulators promulgate a regulation that saves 100 lives at a cost of $1 billion, they are "in effect" valuing a life at (a minimum of) $10 million, whether or not they say that they are willing to place a dollar value on a human life. Since, in this view, cost-benefit analysis "in effect" is inevitable, it might as well be made specific.

This argument misconstrues the real difference in the reasoning processes involved. In cost-benefit analysis, equivalencies are established *in advance* as one of the raw materials for the calculation. One determines costs and benefits, one determines equivalencies (to be able to put various costs and benefits into a common measure), and then one sets to toting things up—waiting, as it were, with bated breath for the results of the calculation to come out. The outcome is determined by the arithmetic; if the outcome is a close call or if one is not good at long division, one does not know how it will turn out until the calculation is finished. In the kind of deliberative judgment that is performed without a common measure, no establishment of equivalencies occurs in advance. Equivalencies are not aids to the decision process. In fact, the decision-maker might not even be aware of what the "in effect" equivalencies were, at least before they are revealed to him afterwards by someone pointing out what he had "in effect" done. The decision-maker would see himself as simply having made a deliberative judgment; the "in effect" equivalency number did not play a causal role in the decision but at most merely reflects it. Given this, the argument against making the process explicit is the one discussed earlier in the discussion of problems with putting specific quantified values on things that are not normally quantified—that the very act of doing so may serve to reduce the value of those things.

My own judgment is that modest efforts to assess levels of benefits and costs are justified, although I do not believe that government agencies ought to sponsor efforts to put dollar prices on non-market things. I also do not believe that the cry for more cost-benefit analysis in regulation is, on the whole, justified. If regulatory officials were so insensitive about regulatory costs that they did not provide acceptable raw material for deliberative judgments (even if not of a strictly cost-benefit nature), my conclusion might be different. But a good deal of research into costs and benefits already occurs—actually, far more in the U.S. regulatory process than in that of any other industrial society. The danger now would seem to come more from the other side.

Notes and Questions

1. Does the rejection of the no-risk, and risk threshold approaches in the excerpt by Fischhoff, Lichtenstein, Slovic, Derby, and Keeney lead inevitably in the direction of cost-benefit analysis? Are there circumstances in which the level of costs, or the cost-risk trade-off, should just not matter?

2. Does the rejection by Fischhoff and colleagues of fixed value trade-offs between cost and risk imply a rejection of cost-benefit analysis? What decision framework do the authors appear to favor?

3. Consider the following instances of market regulation:
    (a) buying a more expensive house rather than one that is cheaper but closer to a hazardous waste site, to a highway, or to a mental hospital
    (b) buying a more expensive car that provides more protection in the event of a crash
    (c) paying more to avoid flying on a commuter airline
    (d) receiving a higher wage to take a job with higher workplace risks
Which, if any, of these forms of market regulation are legitimate? Which, if any, are illegitimate?

4. A standard limiting to one-in-a-million the lifetime probability of getting cancer from exposure to an environmental risk is generally considered stringent and is often advocated by environmentalist groups, whereas a one-in-ten-thousand standard is generally considered lax and is often advocated by industry groups. Why do these numbers so often frame the range for mainstream debate? For further discussion, see Rosenthal, Gray, and Graham, "Legislating Acceptable Cancer Risk from Exposure to Toxic Chemicals," 19 *Ecology Law Quarterly* 269 (1992).

5. In general, technology-based standards under the environmental statutes are set by reference to the best available technology that has been adequately demonstrated, taking costs into account. The consideration of costs has generally been taken to mean costs that a particular industry as a whole could bear and survive, even though particular firms might not survive. Is it sensible to regulate up to the point at which the industry as a whole is about to disappear? If not, what level of profit should the industry be able to keep? How should this malleability affect the assessment of technology-based standards?

6. Consider the following features of using technology-based standards as the means for determining the level of environmental protection:
    (a) More stringent standards would be imposed on more profitable industries.
    (b) More stringent standards would be imposed on industries that faced less foreign competition.
    (c) More stringent standards would be imposed on industries that manufactured products for which there were no substitutes.
Are these features desirable?

7. Compare the relative environmental protection that results from technology-based standards on the one hand and from cost-benefit analysis on the other hand. Which should environmentalists prefer? Which should labor unions prefer? Are the answers case specific? Why do environmentalists generally prefer technology-based standards?

8. If technology-based standards are the sole mechanism for determining the level of environmental quality, this level will vary depending on the rate of entry of new firms. Is this desirable? It is true that when new firms enter the market (or firms exit from the market), the technology-based standards could be altered. Frequent changes, however, would be undesirable because they could render worthless substantial investments in pollution control technology. Given this feature, should the setting of technology-based standards be regarded as a form of risk management, if risk management is understood to be the choice of the level of risk that society chooses to bear?

9. Under the various environmental statutes, technology-based standards are sometimes used to determine the level of environmental protection, as discussed by Lave, but in other instances are used as the means to allocate among various polluters the pollution control burden that is determined through the use of a different framework. For example, under the Clean Air Act, technology-based standards for automobiles, new stationary sources, and certain types of existing sources are used as the means of meeting the National Ambient Air Quality standards, which are nominally set by reference to the no-risk framework (these matters are discussed at greater length in Chapter IX). Is the evaluation of technology-based standards different in this latter context?

10. One commentator draws a distiction between media-quality-based approaches on the one hand, such as ambient air quality standards, and technology-based approaches on the other. See McGarity, "Media-Quality, Technology, and Cost-Benefit Balancing Strategies for Health and Environmental Regulation," *Law and Contemporary Problems,* Summer 1983, at 159. He suggests (p. 161) that

> the general preference of Congress and especially of implementing agencies for the technology-based approach . . . is warranted by an almost universal recognition that citizens of this country have a "right" to a healthy environment and workplace, at least insofar as the societal pursuit of that right is not technologically impossible or prohibitively expensive.

Is this position compelling?

11. Under a risk-risk framework, should the regulatory decision be simply to minimize the aggregate risk (in the Lave example, the sum of the risk of the carcinogenic food additive and the risk of botulism)? Alternatively, should man-made risks be weighted differently from natural risks? In this connection, consider the discussion of risk perceptions in Chapter III. Is this example one in which market regulation, coupled with informational requirements, would do best? For example, the food product could be classified as organic or nonorganic and the relative risks could be disclosed. For an illuminating discussion of this framework, see John D. Graham and Jonathan Baert Wiener, eds., *Risk vs. Risk: Tradeoffs in Protecting Health and the Environment* (Cambridge, Mass.: Harvard University Press, 1995).

12. Under an indirect risk-risk framework, how should the health of workers be traded off against the health of consumers? Consider the relevance of the following factors:
   (a) whether workers are receiving a higher wage to reflect the additional risk to which they are exposed
   (b) whether consumers are sufficiently informed of the risks to make informed decisions
   (c) whether there are less risky substitutes for the product

13. The risk-benefit framework discussed by Lave typically balances the risks that remain under different levels of regulation with the resulting costs. Therefore risk-cost, or

cost-risk, might be a more appropriate term. This term would also be more parallel to cost-benefit, which is simply a special case in which the benefits of regulation are quantified. Under the risk-benefit framework, how should one decide whether it is worth spending $100 million to save thirty lives, or to restore fish to a polluted river?

14. Are the cost effectiveness and regulatory budget approaches really risk management frameworks, if risk management is understood to be the process by which society chooses the risk that it wishes to bear? Under the cost effectiveness approach, on what basis should the underlying objective be chosen? Under a regulatory budget approach, how should the budget be set?

15. Why does Lave believe that benefit-cost analysis is a tool for defending the status quo? If this criticism is valid, how could this bias be eliminated?

16. Evaluate the four criteria advanced by Lave to make comparative decisions among the different frameworks. Are other criteria relevant as well?

17. Compare the various risk management frameworks using Lave's criteria. Is one framework always favored? Is one framework always disfavored? For further discussion of alternative decisional frameworks, see Smith, "A Conceptual Overview of the Foundations of Benefit-Cost Analysis," in Judith D. Bentkover, Vincent T. Covello, and Jeryl Mumpower, eds., *Benefits Assessment: The State of the Art* (Dordrecht: Kluwer, 1986).

18. Consider the hybrid decisional frameworks of the federal environmental statutes, which are discussed in the introduction to this chapter. What is the justification for the approach of the Safe Drinking Water Act? Is this hybrid more or less stringent than what would result from a pure technology-based standard with the same definition of feasibility?

19. Under the hazardous air pollutant provisions of the Clean Air Act, if a one-in-a-million probability is considered the desirable goal, what is the justification for not weakening the technology-based standards if they would result in a lower probability of cancer?

20. Compare the Clean Air Act's approach to hazardous air pollutants to that of the Safe Drinking Water Act. Under the former, a technology-based standard constitutes the first stage in the regulatory process; a health-based standard is the second step, which does not become effective until years later. In contrast, under the latter, the starting point is the health-based standard; at the second stage in the regulatory process, the enforceable, technology-based standard is promulgated. What factors might account for this different structure? For further discussion of risk management standards under the environmental statutes, see Green, "The Role of Congress in Risk Management," 16 *Environmental Law Reporter* 10,220 (1986); Rosenthal, Gray, and Graham, supra.

21. What might explain the multiplicity of risk management frameworks under the federal environmental laws? Would some standardization be desirable?

22. Assess the strength of Kelman's critique of cost-benefit analysis. What decisional framework do you think that he favors? Other critiques of cost-benefit analysis, from different perspectives, include Kennedy, Cost-Benefit Analysis of Entitlement Problems: A Critique," 33 *Stanford Law Review* 387 (1981); Stewart, "Regulation in a Liberal State: The Role of Non-Commodity Values," 92 *Yale Law Journal* 1537 (1983).

23. Is Kelman's primary objection to the monetization of environmental benefits? Is so, should he favor the risk-benefit (risk-cost) approach? Under this approach, how would he decide whether a particular expenditure is worthwhile if it saves a certain number of lives?

24. Should a regulatory system attempt to value human life? If not, on what basis should it decide to regulate an environmental risk?

Even if human life is not explicitly valued in the regulatory process, one can determine what value was "imputed" to life, at least in cases in which there are no additional benefits to the environment, by dividing the cost of the regulation by the number of lives that it expects to save. Studies show that there is a wide disparity across regulatory programs, ranging from about $10,0000 to over $100 million. See Stephen Breyer, *Breaking the Vicious Circle: Toward Effective Risk Regulation* (Cambridge, Mass.: Harvard University Press, 1993). In light of these results, is it appropriate to insist on more stringent regulation in areas in which the imputed valuation is low and less stringent regulation in areas in which it is high? Is such an inquiry a disguised form of cost-benefit analysis? Is consistency desirable? In this connection, consider the factors, discussed in Chapter III, that affect risk perceptions. See Kornhauser, "The Value of Life," 38 *Cleveland State Law Review* 209 (1990). Why is Kelman not critical of implicit valuations?

25. Consider the willingness-to-pay approach to valuing lives. How does one determine that two jobs are identical in al respects except for the risk level? How does one determine the additional risk of one of the jobs? Is it likely to be an expert or lay assessment? Why is this issue significant? Studies show that individuals undervalue voluntary risks relative to involuntary risks (see the discussion in Chapter III). How does this phenomenon affect the valuations? Consider Kelman's criticism that individuals who take risky jobs are likely to have unusually low valuations of risk, either because of unusually weak aversions to risk or unusually strong constraints on their choices.

For discussion of the willingness-to-pay procedure, see Viscusi, *The Value of Risks to Life and Health*, 31 *Journal of Economic Literature* 1912 (1993); Zeckhauser, "Procedures for Valuing Lives," 23 *Public Policy* 419 (1975).

26. While the concept of existence values is now well established in the economic literature, consider the following attack:

> Perhaps the greatest conceptual problem with existence value is deciding which goods and services . . . of the world have an existence value. For example, why not expand the concept to continued . . . existence of the great American farmer. . . .
>
> The family farm is also just the tip of the iceberg for an expanded list of possible existence values. Other candidates include . . . coal mining jobs in West Virginia, automobile manufacturing jobs in Detroit, or jobs anywhere. Because of existence values, trade barriers may start to look more efficient as they generate large economic values by preserving the existence of readily identifiable American jobs. But then, of course, there is also an existence value to reducing poverty in Third World nations that may offset the existence value gains closer to home. (Rosenthal and Nelson, "Why Existence Values Should *Not* Be Used in Cost-Benefit Analysis," 11 *Journal of Policy Analysis and Management* 116, 118 [1992])

How would you respond? See Kopp, "Why Existence Values *Should* be Used in Cost-Benefit Analysis," 11 *Journal of Policy Analysis and Management* 123 (1992). If people really value jobs or the reduction of poverty, why should these factors not enter a cost-benefit analysis?

The procedure generally employed for valuing existence values, the contingent valuation methodology (CVM), is itself highly controversial. The issues are summarized in Portney, "The Contingent Valuation Debate: Why Economists Should Care," 8 *Journal of Economic Perspectives* 3 (1994).

V

Distributional Consequences of Environmental Policy

The preceding chapters have concerned themselves with the aggregate benefits of environmental policy rather than with the distribution of those benefits. Here, we turn to the analysis of distributional issues. The distributional question is posed most starkly by the use of cost-benefit analysis. A policy that maximizes net benefits across the whole population might nonetheless impose significant net costs on a subset of this population. The traditional economic perspective would not advocate modifying such a policy to reduce these inequities, preferring, instead, to deal with distributional questions through other policy instruments such as the tax system. For other perspectives, distributional issues are far more salient and would play an important role in the choice of the preferred policy.

The article by Robert Bullard sets forth the claims of the environmental justice movement, which is an increasingly strong and effective voice in the debates over the shape of environmental policy. Bullard claims that the poor and, primarily, persons of color are disproportionately affected by inadequate environmental quality. He asserts that "racism plays a key factor in environmental planning and decision-making," and that, as a result, communities of color in urban locations "face some of the worst environmental devastation in the nation." Bullard claims that the problem manifests itself in a variety of contexts, including the siting of waste disposal sites, the exposure of children in inner cities to high levels of lead, and the risks faced by workers in the oil, chemical, and nuclear industries. Bullard argues that the disparities in access to environmental quality cannot be explained solely by reference to income differences and that race plays an independent role.

The environmental justice movement has concentrated considerable efforts on attacking the siting of locally undesirable land uses (LULUs), particularly hazardous waste sites. In the 1980s, two influential studies by the United States General Accounting Office (GAO) and the United Church of Christ's Committee for Racial Justice (CRJ) documented that hazardous waste facilities are disproportionately located in areas in which the surrounding communities have a high proportion of people of color and the poor. The article by Vicki Been surveys those studies and questions whether the disparity is the product of disproportionate siting or of market dynamics. She notes that the GAO and CRJ studies both looked at the current demographic characteristics of the surrounding communities, rather than the characteristics at the time that the facilities were sited. It may be that the location of the LULU makes housing in the surrounding community less desirable and depresses housing markets. A disproportionate number of those who can afford to leave the neighborhood will then do so and will be replaced by individuals searching for relatively inexpensive housing. Thus, it is possible that, even if there was no disparity in the racial and economic composition of the community at the time of the siting, market dynamics will subsequently make the community disproportionately composed of people of color and the poor.

Another article by Vicki Been explores the meaning of fairness in connection with the siting of LULUs. She examines three different notions of fairness: fairness in the pattern of distribution, fairness as cost-internalization, and fairness as process. Fairness in the pattern of distribution could involve spreading the burdens of LULUs on a proportional basis over society as a whole by physically spreading the LULUs themselves among the different communities, by equalizing the probability that any community would be chosen as a location for the LULU, or by ensuring that any affected community is compensated. Alternatively, it could involve progressive siting, under which advantaged communities get more than their share of the burden. Fairness as cost-internalization requires that the generators of waste pay the full social cost that results from the disposal of this waste. Finally, fairness as process requires acceptable procedures for the distribution of burdens, but does not inquire about the substance of the allocations.

Finally, an article by Henry Peskin examines the distributional impact of the Clean Air Act (CAA) of 1970. The study allocates the national benefits of the legislation to regions; within a region, it assumes that each individual receives an equal share of the benefit. With respect to costs, the study initially allocates them to industries, governments, or households. It assumes that industrial costs are borne by families in proportion to their consumption of the product (the production of which caused the pollution that is regulated by the CAA); that government costs are allocated on the basis of each family's relative tax burden; and that household costs, which are primarily the costs of pollution controls for automobiles, are allocated proportionately to the number of vehicles that each family owns. Peskin shows that the costs of pollution control are spread out relatively evenly throughout the country, but that benefits are concentrated in highly polluted areas. Aggregating benefits and costs, Peskin finds that people of color are the only net gainers from the legislation, as a result of their lower average vehicle ownership, larger average family size, and disproportionate concentration in urban locations. With respect to income, he concludes that the net benefits of the CAA, which are negative for all groups, are neither progressively nor regressively distributed.

Anatomy of Environmental Racism
and the Environmental Justice Movement

ROBERT D. BULLARD

Communities are not all created equal. In the United States, for example, some communities are routinely poisoned while the government looks the other way. Environmental regulations have not uniformly benefited all segments of society. People of color (African Americans, Latinos, Asians, Pacific Islanders, and Native Americans) are disproportionately harmed by industrial toxins on their jobs and in their neighborhoods. These groups must contend with dirty air and drinking water—the byproducts of municipal landfills, incinerators, polluting industries, and hazardous waste treatment, storage, and disposal facilities.

Why do some communities get "dumped on" while others escape? Why are environmental regulations vigorously enforced in some communities and not in others? Why are some workers protected from environmental threats to their health while others (such as migrant farmworkers) are still being poisoned? How can environmental justice be incorporated into the campaign for environmental protection? What institutional changes would enable the United States to become a just and sustainable society? What community organizing strategies are effective against environmental racism? . . .

Environmental Racism

Racism plays a key factor in environmental planning and decisionmaking. Indeed, environmental racism is reinforced by government, legal, economic, political, and military institutions. It is a fact of life in the United States that the mainstream environmental movement is only beginning to wake up to. Yet, without a doubt, racism influences the likelihood of exposure to environmental and health risks and the accessibility to health care. Racism provides whites of all class levels with an "edge" in gaining access to a healthy physical environment This has been documented again and again.

Whether by conscious design or institutional neglect, communities of color in urban ghettos, in rural "poverty pockets," or on economically impoverished Native-American reservations face some of the worst environmental devastation in the nation. Clearly, racial discrimination was not legislated out of existence in the 1960s. While some significant progress was made during this decade, people of color continue to struggle for equal treatment in many areas, including environmental justice. Agencies at all levels of government, including the federal EPA, have done a poor job protecting people of color from the ravages of pollution and industrial encroachment. It has thus been an up-hill battle convincing white judges, juries, government officials, and policymakers that racism exists in environmental protection, enforcement, and policy formulation.

The most polluted urban communities are those with crumbling infrastructure, ongoing economic disinvestment, deteriorating housing, inadequate schools, chronic unemployment, a high poverty rate, and an overloaded health-care system. Riot-torn South Central Los Angeles typifies this urban neglect. It is not surprising that the "dirtiest" zip code in California belongs to the mostly African-American and Latino neighborhood in that part of the city. In the Los Angeles basin, over 71 percent of the African Americans and 50 percent of the Latinos live in areas with the most polluted air, while only 34 percent of the white population does. This pattern exists nationally as well. As researchers Wernette and Nieves note:

> In 1990, 437 of the 3,109 counties and independent cities failed to meet at least one of the EPA ambient air quality standards . . . 57 percent of whites, 65 percent of African Americans, and 80 percent of Hispanics live in 437 counties with substandard air quality. Out of the whole population, a total of 33 percent of whites, 50 percent of African Americans, and 60 percent of Hispanics live in the 136 counties in which two or more air pollutants exceed standards. The percentage living in the 29 counties designated as nonattainment areas for three or more pollutants are 12 percent of whites, 20 percent of African Americans, and 31 percent of Hispanics.

Income alone does not account for these above-average percentages. Housing segregation and development patterns play a key role in determining where people live. Moreover, urban development and the "spatial configuration" of communities flow from the forces and relationships of industrial production which, in turn, are influenced and subsidized by government policy. There is widespread agreement that vestiges of race-based decisionmaking still influence housing, education, employment, and criminal justice. The same is true for municipal services such as garbage pickup and disposal, neighborhood sanitation, fire and police protection, and library services. Institutional racism influences decisions on local land use, enforcement of environmental regulations, industrial facility siting, management of economic vulnerability, and the paths of freeways and highways.

People skeptical of the assertion that poor people and people of color are targeted for waste-disposal sites should consider the report the Cerrell Associates provided the California Waste Management Board. In their 1984 report, *Political Difficulties Facing Waste-to-Energy Conversion Plant Siting*, they offered a detailed profile of those neighborhoods most likely to organize effective resistance against incinerators. The policy conclusion based on this analysis is clear. As the report states: "All socioeconomic groupings tend to resent the nearby siting of major facilities, but middle and upper socioeconomic strata possess better resources to effectuate their opposition. Middle and higher socioeconomic strata neighborhoods should not fall within the one-mile and five-mile radius of the proposed site."

Where then will incinerators or other polluting facilities be sited? For Cerrell Associates, the answer is low-income, disempowered neighborhoods with a high concentration of nonvoters. The ideal site, according their report, has nothing to do with environmental soundness but everything to do with lack of social power. Communities of color in California are far more likely to fit this profile than are their white counterparts.

Those still skeptical of the existence of environmental racism should also consider the fact that zoning boards and planning commissions are typically stacked with white developers. Generally, the decisions of these bodies reflect the special interests of the individuals who sit on these boards. People of color have been systematically excluded from these decisionmaking boards, commissions, and governmental agencies (or allowed only token representation). Grassroots leaders are now demanding a shared role in all the decisions that shape their communities. They are challenging the intended or unintended racist assumptions underlying environmental and industrial policies. . . .

Beyond the Race Versus Class Trap

Whether at home or abroad, the question of who *pays* and who *benefits* from current industrial and development policies is central to any analysis of environmental racism. In the United States, race interacts with class to create special environmental and health vulnerabilities. People of color, however, face elevated toxic exposure levels even when social class variables (income, education, and occupational status) are held constant. Race has been found to be an independent factor, not reducible to class, in predicting the distribution of (1) air pollution in our society; (2) contaminated fish consumption; (3) the location of municipal landfills and incinerators; (4) the location of abandoned toxic waste dumps; and (5) lead poisoning in children.

Lead poisoning is a classic case in which race, not just class, determines exposure. It affects between three and four million children in the United States—most of whom are African Americans and Latinos living in urban areas. Among children five years old and younger, the percentage of African Americans who have excessive levels of lead in their blood far exceeds the percentage of whites at all income levels.

The federal Agency for Toxic Substances and Disease Registry found that for families earning less than $6,000 annually an estimated 68 percent of African-American children had lead poisoning, compared with 36 percent for white children. For families with incomes exceeding $15,000, more than 38 percent of African-American children have been poisoned, compared with 12 percent of white children. African-American children are two to three times more likely than their white counterparts to suffer from lead poisoning independent of class factors.

One reason for this is that African Americans and whites do not have the same opportunities to "vote with their feet" by leaving unhealthy physical environments. The ability of an individual to escape a health-threatening environment is usually correlated with income. However, racial barriers make it even harder for millions of African Americans, Latinos, Asians, Pacific Islanders, and Native Americans to relocate. Housing discrimination, redlining, and other market forces make it difficult for millions of households to buy their way out of polluted environments. For example, an affluent African-American family (with an income of $50,000 or more) is as segregated as an African-American family with an annual income of $5,000. Thus, lead poisoning of African-American children is not just a "poverty thing."

White racism helped create our current separate and unequal communities. It defines the boundaries of the urban ghetto, *barrio*, and reservation, and influences the provision of environmental protection and other public services. Apartheid-type housing and development policies reduce neighborhood options, limit mobility, diminish job opportunities, and decrease environmental choices for millions of Americans. It is unlikely that this nation will ever achieve lasting solutions to its environmental problems unless it also addresses the system of racial injustice that helps sustain the existence of powerless communities forced to bear disproportionate environmental costs.

Locally Undesirable Land Uses in Minority Neighborhoods: Disproportionate Siting or Market Dynamics?
VICKI BEEN

The environmental justice movement contends that people of color and the poor are exposed to greater environmental risks than are whites and wealthier individuals. The movement charges that this disparity is due in part to racism and classism in the siting of environmental risks, the promulgation of environmental laws and regulations, the enforcement of environmental laws, and the attention given to the cleanup of polluted areas. To support the first charge—that the siting of waste dumps, polluting factories, and other locally undesirable land uses (LULUs) has been racist and classist—advocates for environmental justice have cited more than a dozen studies analyzing the relationship between neighborhoods' socioeconomic characteristics and the number of LULUs they host. The studies demonstrate that those neighborhoods in which LULUs are located have, on average, a higher percentage of racial minorities and are poorer than non-host communities.

That research does not, however, establish that the host communities were disproportionately minority or poor at the time the sites were selected. Most of the studies compare the *current* socioeconomic characteristics of communities that host various LULUs to those of communities that do not host such LULUs. This approach leaves open the possibility that the sites for LULUs were chosen fairly, but that subsequent events produced the current disproportion in the distribution of LULUs. In other words, the research fails to prove environmental justice advocates' claim that the disproportionate burden poor and minority communities now bear in hosting LULUs is the result of racism and classism in the *siting process* itself.

In addition, the research fails to explore an alternative or additional explanation for the proven correlation between the current demographics of communities and the likelihood that they host LULUs. Regardless of whether the LULUs originally were sited fairly, it could well be that neighborhoods surrounding LULUs became poorer and became home to a greater percentage of people of color over the years following the sitings. Such factors as poverty, housing discrimination, and the location of jobs, transportation, and other public services may have led the poor and racial minorities to "come to the nuisance"—to move to neighborhoods that host LULUs—because

Reprinted by permission of the author, The Yale Law Journal Company, and Fred B. Rothman & Company from *The Yale Law Journal*, vol. 103, pp. 1383–1422 (1994).

those neighborhoods offered the cheapest available housing. Despite the plausibility of that scenario, none of the existing research on environmental justice has examined how the siting of undesirable land uses has subsequently affected the socioeconomic characteristics of host communities. Because the research fails to prove that the siting process causes any of the disproportionate burden the poor and minorities now bear, and because the research has ignored the possibility that market dynamics may have played some role in the distribution of that burden, policymakers now have no way of knowing whether the siting process is "broke" and needs fixing. Nor can they know whether even an ideal siting system that ensured a perfectly fair initial distribution of LULUs would result in any long-term benefit to the poor or to people of color. . . .

Market Dynamics and the Distribution of LULUs

The residential housing market in the United States is extremely dynamic. Every year, approximately 17 percent to 20 percent of U.S. households move to a new home. Some of those people stay within the same neighborhood, but many move to different neighborhoods in the same city, or to different cities. Some people decide to move, at least in part, because they are dissatisfied with the quality of their current neighborhoods. Once a household decides to move, its choice of a new neighborhood usually depends somewhat on the cost of housing and the characteristics of the neighborhood. Those two factors are interrelated because the quality of the neighborhood affects the price of housing.

The siting of a LULU can influence the characteristics of the surrounding neighborhood in two ways. First, an undesirable land use may cause those who can afford to move to become dissatisfied and leave the neighborhood. Second, by making the neighborhood less desirable, the LULU may decrease the value of the neighborhood's property, making the housing more available to lower income households and less attractive to higher income households. The end result of both influences is likely to be that the neighborhood becomes poorer than it was before the siting of the LULU.

The neighborhood also is likely to become home to more people of color. Racial discrimination in the sale and rental of housing relegates people of color (especially African-Americans) to the least desirable neighborhoods, regardless of their income level. Moreover, once a neighborhood becomes a community of color, racial discrimination in the promulgation and enforcement of zoning and environmental protection laws, the provision of municipal services, and the lending practices of banks may cause neighborhood quality to decline further. That additional decline, in turn, will induce those who can leave the neighborhood—the least poor and those least subject to discrimination—to do so.

The dynamics of the housing market therefore are likely to cause the poor and people of color to move to or remain in the neighborhoods in which LULUs are located, regardless of the demographics of the communities when the LULUs were first sited. As long as the market allows the existing distribution of wealth to allocate goods and services, it would be surprising indeed if, over the long run, LULUs did not impose a disproportionate burden upon the poor. And as long as the market

discriminates on the basis of race, it would be remarkable if LULUs did not eventually impose a disproportionate burden upon people of color.

By failing to address how LULUs have affected the demographics of their host
communities, the current research has ignored the possibility that the correlation
between the location of LULUs and the socioeconomic characteristics of neighborhoods may be a function of aspects of our free market system other than, or in addition to, the siting process. It is crucial to examine that possibility. Both the justice of
the distribution of LULUs and the remedy for any injustice may differ if market
dynamics play a significant role in the distribution.

If the siting process is primarily responsible for the correlation between the
location of LULUs and the demographics of host neighborhoods, the process may be
unjust under current constitutional doctrine, at least as to people of color. Siting
processes that result in the selection of host neighborhoods that are disproportionately poor (but not disproportionately composed of people of color) would not be
unconstitutional because the Supreme Court has been reluctant to recognize poverty
as a suspect classification. A siting process motivated by racial prejudice, however,
would be unconstitutional. A process that disproportionately affects people of color
also would be unfair under some statutory schemes and some constitutional theories
of discrimination.

On the other hand, if the disproportionate distribution of LULUs results from
market forces which drive the poor, regardless of their race, to live in neighborhoods
that offer cheaper housing because they host LULUs, then the fairness of the distribution becomes a question about the fairness of our market economy. Some might
argue that the disproportionate burden is part and parcel of a free market economy
that is, overall, fairer than alternative schemes, and that the costs of regulating the
market to reduce the disproportionate burden outweigh the benefits of doing so.
Others might argue that those moving to a host neighborhood are compensated
through the market for the disproportionate burden they bear by lower housing costs,
and therefore that the situation is just. Similarly, some might contend that while the
poor suffer lower quality neighborhoods, they also suffer lower quality food, housing, and medical care, and that the systemic problem of poverty is better addressed
through income redistribution programs than through changes in siting processes.

Even if decisionmakers were to agree that it is unfair to allow post-siting market
dynamics to create disproportionate environmental risk for the poor or minorities,
the remedy for that injustice would have to be much more fundamental than the
remedy for unjust siting *decisions*. Indeed, if market forces are the primary cause of
the correlation between the presence of LULUs and the current socioeconomic characteristics of a neighborhood, even a siting process radically revised to ensure that
LULUs are distributed equally among all neighborhoods may have only a short-
term effect. The areas surrounding LULUs distributed equitably will become less
desirable neighborhoods, and thus may soon be left to people of color or the poor,
recreating the pattern of inequitable siting. Accordingly, if a disproportionate burden
results from or is exacerbated by market dynamics, an effective remedy might
require such reforms as stricter enforcement of laws against housing discrimination,
more serious efforts to achieve residential integration, changes in the processes of
siting low and moderate income housing, changes in programs designed to aid the

poor in securing decent housing, greater regulatory protection for those neighborhoods that are chosen to host LULUs, and changes in production and consumption
processes to reduce the number of LULUs needed.

Information about the role market dynamics play in the distribution of LULUs
would promote a better understanding of the nature of the problem of environmental injustice and help point the way to appropriate solutions for the problem.
Nonetheless, market dynamics have been largely ignored by the current research on
environmental justice.

The Evidence of Disproportionate Siting

Several recent studies have attempted to assess whether locally undesirable land uses
are disproportionately located in neighborhoods that are populated by more people of
color or are more poor than is normal. The most important of the studies was published in 1987 by the United Church of Christ Commission for Racial Justice (CRJ).
The CRJ conducted a cross-sectional study of the racial and socioeconomic characteristics of residents of the zip code areas surrounding 415 commercial hazardous
waste facilities and compared those characteristics to those of zip code areas which
did not have such facilities. The study revealed a correlation between the number of
commercial hazardous waste facilities in an area and the percentage of the "non-
white" population in the area. Areas that had one operating commercial hazardous
waste facility, other than a landfill, had about twice as many people of color as a percentage of the population as those that had no such facility. Areas that had more than
one operating facility, or had one of the five largest landfills, had more than three
times the percentage of minority residents as areas that had no such facilities.

Several regional and local studies buttress the findings of the nationwide CRJ
study. The most frequently cited of those studies, which is often credited for first
giving the issue of environmental justice visibility, was conducted by the United
States General Accounting Office (GAO). The GAO examined the racial and
socioeconomic characteristics of the communities surrounding four hazardous waste
landfills in the eight southeastern states that make up EPA's Region IV. The sites
studied include some of the largest landfills in the United States.

The results of the study are summarized in Table V.1. In short, three of the four
communities where such landfills were sited were majority African-American in
1980; African-Americans made up 52 percent, 66 percent, and 90 percent of the
population in those three communities. In contrast, African-Americans made up
between 22 percent and 30 percent of the host states' populations. The host communities were all disproportionately poor, with between 26 percent and 42 percent of
the population living below the poverty level. In comparison, the host states'
poverty rates ranged from 14 percent to 19 percent.

Another frequently cited local study was conducted by sociologist Robert Bullard
and formed important parts of his books, *Invisible Houston* and *Dumping in Dixie*.
Professor Bullard found that although African-Americans made up only 28 percent
of the Houston population in 1980, six of Houston's eight incinerators and mini-
incinerators and fifteen of seventeen landfills were located in predominantly African-
American neighborhoods. . . .

Table V.1.   Summary of GAO's Findings

	Population, % African- American	Mean Family Income		Population Below Poverty Level, %
Landfill		All Races	African- Americans	
Chemical Waste	90	$11,198	$10,752	42
SCA Services	38	16,371	6,781	31
Ind. Chem.	52	18,996	12,941	26
Warren Cty. PCB	66	10,367	9,285	32

[N]one of the . . . studies addressed the question of which came first—the people of color and the poor, or the LULU. As noted by the CRJ, the studies "were not designed to show cause and effect," but only to explore the relationship between the current distribution of LULUs and host communities' demographics. The evidence of disproportionate siting is thus incomplete: it does not establish that *the siting process* had a disproportionate effect upon minorities or the poor. . . .

Did the Siting Disparities Revealed by the GAO and Professor Bullard Result from Siting Practices, Market Dynamics or Both?

To begin to fill the gaps in the literature, this Part expands the GAO and Bullard studies described above. First, it adds to those studies data regarding the socioeconomic characteristics of the host communities at the time the siting decisions were made. Second, it traces changes in the demographics of the host communities since the sitings . . . .

The extensions of the GAO and Bullard studies, . . . show the effect of using demographic data from the census closest to the actual siting or capacity change decision (rather than the latest census data). Tracing changes in the demographics from this baseline reveals a significant difference in the evidence the studies provide regarding the burden LULUs impose on minorities and the poor. These studies suggest that the siting process bears some responsibility for the disproportionate burden waste facilities now impose upon the poor and people of color. The extension of the GAO study suggests that market dynamics play no role in the distribution of the burden. The extension of the Bullard study, on the other hand, suggests that market dynamics do play a significant role in that distribution.

The different results obtained by the two extensions may be attributable to the generally slower rate of residential mobility in rural areas, such as those hosting the GAO sites, versus urban areas, such as those hosting the Houston sites. The difference also may be attributable to the size and nature of the facilities studied in the two extensions. The sites studied in the GAO report are quite large, and provide a substantial number of jobs to residents of the host counties. Persons moving to the area to take those jobs may have displaced the African-Americans who previously lived in the community. The sites at issue in Professor Bullard's study, on the other hand, were unlikely to have created many new jobs, and those jobs that were created would have been much less likely than the jobs at the GAO sites to induce people to move nearby in order to take them.

Conclusion

Significant evidence suggests that LULUs are disproportionately located in neighborhoods that are now home to more of the nation's people of color and poor than
other neighborhoods. Efforts to address that disparity are hampered, however, by the
lack of data about which came first—the people of color and poor or the LULU. If
the neighborhoods were disproportionately populated by people of color or the poor
at the time the siting decisions were made, a reasonable inference can be drawn that
the siting process had a disproportionate effect upon the poor and people of color. In
that case, changes in the siting process may be required.

On the other hand, if, after the LULU was built, the neighborhoods in which
LULUs were sited became increasingly poor, or became home to an increasing percentage of people of color, the cure for the problem of disproportionate siting is
likely to be much more complicated and difficult. The distribution of LULUs would
then look more like a confluence of the forces of housing discrimination, poverty,
and free market economics. Remedies would have to take those forces into account.

The preliminary evidence derived from this extension of two of the leading studies of environmental justice . . . shows that research examining the socioeconomic
characteristics of host neighborhoods at the time they were selected, then tracing
changes in those characteristics following the siting, would go a long way toward
answering the question of which came first—the LULU or its minority or poor
neighbors. Until that research is complete, proposed "solutions" to the problem of
disproportionate siting run a substantial risk of missing the mark.

What's Fairness Got to Do With It?
Environmental Justice and the Siting of
Locally Undesirable Land Uses
VICKI BEEN

Fairness in the Pattern of Distribution

Fairness Requires Equal Division—A Per Capita or Proportional Distribution of the Burdens of LULUs

A broad conception of fairness in siting would require that a LULU's burdens be
spread on a per capita or proportional basis over society as a whole. This fairness
concept is implicit in the contention that LULUs are inequitably sited if the percentage of LULUs in minority neighborhoods is disproportionate to the percentage

of minorities in the nation's population. It is also inherent in the calls by the environmental justice movement demanding that people of color receive an "equitable distribution of 'healthy' physical environments" and that no neighborhood bear more than its proportionate share of LULUs. . . .

Several means of distribution are plausible under the proportional distribution of burden theory. One scheme would impose a *physical* proportional distribution: LULUs themselves would be distributed equally among neighborhoods. This distribution could be either equal *ex post* or equal *ex ante*. In an *ex post* scheme, the facilities and the harms that they pose would be distributed proportionately among neighborhoods. For example, if New York City requires facilities for 10,000 homeless individuals and has 100 neighborhoods, all holding some land suitable for a facility, each neighborhood would receive one facility housing 100 individuals. In an *ex ante* scheme, each neighborhood has an equal chance of being selected for the site through a lottery process. For example, if New York City requires a sewage sludge treatment plant, each of the 100 neighborhoods would have a 1/100 chance of being selected for the site. The *ex ante* physical distribution scheme is particularly well suited to situations in which there are economies of scale in building and operating fewer but larger LULUs. Some types of hazardous waste, for example, are stored most efficiently in large, centralized facilities. To accommodate such efficiency considerations, most neighborhoods should be spared the burden of having the facility nearby. A lottery procedure can ensure that although most neighborhoods will not have to host the site, all have an equal chance of being selected as the host site. The lottery accordingly achieves equality of opportunity before the actual distribution.

Instead of either *ex ante* or a *ex post* physical equality, a distribution might seek "compensated" equality. In this distribution scheme, all individuals or communities that gain a net benefit from a particular LULU must compensate those who suffer a net loss. For example, if a sludge treatment plant imposed costs upon a neighborhood, each of the neighborhoods that benefitted from the plant, but did not suffer the detriment of close proximity, would have to pay a proportionate share of the costs. Compensated equality can operate either *ex ante* or *ex post*. In an *ex ante* scheme, the siting neighborhood would be compensated for the expected loss that the site might inflict, even though the loss might never occur. In an *ex post* scheme, the siting neighborhood would be compensated only as injuries occurred. Compensation could be in the form of cash, neighborhood amenities, insurance, or indemnification. Compensation also could include "risk substitution," involving commitments to reduce some other burden borne by the community, such as a landfill developer's promise to clean up existing waste dumps. The amount and nature of the compensation would be determined either by a government authority, such as an administrative agency, or by the affected neighborhood itself. . . .

Fairness Requires Progressive Siting

One could argue that a fair distribution of LULUs would require advantaged neighborhoods to bear more of the burden that LULUs impose than poor and minority neighborhoods. Such a distribution could involve either a physical siting scheme in which advantaged neighborhoods receive a disproportionately greater number of LULUs or a compensated siting scheme in which advantaged neighborhoods pay a greater share of the cost of LULUs. One rationale for such "progressive" siting would be compensatory justice: advantaged neighborhoods should bear more of the LULU burden in order to redress or remedy past discrimination against poor and minority neighborhoods.

Although the compensatory argument for progressive siting is backward-looking, at least four forward-looking justifications exist. First, progressive siting may be necessary to achieve equality of results, or equal impact of the burdens of LULUs. Because residents of poor and minority neighborhoods suffer from numerous disadvantages, such as poor health and barriers to mobility, a LULU in a disadvantaged neighborhood will have a greater impact than one in a more advantaged community. Thus, to achieve the same level of impact, advantaged communities must bear a greater share of the burden of LULUs. The environmental justice movement's focus on tailoring risk assessments and regulatory activities to account for the special health risks faced by low income or minority communities reflects this argument.

Second, if the marginal utility of environmental quality declines as neighborhoods receive more environmental amenities, LULUs would impose less disutility upon advantaged neighborhoods than upon poor and minority neighborhoods. Thus, progressive siting would induce equal sacrifice from all neighborhoods and impose the least damage to society as a whole. Third, progressive siting could maximize total utility if putting more LULUs in advantaged neighborhoods encouraged society to reduce the number of LULUs it requires.

Fourth, John Rawls' difference principle might justify progressive siting. Rawls' theory of justice does not directly apply to siting controversies because it addresses the design of fair institutional structures, not the fairness of individual distributional choices. Nevertheless, on a micro-level the difference principle requires the siting process to yield the greatest benefit, or the least burden, to the least advantaged. A progressive siting scheme is justified, in other words, if such siting would be more like to improve the condition of the poorest members of society than one that either distributed the burden of LULUs equally or imposed the burden disproportionately upon the poor and minorities. . . .

Fairness Requires an Equal Initial Split and Competitive Bidding

A much different conception of fairness can be drawn from Ronald Dworkin's work on the meaning of equality. In exploring the ideal of equality of resources, Dworkin asks how resources should be distributed among shipwreck survivors washed up on a desert island. Dworkin's thought experiment is far removed from the problems of siting LULUs, but his analysis helps illuminate the notions of fairness inherent in several recent proposals that LULUs be "auctioned" among communities. Dworkin posits that equality of resources would result if each shipwreck survivor were assigned initially an equal share of clamshells that she could use to bid competitively for resources. He argues that a distribution is fair if no individual prefers the distribution that some other individual obtains. Competitive bidding among those with initially equal bidding currency will produce such a distribution.

Applied to the siting context, Dworkin's scheme requires that communities be given an equal number of bargaining chips with which to bid against LULUs. Society would decide which LULUs it would need for some period of time and put them on the auction block. Each community would then receive chips to buy its way out of particular LULUs—essentially a currency of vetos.

For example, suppose there are five communities of equal population and land area and there are fifteen LULUs, ranging from a home for juvenile delinquents to a low-level radioactive waste dump. Each community would be allocated ten veto chips. The auctioneer would begin by announcing that a particular LULU would be randomly distributed among the communities, but the highest bidder could remove itself from the eligible pool. If community A bids two chips to avoid it, B bids three, and C bids four, C "wins" the veto, paying four of its chips to disqualify itself from the selection process. Communities A, B, D, and E are still eligible, unless one wants to bid again. Suppose that A and E each again bid two, and B bids three; B is then eliminated, and so on, until none of the eligible communities wishes to veto the LULU. It is then randomly distributed among them. The process repeats for the next LULU. Once a community spends its ten veto chips, it is eligible for all remaining LULUs. The auction ends once all LULUs have been distributed.

If communities change their minds about the LULUs they prefer, they may trade their LULUs with other willing communities. Suppose that through the auction community A receives a large oil refinery and a prison, while community B receives a low-level radioactive waste storage facility. If community B believes that its geographical characteristics and the qualifications of its work force would enable it to host the oil refinery and prison more efficiently than the radioactive waste facility, community B could seek to trade with community A.

Under this conception of fairness, an auction for LULUs involves an equitable distribution of veto power. Proposals to hold reverse auctions, in which the siting agency or developer offers to pay a specific sum of money to the first community that agrees to accept the LULU, then increases the amount until some community steps forward with an acceptable site, are not fair under this conception because they take advantage of any inequalities of wealth that existed prior to the auction. . . .

Fairness as Cost-Internalization

Many environmental justice advocates argue that fairness requires those who benefit from LULUs to bear the cost of the LULUs. Forcing the internalization of costs leads to greater fairness in two ways. First, it is fairer to hold individuals responsible for their actions than to let costs fall on innocent bystanders. Second, forcing the internalization of costs results in greater efficiency, and greater efficiency is likely to mean fewer LULUs. Purchasers of products that generate waste will reduce consumption once the prices of the products reflect the true cost of waste facilities. In turn, producers will develop more efficient means of production, given the cost of disposing the waste generated. The number of LULUs will thereby decrease to the socially optimal level—the level at which the marginal utility of the product necessitating the LULU equals its costs. . . .

The cost-internalization conception of fairness requires that those who consume a product or make an allocation decision bear the product's or decision's full costs. To attain that goal, the manufacturer whose product generates the need for a LULU could compensate the host neighborhood for the full damages that the LULU inflicts, and then pass that cost forward to the consumers of the product. However, those who advocate cost-internalization as a means of combatting disproportionate siting generally oppose compensatory schemes. Instead, they either support programs to physically distribute LULUs more evenly across all neighborhoods or favor blocking all LULUs to encourage pollution prevention that would make LULUs unnecessary.

A physical distribution scheme would be less effective than compensation, however, at forcing the full internalization of the product's costs. It would be nonsensical to distribute one LULU to each consumer; therefore, a physical distribution of sites would always allow most consumers to avoid the full costs of their consumption decisions. Additionally, a physical distribution scheme would usually be both overinclusive and underinclusive: some residents of the host neighborhood might never consume the product in question, yet bear its burdens, while heavy users of the product outside the host neighborhood would bear no burden at all. . . .

Fairness as Process

Rather than focusing on the distribution of burdens to determine whether the siting process is equitable, the fairness as process theory focuses on the procedures by which the burden is distributed. The most obvious theory of fairness as process would assert that a distribution is fair as long as it results from a process that was agreed upon in advance by all those potentially affected. Although there are examples of interstate siting compacts and regional intrastate siting agreements, in which all participants voluntarily agree to a particular siting process, most LULUs are sited in communities that had no opportunity to remove themselves from the selection process. Therefore, this Section focuses on theories of fairness as process that do not rest upon voluntary agreement for their legitimacy.

Fairness Requires a Lack of Intentional Discrimination

A siting decision motivated by hostility toward people of a particular race is unfair under almost any theory of justice, and would not be considered fair under the Constitution. Under the intentional discrimination theory, fairness requires that a decision to site a LULU be made without any intent to disadvantage people of color. . . .

Fairness Requires Treatment as Equals

Even if discrimination is unintentional or based upon characteristics that do not trigger strict scrutiny under the Equal Protection Clause, disproportionate siting arguably would be inappropriate if it stemmed from a siting process that failed to treat people with "equal concern and respect," instead valuing certain people less than others. Under this theory, if a siting process is more attentive to the interests of wealthier or white neighborhoods than to the interests of poor or minority neighborhoods, that process illegitimately treats the poor and people of color as unequal.

Thus, if two potential sites were otherwise identical but one was in a poor neighborhood and one was in a wealthier neighborhood, society could not take note of the costs that the siting would impose on the wealthy, ignore the costs it would impose on the poor, and consequently site the LULU in the poor neighborhood. Nor could the siting decision consider the impact that the LULU would have on the poor, but discount that impact on the ground that the value of being free from certain kinds of risks or harms is worth less to poor people. Instead, the siting decision would have to consider the interests of the poor just as fully and sympathetically as it considered the interests of the more wealthy. If the decision-maker then concluded that both neighborhoods faced equal risk or loss, the choice between the two neighborhoods would have to be made with the flip of a coin or some other lottery mechanism.

If the two potential sites were not identical, treatment as equals would require only that the harm that a site would cause to the poor be considered in exactly the same manner as the harm that a site would cause to the more affluent. Thus, if siting the LULU in the poor neighborhood would expose five neighbors to a particular risk while siting it in the wealthier area would expose twenty-five neighbors to risk, society would be justified under this theory in choosing the site in the poor neighborhood. . . .

Environmental Policy and the Distribution of Benefits and Costs

HENRY M. PESKIN

It is hard to conceive of any federal environmental policy—or, indeed, any federal policy—that does not affect various people differently. If a policy were designed in such a way that all affected parties were made better off and none worse off, then the fact that some gained more than others would probably not be of great concern to the designers of policy. At least, this is what one might surmise from the policy designer's frequent emphasis on total benefits and costs. The fact that policies with benefit-cost ratios greater than unity have the *potential* of making everyone better off has apparently eased the conscience of many a policy maker. In the real world, however, such potential outcomes are rarely realized. Regardless of the total of benefits and costs, the usual state of affairs is that some parties gain while others lose.

Federal environmental policy is not exceptional in this respect. What may be exceptional about it is that there seems to be widespread political acceptance for the environmental policies we have adopted even though there is evidence that the number of losers may exceed the number of gainers. In this chapter we will first discuss why this disparity between gainers and losers is an expected consequence of the design of the

Reprinted by permission from Paul R. Portney, ed., *Current Issues in Environmental Policy*, Copyright © 1978, Resources for the Future.

policy. Then, using data from a study of the distributional consequences of the Clean Air Amendments of 1970, we provide evidence that substantiates these expectations. . . .

Uneven Benefits and Uneven Burdens

Our task of identifying who will benefit from and who will bear the costs of pollution control is difficult because there is no general agreement as to who are the actual beneficiaries. This lack of agreement is due to conflicting perceptions of the nature of pollution and environmental quality.

There are those who believe that environmental quality is a commodity provided equally to everyone and enjoyed by all members of society. According to this view, no individual member would be able to purchase more environmental quality than his neighbor even if he wished. If this view were adopted, one might conclude that environmental quality improvements would benefit everyone equally. Yet this conclusion does not necessarily follow. For, even if the *physical* quality of the environment had improved identically for all citizens, these physical improvements might not be valued equally by all of them. . . . [I]t is likely that those with more money might place a higher value on environmental quality improvements if only by virtue of the fact that they can pay more for them. For this reason, if one accepted the view that environmental quality is a purely public commodity, then it could be argued that the rich would benefit the most from environmental cleanup.

On the other hand, there is another view that emphasizes the variability of pollution across geographical locations. According to this view, households can determine the pollution to which they are exposed by choosing their residence. Households can thus, "purchase" environmental quality like any other private commodity by buying a home in an unpolluted neighborhood. However, since poorer households often have very little choice of residential location, they often end up in polluted neighborhoods where rents or property values are depressed. Therefore, under this latter view, while environmental quality is perceived to be more nearly a private commodity, adjusting the consumption of this commodity is difficult for many individuals, especially for those with low incomes.

Consequently, while the rich would appear to benefit most under the former, public-commodity view of environmental quality, the urban poor would clearly benefit most if environmental quality were viewed as "purchasable" through location. Therefore, conclusions regarding the distribution of environmental quality improvement depend critically on which view of environmental quality is adopted.

When one considers the costs of cleaning up pollution as well as the benefits, it becomes even more difficult to theorize about the distributional consequences. There are many ways to finance a pollution reduction program and each way can present its own pattern of financial burdens to members of society. A federal program financed with the federal income tax structure is likely to place a heavier burden on the rich (that is, be more progressive) than if the program were administered and financed by the states and local governments with their generally less progressive tax structures.

Actually, the more prominent environmental programs . . . are characterized by a combination of control strategies, with the initial financial burdens shared by several layers of government, industrial sectors, and households. The ultimate distributional consequences of these complex programs, however, depend on the way in which their initial financial burdens are shifted to the general population. In principle, this shifting depends on the tax structures of the various levels of government, the structure of the markets within which the industrial sectors operate, and the ownership of governmental debt and private capital.

One clear implication of these observations is that no realistic environmental policy is likely to have *uniform* distributional effects—that is, no policy will affect each and every person, family, or household in the same way. Moreover, the ultimate impact may be very difficult to predict. As a result, while a particular policy approach may at first appear equitable to the general public, its final distributional consequences might not be judged fair at all. For example, while it may seem fair (and it may make good economic sense) to require the steel industry to pay for its own cleanup, it is likely that these costs can easily be shifted to steel customers and ultimately to final consumers. To the extent that such a shifting is possible, the financing of the cleanup will take on the aspects of a general sales tax because steel is a major component of a large number of products destined for final consumption. What at first seemed fair ultimately may appear regressive and unfair, since the poor will bear a proportionately greater burden of the cost than the rich.

Of course, if it were also the case that the poor were the primary victims of the steel industry's pollution, they might receive proportionately larger benefits from the policy. Thus, when *net* benefits are considered, the total impact of the policy could be progressive and, again, judged fair.

To make final assessments of the distribution of benefits and costs requires a difficult type of empirical analysis. There are perhaps two unsatisfactory characteristics of such analysis. First, it will be almost impossible to draw general conclusions about all environmental policies since the analysis must be tailored to specific approaches, and the distributional effects of one pollution control strategy may easily differ from those of another. Second, the factors that determine the policy's ultimate effect on families are so complex that it is impossible to undertake the analysis without making a number of strong assumptions, many of which are disputable. . . .

Analysis of the Clean Air Amendments

. . . Because of the complexities of the legislation, an analysis of likely distributional consequences makes it necessary to estimate what will, in fact, be done (that is, what particular strategies will be adopted), as well as the costs and benefits of so doing. Fortunately this speculative exercise was largely accomplished by the Environmental Protection Agency (EPA) in their 1970 report to Congress. This report assumes that in addition to mobile-source emissions requirements, the ambient standards in the act will be attained by certain technological responses on the part of industry. EPA has estimated the costs of these responses as well as the dollar value of the likely benefits from air pollution control at the national level. . . .

Before discussing the details of our distribution techniques, we should point out the basic methodological approach. It is impossible to observe the actual impacts on millions of people of a policy that has yet to be fully implemented. Therefore, instead we "observe" the impacts on a random sample of the population of a policy that is *assumed* to be fully implemented. The random sample is the one-in-a-thousand 1970 Census Public Use File, a self-weighting sample of household and person records of the 1970 population census. Our approach is an example of a class of approaches known as microsimulation techniques. These techniques have become feasible only because of the ability of the modern computer to process rapidly extremely large data sets. In our case, we have had to process over 60,000 household and 500,000 individual records.

Benefit Estimates

National benefit totals were first distributed to regions known as Census County Groups and to subportions of these groups, known as Standard Metropolitan Statistical Areas (SMSAs). Each individual within these regions was assumed to receive an even share of a region's estimated benefit total. Thus, within a region, a family's benefit was determined solely by the number of its members.

The procedure for distributing the national benefit totals to regions, the assumptions behind the procedure, and its weaknesses are fully discussed in two recent studies. Here, we make only the following observations about our procedure. We basically assume that pollution damage (and, hence, the benefits from alleviating such damage) is proportional to local physical air quality or pollution concentrations. Hence, we reject the pure public-commodity view of environmental quality. Two factors that may have nonproportional influences on damage are also neglected. The first is the income of the damaged party. . . . Baumol and others have argued that the dollar valuation a person places on physical environmental damage can vary with the person's income; furthermore, there is reason to believe that the valuation increases as the individual becomes wealthier. Our analysis does not account for this. The second neglected factor is the possibility of "increasing returns" to bad health from pollution concentrations. There is reason to believe that when the air is rather clean, small increments in pollution concentrations will have small detrimental effects. However, increments of the same size can have very large detrimental effects when the air is very dirty.

Since we distributed a fixed national benefit total, neglect of these two factors may work in opposite directions. Ignoring income differences may imply that we underestimate the policy benefits to wealthier people relative to poor people. However, to the extent that the poor live in more highly polluted areas, and would benefit the most from improvements in very polluted air, the opposite is the case. Thus, the proportionality assumption may provide a good approximation of relative damages in spite of the neglect of the nonproportional influences of income and the potentially cumulative health–pollution concentration relationships. . . .

While it is not surprising that the residents of the heavily industrialized northeastern areas have the most to gain from pollution control, it is perhaps surprising that these benefits are concentrated in so few regions. Our analysis suggests that over 30 percent of total national benefits go to the residents of the five dirtiest SMSAs, although they account for but 8 percent of the U.S. population. As we shall see, costs appear to be much more widely distributed. Therefore, the concentration of benefits in a very few areas is the major determinant of all our subsequent results concerning the ultimate distributional consequences of the 1970 Clean Air Amendments on income classes and racial groups.

Cost Estimates

The costs of air pollution abatement . . . were distributed to families in one of three ways, depending on whether the initial incidence of costs fell on industries, governments, or households. Industrial costs—which we assume to be passed on in the form of higher product prices—were apportioned to families according to an estimate of their total consumption (which, in turn, was based on their income). The apportionment of governmental costs—assumed to be paid for by tax increases rather than reductions in other expenditures—was based on estimates of family tax burden. Household costs, which for the most part, are automobile pollution control costs, were apportioned on a per-vehicle-owned basis, one of the many household characteristics reported in the Public Use Sample.

No region of the United States can claim an overwhelming preponderance of vehicle ownership or families residing in a given income class. Therefore, average per-family costs are spread much more narrowly between the ten highest cost areas and the ten lowest cost areas than are the per-family benefits. . . . What differences do exist are explained by the fact that the high cost areas have relatively fewer low income families and high per-family automobile ownership.

Net Benefits

Net benefits are defined as the difference between per-family benefits and costs. When one calculates average net benefits per family for all 274 county groups and SMSAs, it appears that these average net benefits are positive for only a minority of these areas. For the air pollution policy as a whole, there appear to be only twenty-four areas (accounting for about 28 percent of the U.S. population) where average net benefits per family are positive. If the automobile policy is considered in isolation, only four areas (the Jersey City, New York, Paterson, and Newark SMSAs) show positive net benefits.

These results should not be surprising since we have seen that costs seem to be spread, geographically, far more evenly than benefits. The areas enjoying positive net benefits are those that suffer from a disproportionately large share of the nation's air pollution.

These results also imply that only a minority of the nation's population—principally those who live in the highly polluted areas—gain on net from the 1970 Clean Air Amendments. A more exact estimate of gainers and losers can be obtained by inspecting estimated benefits and costs for each family in the Public Use Sample. By this procedure, it is estimated that about 19 million families—or 29 percent of all families—enjoy positive net benefits. When the automobile policy is considered in isolation, only about 16 million families gain on net. It should be noted, however, that while net gainers appear to be in the minority, their gains on average greatly exceed the loss of net losers even though . . . total losses (costs) exceed total gains (benefits) by about $3 billion. . . .

Income Relationships

Table V.2 shows average benefits, costs, and net benefits by family income classes and by race. These data provide another view of the relative benefits and burdens of the air pollution control policy.

Table V.2.   Annual U.S. Per-Family Air Pollution Control Costs and Benefits by Income Class and Race ($)

Costs and Benefits	Income Class									
	Less than 3,000	3,000–3,999	4,000–5,999	6,000–7,999	8,000–9,999	10,000–11,999	12,000–14,999	15,000–19,999	20,000–24,999	More than 25,000
Industry and govt. costs										
Whites	59.79	81.90	102.63	130.47	155.90	177.56	210.49	248.17	299.87	457.29
Non-whites	59.79	81.90	102.63	130.47	155.90	177.56	210.49	248.17	299.87	457.29
Total	59.79	81.90	102.63	130.47	155.90	177.56	210.49	248.17	299.87	457.29
Industry and govt. benefits										
Whites	100.27	123.49	138.94	177.39	228.40	258.04	294.59	334.85	377.25	392.57
Non-whites	166.13	216.42	286.31	337.07	391.29	399.25	490.57	491.77	500.95	495.76
Total	112.57	138.76	160.33	195.48	242.19	267.16	306.99	343.27	382.75	395.22
Industry and govt. net benefits										
Whites	40.48	41.59	36.32	46.92	72.50	80.48	84.10	86.68	77.38	−64.72
Non-whites	106.34	134.52	183.68	206.60	235.39	221.69	280.08	243.60	201.08	38.47
Total	52.78	56.86	57.70	65.01	86.29	89.60	96.50	95.10	82.88	−62.07
Household costs (automobile)										
Whites	92.87	131.01	153.75	179.69	204.27	225.04	245.92	274.29	299.99	303.78
Non-whites	52.75	82.33	100.81	138.89	152.96	175.16	197.35	228.70	250.82	287.24
Total	85.25	123.01	146.03	175.04	199.91	221.82	242.83	271.83	297.81	303.35
Household benefits										
Whites	27.00	33.34	36.21	44.66	58.59	65.61	76.00	88.40	102.58	112.91
Non-whites	44.11	59.07	84.69	89.78	100.49	113.26	129.67	138.16	128.63	146.28
Total	30.25	37.56	43.28	49.79	62.14	68.69	79.41	91.08	103.74	113.77
Household net benefits										
Whites	−65.87	−97.67	−117.54	−135.03	−145.68	−159.43	−169.92	−185.89	−197.41	−190.87
Non-whites	−8.64	−23.26	−16.12	−49.11	−52.47	−61.90	−67.68	−90.54	−122.19	−140.96
Total	−55.00	−85.45	−102.75	−125.25	−137.77	−153.13	−163.42	−180.75	−194.07	−189.58
Total costs										
Whites	152.66	212.91	256.38	310.16	360.17	402.60	456.41	522.46	599.86	761.07
Non-whites	112.54	164.33	203.44	269.36	308.86	352.72	407.84	476.87	550.69	744.53
Total	145.04	204.91	248.66	305.51	355.81	399.38	453.32	520.00	597.68	760.64
Total benefits										
Whites	127.27	156.83	175.15	222.05	286.99	323.65	370.59	423.25	479.83	505.48
Non-whites	210.24	275.49	371.00	426.85	491.78	512.51	620.24	629.93	629.58	642.04
Total	142.82	176.32	203.61	245.27	304.33	335.85	386.40	434.35	486.49	508.99
Total net benefits										
Whites	−25.39	−56.08	−81.23	−88.11	−73.18	−78.95	−85.82	−99.21	−120.03	−255.59
Non-whites	97.70	111.16	167.56	157.49	182.92	159.79	212.40	153.06	78.89	−102.49
Total	−2.22	−28.59	−45.05	−60.24	−51.48	−63.53	−66.92	−85.65	−111.19	−251.65

It is apparent that the industrial air pollution control policy and the household (primarily automobile) policy have markedly different impacts when viewed in relation to income. Nearly all income and racial groups enjoy, on average, positive net benefits from the control of industrial air pollution. In contrast, no racial group or income class appears to gain from the control of automobile emissions. Except for those in the highest income class, non-whites appear to be the only net gainers from the 1970 Clean Air Amendments as a whole—a consequence of their lower average vehicle ownership (which lowers automobile control costs), their larger average family size (which increases per-family benefits), and their tendency to reside in more urbanized locations (areas that receive the greatest pollution reductions).

Total benefits appear to increase as income rises, implying that higher income groups suffer (prior to the policy implementation) greater absolute damage than the poor. This finding . . . may appear strange to those who picture the poor living in highly polluted urban slums. Although this characterization is no doubt valid, it must be kept in mind that the poorest of the poor actually live in relatively clean rural areas. . . .

If a family's share of pollution control costs is viewed as a tax, while the family's benefit is viewed as a subsidy, then the "progressivity" of this tax-subsidy package can be analyzed by dividing the cost and benefit totals of Table V.2 by the average incomes of the various income classes. These results are displayed in Table V.3.

Since average per-family cost as a percentage of income declines with income, the relative burden of the air pollution control policy clearly falls more heavily on the poor. In other words, the cost structure is "regressive." On the other hand, the benefits are "progressive"—the rich enjoy fewer benefits per dollar of income than the poor.

Table V.3. Average Incidence of Air Pollution Control Costs and Benefits by Income Class and Source for the United States (% of income)

	Income Class									
	Less than 3,000	3,000– 3,999	4,000– 5,999	6,000– 7,999	8,000– 9,999	10,000– 11,999	12,000– 14,999	15,000– 19,999	20,000– 24,999	More than 25,000
Costs										
Industry and government	3.4	2.4	2.1	1.9	1.7	1.6	1.6	1.4	1.4	1.1
Household	4.8	3.6	3.0	2.5	2.2	2.0	1.8	1.6	1.3	0.7
Total	8.2	6.0	5.1	4.4	3.9	3.6	3.4	3.0	2.7	1.8
Benefits										
Industry and government	6.3	4.0	3.2	2.8	2.7	2.4	2.3	2.0	1.7	1.0
Household	1.7	1.1	0.9	0.7	0.7	0.6	0.6	0.5	0.5	0.3
Total	8.0	5.1	4.1	3.5	3.4	3.0	2.9	2.5	2.2	1.3
Net benefits										
Industry and government	2.9	1.6	1.1	0.9	1.0	0.8	0.7	0.6	0.3	−0.1
Household	−3.1	−2.5	−2.1	−1.8	−1.5	−1.4	−1.2	−1.1	−0.8	−0.4
Total	−0.2	−0.9	−1.0	−0.9	−0.5	−0.6	−0.5	−0.5	−0.5	−0.5

Overall, net benefits, although negative for all classes, appear neither progressive nor regressive (indeed, they are proportionally distributed for incomes above $10,000). However, there is obviously a big difference between the burdens of the policy with respect to its industrial component as opposed to its household (primarily automobile) component. The net benefits of industrial air pollution control appear not only to be positive for almost all income groups but also progressively distributed. The net benefits of the household component, on the other hand, because of its large costs relative to benefits for all income classes, are both negative and regressive. . . .

Consistency with Distributive Goals

The analysis of the distributive impacts of the 1970 Clean Air Amendments indicates that those who gain when both benefits and costs are taken into account are more likely to be non-white, low income, inner-city residents. The law apparently serves to shift welfare toward a population group that is also a focus of a large body of explicit redistributive policy. It is quite likely that certain pro-environment congressmen were influenced by the image of the inner-city resident living with both the burdens of poverty and dirty air. To these congressmen, the redistributive character of the environmental policy is simply another point in its favor.

If this conjecture is correct, it does not bode well for certain aspects of water quality legislation or for certain suggested amendments to the Clean Air Act. That is, it has been estimated that 70 percent of the benefits of improved water quality will be in the nature of improved recreational opportunities. These benefits will be biased toward the wealthy since those in the higher income classes are the major users of water-based recreation. In a similar fashion, suggested policies to limit air quality degradation in already clean areas cannot help but dampen economic growth, and, regardless of how noble the objectives may be, such growth-limiting policies, unless accompanied by substantial redistribution efforts, will almost certainly harm the poor who have yet to attain high levels of wealth far more than the rich who already have. To the extent that these and other aspects of environmental policy are perceived as primarily benefiting the wealthy, they may lose the support of welfare-oriented congressmen.

Notes and Questions

1. Consider three different explanations for Bullard's claim that persons of color are disproportionately affected by inadequate environmental quality. First, it might be that before 1970, when environmental quality first began to be regulated in a comprehensive manner, the disparities were even greater than they are now and that the process of equalizing access to environmental quality is still ongoing. Second, it might be that environmental regulation has unintentionally exacerbated the disparities (a disparate impact claim). Third, it might be that environmental regulation has intentionally discriminated against poor and minority communities (an intentional discrimination claim). To which of these views does Bullard subscribe? Other representative works by environmental justice advocates include Cole, "Empowerment as the Key to Environmental Protection: The Need for Environmental Poverty Law," 19

Ecology Law Quarterly 619 (1992); Mohai and Bryant, "Environmental Injustice: Weighing Race and Class as Factors in the Distribution of Environmental Hazards," 63 *University of Colorado Law Review* 921 (1992).

2. Are Bullard's claims based primarily on the lack of political power or of economic power on the part of people of color? For discussion of the causes of environmental inequities, see Lazarus, "Pursuing 'Environmental Justice': The Distributional Effects of Environmental Protection," 87 *Northwestern University Law Review* 787 (1993).

3. Are racial and economic disparities in communities surrounding environmentally undesirable facilities problematic only if the risk resulting from the environmental exposure is serious? If so, what should be more relevant: the objective estimates of risk or the perceptions of risk on the part of the surrounding community? For further discussion of this issue, see Chapter III.

4. Does the siting of hazardous waste sites in poor or minority areas raise issues that are analytically different from the siting of highways or of homeless shelters? What are possible differences? Are these differences persuasive? Are the claims of the environmental justice movement simply a subset of claims about unfairness in the siting of any facility that a community might regard as undesirable, even if the effects of the facility are nonenvironmental?

5. Is it appropriate for the price of land to be considered in siting decisions? If so, is it inevitable that facilities will be disproportionately sited where land is cheapest? Are these areas likely to be disproportionately minority or poor?

6. Does the disparate siting of undesirable uses (environmental or nonenvironmental) raise issues that are analytically different from the disparate allocation of public amenities, such as higher quality schools or parks?

7. It is inevitable that income disparities will result in access to different qualities of housing. The following factors contribute to such disparities:
   (a) differences in the natural attractiveness of the location
   (b) differences in publicly provided amenities
   (c) differences in the allocation of nonenvironmentally undesirable uses
   (d) differences in the allocation of environmentally undesirable uses
Which of these differences in housing quality ought to be permissible?

8. Some facilities may bring economic benefits in the form of jobs for the surrounding community. What is the argument for basing environmental justice claims solely on costs without considering possible benefits?

9. Communities might also get compensated for accepting the siting of a facility. Should they be prevented from striking a deal that they consider desirable? If so, why? Who should represent the interests of the affected community in the negotiations? How should environmental justice statistics be presented in light of the possibility of compensation? By what means should one determine whether the community in fact got compensated?

10. Similarly, as was discussed in Chapter IV, workers generally get wage premiums for accepting more risky occupations. Does this form of compensation cure the environmental justice problem that would otherwise exist? Should environmental policy seek to reduce the disparities in exposure to risk even if the affected workers prefer the additional income? In a portion of his book that is not excerpted (p. 23), Bullard makes the following argument:

> Workers of color are especially vulnerable to job blackmail because of the greater threat of unemployment they face compared to whites and because of their concentration in low-paying, unskilled nonunionized occupations. . . . Fear of unemployment

acts as a potent incentive for many African-American workers to accept and keep jobs they know are health threatening.

How does this perspective affect the assessment of the adequacy of wage premiums or other forms of compensation?

11. The Commission of Racial Justice (CRJ) study found that race was more significant than income as a predictor of the location of hazardous waste sites. What factors might explain this phenomenon? Consider the following explanations:
   (a) differential political power
   (b) differential information
   (c) differential mobility
   (d) discrimination in housing

12. A recent study conducted by the Social and Demographic Research Institute (SADRI) at the University of Massachusetts-Amherst reached quite different conclusions. Examining the siting of the same facilities as CRJ, it found no significant difference in the percentage of African Americans in census tracts with commercial hazardous waste facilities, as compared to tracts without such facilities. There was, however, a marginally significant difference between the percentage of Latinos in host and non-host tracts (Anderton, Anderson, Oakes, and Fraser, "Environmental Equity: The Demographics of Dumping," 31 *Demography* 229 [1994]). Whereas the CRJ study had used zip codes as its unit of analysis, the SADRI study used census tracts, reasoning that they are intended to have a relatively homogeneous population. The SADRI study found, however, that while the host census tracts and the immediately adjoining census tracts did not have important racial disparities, tracts further from the facility but within a two-and-a-half mile radius contained significantly higher percentages of people of color and the poor than the remainder of the metropolitan area. What are possible explanations for this pattern? For a detailed overview of the evidence of disparate siting, see Been, "Environmental Justice and Equity Issues," in Patrick J. Rohan, *Zoning and Land Use Controls*, chap. 25D (New York: Matthew Bender, 1995). For analyses of how to define the affected neighborhood, see Fahsbender, Note, "An Analytical Approach to Defining the Affected Neighborhood in the Environmental Justice Context," 5 *New York University Environmental Law Review* (forthcoming 1996); Zimmerman, "Issues of Classification in Environmental Equity: How We Manage Is How We Measure," 21 *Fordham Urban Law Journal* 633 (1994).

13. There is also debate about whether there are racial and income disparities in the enforcement of the environmental laws. A study by the National Law Journal (NLJ) conducted in 1992 found significant disparities. For example, with respect to cleanups of hazardous waste sites under the Superfund statute, it found that sites in minority neighborhoods took longer to be listed on the National Priorities List (NPL), which would make them eligible for the expenditure of federal funds for remedial action, and that the cleanups of these sites were less extensive. See Lavelle and Coyle, "Unequal Protection: The Racial Divide in Environmental Law," *National Law Journal*, Sept. 21, 1992, at S2. In contrast, a more recent study, conducted by researchers at the University of Maryland, found no statistically significant racial disparities in the choice of cleanup levels. Unlike the NLJ study, the University of Maryland study controlled for other variables, such as whether the site is located in an urban or rural area, that might influence the choice of cleanup remedy (Gupta, Van Houtven, and Cropper, "Do Benefits and Costs Matter in Environmental Regulation? An Analysis of EPA Decisions Under Superfund," in Richard L. Revesz and Richard B. Stewart, eds., *Analyzing Superfund: Economics, Science, and Law* [Washington, D.C.: Resources for the Future, 1995], at 83). What role should environmental justice concerns play in the setting of enforcement priorities?

14. The environmental justice movement also contends that risk assessments systematically understate the risks that affect people of color and the poor. It makes two different types of claims: (1) that these individuals are more likely to be exposed to risk, principally as a result of multiple exposures and synergistic effects; and (2) that they are more susceptible to risk, in part as a result of genetic differences and social inequalities. For discussion, see Israel, "An Environmental Justice Critique of Risk Assessment," 3 *New York University Environmental Law Journal* 469 (1995). In what ways, if any, should the risk assessment process discussed in Chapter III be modified to address these concerns?

15. If the market dynamics hypothesis presented by Been were to explain an important part of the *current* disparities in the socioeconomic characteristics of host communities, what solutions are possible to reduce or eliminate these disparities? Would anything short of massive economic redistribution work?

16. Does the market dynamics hypothesis explain why communities surrounding hazardous waste sites are not only disproportionately poor but also, even after adjusting for income, disproportionately composed of people of color?

17. How does the market dynamics hypothesis affect the evaluation of compensation schemes? Who should get the compensation?
   (a) current owners of housing
   (b) current renters of housing
   (c) future owners of housing
   (d) future renters of housing
   (e) community groups

18. On February 11, 1994, President Clinton issued Executive Order No. 12,898, dealing with environmental justice. Most pertinently, this order provides:

> [E]ach agency shall develop an agency-wide environmental strategy . . . that identifies and addresses disproportionately high and adverse human health or environmental effects of its programs, policies, and activities on minority populations and low-income populations. The environmental justice strategy shall list programs, policies, planning and public participation processes, enforcement, and/or rulemaking related to human health or the environment that should be revised to, at a minimum: (1) promote enforcement of all health and environmental statutes in areas with minority populations and low-income populations; (2) ensure greater public participation; (3) improve research and data collection relating to the health of and environment of minority populations and low-income populations; and (4) identify differential patterns of consumption of natural resources among minority populations and low-income populations.

To what extent is the order likely to make a difference?

19. Several bills designed to promote environmental justice were introduced in Congress in the early 1990s, though none were passed. For example, the Environmental Justice Act, initially introduced by then Senator Albert Gore, defines Environmental High Impact Areas (EHIA) as "the 100 counties or appropriate geographical units with the highest total weight of toxic chemicals present during the course of the most recent 5-year period for which data is available." Any such area in which there are "significant adverse impacts of environmental pollution on human health," would benefit, with certain exceptions, from a moratorium on the siting of any new toxic chemical facility "shown to emit toxic chemicals in quantities found to cause significant adverse impacts on human health" (H.R. 2105, 103d Cong., 1st Sess. [1993]). How would you assess this approach? Consider the burden to the Environmental

Protection Agency (EPA) of identifying the 100 EHIAs. What challenges to these determinations do you foresee? Should the size and population of the county matter?

20. In contrast, the Environmental Equal Rights Act defines an "environmentally disadvantaged community" as an area within two miles of a site on which a solid waste facility is proposed in which two conditions are met: (1) the percentage of residents of color is greater than their percentage in that state or the United States as a whole, twenty percent or more of the population lives below the poverty line, *or* the per capita income is less than 80 percent of the national average; and (2) there already is a solid or hazardous waste facility, or certain other facilities that have an adverse impact on human health. The bill would allow any citizen of the state in which the proposed facility would be located to petition to prevent the issuance of a permit in an environmentally disadvantaged community. A permit would be denied if the facility "may adversely affect" human health or the environment (H.R. 1924, 103d Cong., 1st Sess. [1993]). How would you assess this approach? Does the bill contain a sensible definition of the affected class? How does the definition affect the siting of facilities in states that have percentages of residents of color that are far higher than the national average? Why is the bill concerned only with the siting of waste facilities, rather than with other polluting activities? Should the population density of the affected community matter? If the affected community favors a facility (perhaps because it will receive compensation), is it sensible to allow any citizen of the state (even one outside the community) to file an action? Is the problem, if any, cured by the requirements of Article III standing?

21. Compare the approaches of the Environmental Justice Act to the Environmental Equal Rights Act. In particular, focus on the following:
   (a) the geographic definition of affected area
   (b) the definition of the disadvantaged population
   (c) the procedures for keeping out proposed facilities

22. The Community Information Statement Act would require, as part of the permitting process for any offsite hazardous waste treatment or disposal facility, the preparation of a statement detailing the demographic characteristics of the host community and the impact of the facility on the community. The statement must be prepared by a contractor chosen jointly by the applicant for the permit and the chief elected official of the affected community. Assess the likely effectiveness of this approach. The bill requires that "the permitting authority . . . take the statement into account." In a judicial challenge to a permit decision, how should the court determine compliance with this requirement?

23. Evaluate the competing notions of fairness advanced by Been. Do these notions aid in the evaluation of changes brought about by market dynamics in the demographic composition of communities hosting LULUs?

24. Been writes, "[I]f New York City requires facilities for 10,000 homeless individuals and has 100 neighborhoods, all holding some land suitable for a facility, each neighborhood would receive one facility housing 100 individuals." Consider the relevance of the following:
   (a) Land in certain communities is a great deal cheaper than in others.
   (b) Certain communities offer better social services for the homeless.
   (c) The optimal size of each facility is 200 individuals.
   (d) The facilities impose less disutility in some communities than in others.
More generally, to what extent should fairness notions trump economic considerations?

25. How would one define what counts as a community for the purposes of an auction to allocate LULUs? How should a community's auction strategy be determined? Are the interests of all members of the community likely to be coextensive?

26. The internalization of the costs associated with waste disposal is achieved if the disposal charge reflects the expected harm of the waste. Is cost-internalization fair even if the community surrounding the site is not compensated for this harm? In contrast, if the community is compensated for this harm, does it matter if the funds are being disbursed by the disposer of the wastes? Should cost-internalization be seen as an independent notion of fairness?

27. Evaluate the fairness of a process that allocates LULUs on the basis of where they will decrease property values the least and under which, on the basis of unimpeachable studies, the LULUs always get allocated to poor communities.

28. The Peskin study shows that the benefits of pollution control increase with increasing income but that the corresponding costs increase even more. Is the environmental justice movement most interested in the distribution of benefits or net benefits? Which measure is more relevant?

29. The Peskin study shows that the greatest benefits of the Clean Air Act in its early years accrued to the dirtiest areas. It is likely, however, that, despite the improvement, these areas continue to be dirtier than the norm. In evaluating the environmental laws from a justice perspective, should the focus be on the changes that have taken place since their adoption or on the inequities that remain? What approach is taken by the Executive Order promulgated by President Clinton?

30. Peskin suggests that people of color are the greatest net beneficiaries of the Clean Air Act. They may, however, be disproportionately disadvantaged by other environmental programs. Is it appropriate to disaggregate the effects of the environmental laws and urge a massive restructuring of any program that has disproportionately negative effects on people of color or the poor? Why should the effects not be aggregated? Might it be appropriate to look even more broadly at the aggregate distributional consequences of all governmental programs, including taxes and transfer payments? For further discussion of the distributional effects of the Clean Air Act, see Gianessi, Peskin, and Wolff, "The Distributional Effects of Uniform Air Pollution Policy in the United States," 93 *Quarterly Journal of Economics* 281 (1979).

VI

The Choice of Regulatory Tools

This chapter explores the choice among different tools for improving environmental quality. Since the 1970s, command-and-control regulation has been the tool most used by the federal environmental statutes. Under this technique, polluting sources must meet a regulatory standard, typically set for categories of similar sources (for example, cement plants), by reference to what can be achieved through the use of "best available technology" (BAT). Thus, for example, under the Clean Air Act, BAT standards are set for categories of new stationary sources. Under the Clean Water Act, similar standards apply to effluent discharges by both new and existing sources.

The article by Bruce Ackerman and Richard Stewart argues forcefully that the emphasis on command-and-control regulation is misplaced. The authors raise two principal objections. First, firms that meet the command-and-control standard have no incentives to reduce their pollution further. Second, command-and-control regulation is generally insensitive to differences in the costs of pollution abatement. Thus, the same standard might apply to a source that can reduce its emissions cheaply as to one that can do so only at great expense. The cost of meeting a given level of environmental quality would be lower if the former source were regulated more stringently than the latter.

Ackerman and Stewart advocate the establishment of a system of marketable permits. Under such a scheme, the regulator would determine only the aggregate amount of permissible pollution and, by reference to this amount, define the number of permits to be allocated. Individual sources would have to possess a permit for each unit of pollution that they emitted. Initially, the permits would be auctioned off by the regulator, and subsequently, they would be traded in a market. The authors show that such a scheme satisfies the cost-effectiveness criterion: it leads to the achievement of a given level of environmental quality at least cost. This property holds regardless of whether the level of environmental quality is chosen to maximize social welfare or, instead, by reference to some other criterion. In the latter scenario, the use of a marketable permit scheme is "second best" in that, even though it does not lead to the maximization of social welfare, it provides the cheapest method for achieving the desired level of environmental quality.

The excerpt from the book by William Baumol and Wallace Oates explains that effluent fees (taxes per unit of pollution), like marketable permit schemes, can lead to the achievement of environmental standards at least cost. The authors compare the properties of these two tools. For example, marketable permits reduce the uncertainty about whether a given level of environmental quality will be met: with effective enforcement, that level is determined by the number of permits. In contrast, the environmental quality that results from a system of effluent fees depends on the extent to which individual firms reduce their pollution as a result of the imposition of the fees. Similarly, the effectiveness of marketable permit schemes is not eroded by economic growth or price inflation.

The excerpt from the book by Peter Bohm analyzes the use of deposit-refund systems, under which purchasers of products that are harmful to the environment must pay a deposit, which is refunded to them when the product is properly disposed. These schemes are a combination of a tax at the time of purchase and a subsidy at the time of return. Bohm explains that in certain instances deposit-refund systems are the only tool likely to be effective. For example, he notes that it is difficult to detect the improper disposal of small items that damage the environment, such as cadmium batteries. Thus, a ban or tax on improper disposal are unlikely to be effective tools. In contrast, if the tax and subsidy are set at a sufficiently high level, users of such batteries will have an economic incentive to dispose of them properly.

The article by Steven Shavell compares regulation and liability as means of controlling environmental risk. Regulation operates prospectively (or *ex ante*): a polluter must meet a given standard or face fines and other penalties. In contrast, liability operates retrospectively (or *ex post*): the polluter can engage in the activity in whatever way it wants but may be liable for any resulting harm. Regulation controls the activity directly, at least as long as enforcement is sufficiently vigorous and the penalties for noncompliance are sufficiently high. In contrast, liability controls the activity indirectly: Because an actor's expected liability is a function of the risk of its activities, liability creates incentives for the reduction of risk. Shavell identifies four factors as relevant for the choice between regulation and liability. First, liability is preferable if private actors have better information than regulatory authorities about an activity's risk. Second, regulation is preferable if the actors causing risk might have insufficient assets to pay for the resulting harm. Third, regulation is preferable if there is a sufficient possibility that suits will not be brought in the event of damage from the risky activity. Fourth, the relative administrative costs of the two schemes are relevant to the choice between them.

Finally, the excerpt from the book by Wesley Magat and W. Kip Viscusi analyzes hazard-warning schemes. The authors describe instances in which the provision of information might be more effective than direct regulation. For example, certain health-and-safety risks are a function of the decisions of both the manufacturer of a product, who decides how much to spend in order to reduce risk; and of the product's consumer, who decides how to operate the product. Thus, safety is likely to be increased if the consumer understands how to operate the product properly. Advocates of informational approaches to regulation also note that individuals have different preferences for risk. For example, some workers might prefer to take a relatively more risky job if it pays a sufficiently high wage premium, whereas others might make the opposite decision. Similarly, some consumers might be willing to pay a premium for a safer product whereas others might not. With accurate information about risk, workers and consumers can make the decisions that are best in light of their preferences.

Reforming Environmental Law

BRUCE A. ACKERMAN AND RICHARD B. STEWART

The Existing System

The existing system of pollution regulation. . . is primarily based on a Best Available Technology (BAT) strategy. If an industrial process or product generates some nontrivial risk, the responsible plant or industry must install whatever technology is available to reduce or eliminate this risk, so long as the costs of doing so will not cause a shutdown of the plant or industry. BAT requirements are largely determined through uniform federal regulations. Under the Clean Water Act's BAT strategy, the EPA adapts nationally uniform effluent limitations for some 500 different industries. A similar BAT strategy is deployed under the Clean Air Act for new industrial sources of air pollution, new automobiles, and industrial sources of toxic air pollutants. BAT strategies are also widely used in many fields of environmental regulation other than air and water pollution. . . .

BAT was embraced by Congress and administrators in the early 1970s in order to impose immediate, readily enforceable federal controls on a relatively few widespread pollutants, while avoiding widespread industrial shutdowns. Subsequent experience and analysis has demonstrated:

1. Uniform BAT requirements waste many billions of dollars annually by ignoring variations among plants and industries in the cost of reducing pollution and by ignoring geographic variations in pollution effects. A more cost-effective strategy of risk reduction could free enormous resources for additional pollution reduction or other purposes.

2. BAT controls, and the litigation they provoke, impose disproportionate penalties on new products and processes. A BAT strategy typically imposes far more stringent controls on new sources because there is no risk of shutdown. Also, new plants and products must run the gauntlet of lengthy regulatory and legal proceedings to win approval; the resulting uncertainty and delay discourage new investment. By contrast, existing sources can use the delays and costs of the legal process to burden regulators and postpone or "water-down" compliance. BAT strategies also impose disproportionate burdens on more productive and profitable industries because these industries can "afford" more stringent controls. This "soak the rich" approach penalizes growth and international competitiveness.

3. BAT controls can ensure that established control technologies are installed. They do not, however, provide strong incentives for the development of new, environmentally superior strategies, and may actually discourage their development. Such innovations are essential for maintaining long-term economic growth without simultaneously increasing pollution and other forms of environmental degradation.

4. BAT involves the centralized determination of complex scientific, engineering, and economic issues regarding the feasibility of controls on hundreds of thousands of pollution sources. Such determinations impose massive information-gathering burdens on administrators, and provide a fertile ground for complex litigation in the form of massive adversary rulemaking proceedings and protracted judicial review. Given the high costs of regulatory compliance and the potential gains from litigation brought to defeat or delay regulatory requirements, it is often more cost-effective for industry to "invest" in such litigation rather than to comply.

5. A BAT strategy is inconsistent with intelligent priority setting. Simply regulating to the hilt whatever pollutants happen to get on the regulatory agenda may preclude an agency from dealing adequately with more serious problems that come to scientific attention later. BAT also tends to reinforce regulatory inertia. Foreseeing that "all or nothing" regulation of a given substance under BAT will involve large administrative and compliance costs, and recognizing that resources are limited, agencies often seek to limit sharply the number of substances on the agenda for regulatory action.

This indictment is not idle speculation, but the product of years of patient study by lawyers, economists, and political scientists. There are, for example, no fewer than fifteen careful efforts to estimate the extra cost burden generated by a wide range of traditional legalistic BAT systems used to control a variety of air and water pollutants in different parts of the country. Of the twelve studies of different air pollutants—ranging from particulates to chlorofluorocarbons—seven indicated that traditional forms of regulation were more than 400 percent more expensive than the least-cost solution; four revealed that they were about 75 percent more expensive; one suggested a modest cost-overrun of 7 percent. Three studies of water pollution control in five different water sheds also indicate the serious inefficiency of traditional forms of command-and-control regulation. These careful studies of selected problems cannot be used to estimate precisely the total amount traditional forms of regulation are annually costing the American people. Nonetheless, very large magnitudes are at stake. Even if a reformed system could cut costs by "only" one-third, it could save more than $15 billion *a year* from the nation's annual expenditure of $50 billion on air and water pollution control alone. . . .

Implementation

A BAT system has an implicit environmental goal: achievement of the environmental quality level that would result if all sources installed BAT controls on their discharges. The usual means for implementing this goal are centralized, industry-uniform regulations that command specific amounts of cleanup from specific polluters. When a polluter receives an air or water permit under existing law, the piece of paper does not content itself, in the manner of Polonius, with the vague advice that he "use the best available technology." Instead, the permit tries to be as quantitatively precise as possible, telling each discharger how much of the regulated pollutants he may discharge.

. . . [O]ur reforms build upon, rather than abandon, this basic permit system. Indeed, we have only two, albeit far-reaching, objections to the existing permit mechanism. First, existing permits are free. This is bad because it gives the polluter no incentive to reduce his wastes below the permitted amount. Second, they are non-transferable. This is bad because polluter A is obliged to cut back his own wastes even if it is cheaper for him to pay his neighbor B to undertake the extra cleanup instead.

Our basic reform would respond to these deficiencies by allowing polluters to buy and sell each other's permits—thereby creating a powerful financial incentive for those who can clean up most cheaply to sell their permits to those whose treatment costs are highest. This reform will, at one stroke, cure many of the basic flaws of the existing command-and-control regulatory systems discussed earlier.

A system of tradeable rights will tend to bring about a least-cost allocation of control burdens, saving many billions of dollars annually. It will eliminate the disproportionate burdens that BAT imposes on new and more productive industries by treating all sources of the same pollutant on the same basis. It will provide positive economic rewards for polluters who develop environmentally superior products and processes. It will, as we show below, reduce the incentives for litigation, simplify the issues in controversy, and facilitate more intelligent setting of priorities.

Would allowing the sale of permits lead to a bureaucratic nightmare? Before proceeding to the new administrative burdens marketability will generate, it is wise to pause . . . to consider marketability's great administrative advantages.

First, marketability would immediately eliminate most of the information-processing tasks that are presently overwhelming the federal and state bureaucracies. No longer would the EPA be required to conduct endless adversary proceedings to determine the best available control technologies in each major industry of the United States, and to defend its determinations before the courts; nor would federal and state officials be required to spend vast amounts of time and energy in adapting these changing national guidelines to the particular conditions of every important pollution source in the United States. Instead of giving the job of economic and technological assessment to bureaucrats, the marketable rights mechanism would put the information-processing burden precisely where it belongs: upon business managers and engineers who are in the best position to figure out how to cut back on their plants' pollution costs. If the managers operating plant A think they can clean up a pollutant more cheaply than those in charge of plant B, they should be expected to sell some of their pollution rights to B at a mutually advantageous price; cleanup will occur at the least cost without the need for constant bureaucratic decisions about the best available technology. . . .

Second, marketable permits would open up enormous financial resources for effective and informed regulation. While polluters would have the right to trade their permits among themselves during the n years they are valid, they would be obliged to buy new ones when their permits expired at an auction held by the EPA in each watershed and air quality control region. These auctions would raise substantial sums of money for the government on a continuing basis. While no study has yet attempted to make global estimates for the United States as a whole, existing work suggests that auction revenues could well *equal* the amount polluters would spend in cost-minimizing control activities. Even if revenues turned out to be a third of this amount, the government would still be collecting more than $6 to 10 billion a year. Moreover, it seems reasonable to suppose that Congress would allow the EPA (and associated state agencies) to retain a substantial share of these revenues. Since the current EPA operating budget is $1.3 billion, using even a fraction of the auction fund to improve regulatory analyses, research, and monitoring would allow a great leap forward in the sophistication of the regulatory effort.... Given its revenue-raising potential, environmental reform is hardly a politically unrealistic pipe dream. To the contrary, it is only a matter of time before the enormous federal deficit forces Congress and the President to consider the revenue-raising potential of an auction scheme.

Third, the auction system would help correct one of the worst weaknesses of the present system: the egregious failure of the EPA and associated state agencies to enforce the laws on the books in a timely and effective way. Part of the problem stems from the ability of existing polluters to delay regulatory implementation by using legal proceedings to challenge the economic and engineering bases of BAT regulations and permit conditions. But agencies also invest so little in monitoring that they must rely on polluters for the bulk of their data on discharges. Since polluters are predictably reluctant to report their own violations, the current system perpetuates a Panglossian view of regulatory reality. For example, a General Accounting Office investigation of 921 major air polluters officially considered to be in compliance revealed 200, or 22 percent, to be violating their permits; in one region, the number not complying was 52 percent. Even when illegal polluters are identified, they are not effectively sanctioned: The EPA's Inspector General in 1984 found that it was a common practice for water pollution officials to respond to violations by issuing administrative orders that effectively legitimized excess discharges. Thus, while the system may, after protracted litigation, eventually "work" to force the slow installation of expensive control machinery, there is no reason to think this machinery will run well when eventually installed. Although there are many reasons for this appalling weakness in enforcement, one stands out above all others: The present system does not put pressure on agency policymakers to make the large investments in monitoring and personnel that are required to make the tedious and unending work of credible enforcement a bureaucratic reality.

The auction system would change existing compliance incentives dramatically. It would reduce the opportunity and incentive of polluters to use the legal system for delay and obstruction by finessing the complex BAT issues, and it would limit dispute to the question of whether a source's discharges exceeded its permits. It would also eliminate the possibility of using the legal system to postpone implementation of regulatory requirements by requiring the polluter that lost its legal challenge to pay for the permits it would have been obliged to buy during the entire intervening period of noncompliance (plus interest).

The marketable permit system would also provide much stronger incentives for effective monitoring and enforcement. If polluters did not expect rigorous enforcement during the term of their permits, this fact would show up at the auction in dramatically lower bids: Why pay a lot for the right to pollute legally when one can pollute illegally without serious risk of detection? Under a marketable permit approach, this problem would be at the center of bureaucratic attention. For if, as we envisage, the size of the budget available to the EPA and state agencies would depend on total auction revenues, the bureaucracy's failure to invest adequately in enforcement would soon show up in a potentially dramatic drop in auction income available for the next budgetary period. This is not a prospect that top EPA administrators will take lightly. Monitoring and enforcement will become agency priorities of the first importance. Moreover, permit holders may themselves support strong enforcement in order to ensure that cheating by others does not depreciate the value of the permit holders' investments. . . .

The reformed system we have described involves the execution of four bureaucratic tasks. First, the agency must estimate how much pollution presently is permitted by law in each watershed and air quality region. Second, it must run a system of fair and efficient auctions in which polluters can regularly buy rights for limited terms. Third, it must run an efficient title registry in each region that will allow buyers and sellers to transfer rights in a legally effective way. Fourth, it must consistently penalize polluters who discharge more than their permitted amounts.

And that's that. So far as the fourth bureaucratic task is concerned, we have already given reasons to believe that the EPA would enforce the law far more effectively under the new regime than it does at present. So far as the first three management functions are concerned, we think that they are, in the aggregate, far *less* demanding than those they displace under the BAT system.

Taking the three functions in reverse order . . . a system of title registration is within the range of bureaucratic possibility. In contrast, the second task—running fair and efficient auctions—is a complicated affair, and it is easy to imagine such a system run incompetently or corruptly. Nonetheless, other agencies seem to have done similar jobs in satisfactory fashions: If the Department of Interior can auction off oil and gas leases competently, we see no reason the EPA could not do the same for pollution rights. Finally, there remains the task of estimating the total allowable wasteload permitted under existing law in each watershed and air control region. If the BAT system functioned properly, these numbers would be easy to obtain. EPA's regional administrators would simply have to add up the allowed amounts appearing in the permits that are in their filing cabinets. We have no illusions, however, about present realities: So much bureaucratic time and energy has been diverted into the counterproductive factfinding tasks generated by the BAT system, and so little attention has been paid to actual discharges, that even the data needed for these simple arithmetic operations may well be incomplete and inadequate. Nonetheless, total permitted emissions in a region can be approximated in order to get a system of permits and auctions started. Surely this start-up effort would be less complex than the unending inquiries into available technologies required by the existing system. . . .

Would a system of marketable rights preclude improvement of environmental quality? By no means. The initial stock of rights can be amortized on a fixed schedule in order to reach a targeted goal, or the government may decide not to reissue existing rights after they expire. Any such reductions will increase the price of rights by reducing supply. Prices will also automatically tend to rise over time as the economy grows and the demand for rights increases. Under a BAT approach, by contrast, regulators must consistently undertake new, difficult, and unpopular initiatives to impose ever more stringent BAT controls on existing sources in order to accommodate economic growth without increased pollution. The prospect of steady increases in the price of rights will be a powerful incentive—far more powerful than the patchwork efforts at "technology forcing" under the BAT system—for businesses to develop cleaner products and processes.

A more serious objection to our proposal is that it ignores the problem of defining the region within which trades are permitted. The short answer is that the EPA and the States have already divided the nation into several hundred air quality control regions; similarly, the states have delineated the watershed boundaries for pollution control and other water management purposes. Rather than starting from scratch, our proposal can proceed on the basis of these existing boundaries. Especially in the area of air pollution, however, we have no doubt that existing regional lines have been drawn in a way that is extremely insensitive to ecological realities. We strongly recommend, therefore, that a reformed statute provide a mechanism for the orderly reexamination of existing regional boundaries— although it may well be wiser to defer this question for five or ten years to allow the EPA to concentrate on the challenges involved in managing the transition to a marketable permit system. . . .

[A] reformed implementation system would not easily solve all foreseeable regulatory problems. In particular, the market system we have described could allow the creation of relatively high concentrations of particular pollutants in small areas within the larger pollution control region. In tolerating "hot spots," of course, our reform proposal shares the defects of the existing BAT system, which also generates risk of "hot spots" by imposing the same controls on sources regardless of their location, the size of the human population affected by their discharges, and the nature and vulnerability of affected ecosystems. Nonetheless, the blindness of both systems to intraregional variation is a serious source of concern. The extensive literature on marketable permits . . . points to a variety of feasible means for dealing with the hot spot problem. We believe that a long-run strategy for institutional reform should strive to take advantage of these more sophisticated market solutions to the problem of intraregional variation. For the present, it will be enough to emphasize . . . that administrative feasibility is an important constraint on the degree of sophistication that we may reasonably expect.

The critical question here, however, is not whether our market reform fails to solve problems that the BAT system also fails to solve. It is whether the reformed implementation system will generate *new* problems that offset its great economic, environmental, and administrative advantages. . . . We can foresee situations in which existing polluters might try to manipulate the rights market to deter entry by new firms in a way that is inconsistent with the antitrust laws, by either monopolizing the pollution rights market itself, or using it to block entry by competitors. There is, however, a considerable literature in which problems like this are discussed, and we ourselves have worried about them.

The Theory of Environmental Policy
WILLIAM J. BAUMOL AND WALLACE E. OATES

Although both effluent fees and systems of marketable permits have the capacity to achieve a set of environmental standards at least cost, they are by no means equivalent policy instruments from the viewpoint of an environmental agency. We shall consider first the grounds on which the environmental authority might prefer such permits to fees and shall then turn to the case for fees.

The first, and a major, advantage of marketable permits over fees is that permits promise to reduce the uncertainty and adjustment costs involved in attaining legally required levels of environmental quality. The environmental authority cannot be completely sure of the response of polluters to a particular magnitude of an effluent charge; in particular, if the authority inadvertently sets the fee too low, environmental standards will not be met. . . . [T]he fee may have to be raised and then altered again to generate an iterative path converging toward the target level of emissions. This means costly adjustments and readjustments by polluters in their levels of waste discharges and the associated abatement technology. The need for repeated changes in the fee is also an unattractive prospect for administrators of the program. In contrast, under a permit scheme, the environmental agency directly sets the total quantity of emissions at the allowable standard; there is, in principle, no problem in achieving the target.

Second, and closely related to the issue just discussed, are the complications that result from economic growth and price inflation. Continuing inflation will erode the real value of a fee; similarly, expanding production of both old and new firms will increase the demand for waste emissions. Both of these will require the fee to be raised periodically if environmental standards are to be maintained. The burden of initiating such corrective action under a system of fees falls necessarily upon environmental officials; they are forced to choose between unpopular fee increases or nonattainment of standards. Under a system of permits, market forces automatically accommodate themselves to inflation and growth with no increase in pollution. The rise in demand for permits, real and nominal, simply translates itself directly into a higher price.

Third, the introduction of a system of effluent fees may involve enormous increases in costs to polluters *relative* to alternative regulatory policies. This point may seem somewhat paradoxical in light of the widespread recognition that systems of pricing incentives promise large savings in aggregate abatement costs. But the two are not inconsistent. Although a system of effluent charges will reduce total abatement costs, it will impose a new financial burden, the tax bill itself, on polluting firms. Although these taxes represent a transfer payment from the viewpoint of society, they are a cost of operation for the firm. Some recent evidence on this issue suggests that the figures can be rather staggering. One such study of the use of pricing incentives to restrict emissions of certain halocarbons into the atmosphere estimates that aggregate abatement costs under a realistic program of direct controls would total about $230 million; a system of fees or of marketable permits would reduce these costs to an estimated $110 million (a saving of roughly 50 percent). However, the cost of the fees or permits to polluters would total about $1,400 million so that, in spite of the substantial savings in abatement costs, a program of pricing incentives would, in this instance, increase the total cost to polluters by a factor of *six* relative to a program of direct controls! Some studies of other pollutants also suggest that fees can be a major source of new costs. It is true that a system of marketable permits *making use of an auction for the initial acquisition of these rights* is subject to the same problem, because sources face high prices for permits. However, there is an alternative that gets around the problem: A permit system can be initiated through a *free* initial distribution of the permits among current polluters. This version of the permit scheme effectively eliminates the added costs for existing firms without any necessarily adverse consequences for the efficiency properties of the program and with some obvious and major advantages for its political acceptability. It is interesting in this regard that existing systems of marketable permits in the United States embody a kind of "grandfathering" scheme involving an initial distribution of emission permits or "rights" among polluters based on historical levels of emissions.

Fourth, . . . there may be instances where geographical distinctions among polluters are important. In fact, for several important air and water pollutants, various studies indicate that it is imperative for the environmental authority to differentiate among polluters according to their location if environmental standards are to be realized in a cost-effective way. Sources at a highly polluted location within an air shed cannot be allowed to increase their emissions on a one-to-one basis in exchange for emissions reductions by other sources at a less-polluted point. As we have indicated, it can be administratively quite cumbersome to deal with the spatial problem under a system of effluent charges, for it will typically require the environmental agency to determine a separate effluent fee for each source, depending upon its location in the air shed or river basin (or alternatively, it will be necessary to introduce a system of zones with different charges). Such discrimination among sources in fee levels may either be explicitly illegal or politically infeasible. In contrast, a system of marketable permits can address these spatial dimensions of the pollution problem in a manner that is less objectionable.

Fifth, marketable permits may well be the more feasible approach on grounds of familiarity. The introduction of a system of effluent fees requires the adoption of a wholly new method of controlling pollution, new both to regulators and polluters. Such sharp departures from established practice are hard to sell; moreover, some real questions have been raised about the legality of charging for pollution. In contrast, permits already exist, and it may be a less-radical step to make these permits effectively marketable.

There is thus a strong case on administrative grounds for favoring marketable permits over effluent fees. But the case is far from ironclad. Where charges are feasible, they represent a most attractive source of revenues for the public sector. Most taxes in the economy have undesired side effects: they distort economic choices in various ways. Income taxes, for example, can induce individuals to choose untaxed leisure activities rather than work; excise taxes shift peoples' purchases away from the taxed goods; and so on. Such taxes generate an "excess burden" on the economy—a cost in addition to the reduced disposable income directly attributable to the revenues. Effluent fees, in contrast, have a beneficial side effect: They tend to correct distortions in the economy while at the same time generating public revenues. Such fees can be said to impose a "negative excess burden." Fees, then, to the extent they are feasible, are a very desirable source of public revenues in terms of economic efficiency. . . .

There is yet another argument favoring effluent fees—one that involves savings in certain transactions costs. A system of marketable emission permits requires an initial distribution of the permits. However, if this initial distribution is based on the grandfathering principle or some other mechanism that does not reflect the relative marginal abatement costs of the different sources, a series of transfers (purchases and sales) of permits will be required if the least-cost allocation is to be attained. The incentives for such transfers exist: Buyers who can reduce emissions only at a higher real cost will be willing to pay more than the reservation price of sellers. But there may well be significant search costs and elements of strategic behavior that impede the transfers of emissions entitlements that are necessary to achieve the least-cost outcome. In contrast, under a system of fees, no such transfers of permits are needed—each source simply responds directly to the incentive provided by the fee. It may thus prove easier in certain circumstances to attain the least-cost allocation of waste emissions under a set of fees than under a system of marketable permits.

Deposit-Refund Systems: Theory and Applications to Environmental, Conservation, and Consumer Policy

PETER BOHM

A deposit-refund system is essentially a combination of a tax and a subsidy. To attain objectives in areas such as environmental, conservation, and consumer policy, a refund (subsidy) is offered to consumers or producers, or both, in order to stimulate activities that otherwise would not have been undertaken or to guide existing activities in time and space. For example, refund offers can be introduced to avoid having used objects such as beverage containers, discarded automobiles, and waste lubricating oil dumped in places where they would be harmful in one way or another. Or refund offers can be used as an instrument for reducing waste management costs to society and for recovering certain materials from waste when there are economies of scale in recovery operations or lags in adjustment of disposal behavior.

To avoid extensive distributional and budgetary effects and to create adequate incentives, the refund offer is coupled with a deposit (tax) that is introduced at an earlier point in the chain of transactions. Thus in a complete deposit-refund system the deposit is refunded if certain conditions are met by the decision maker. A complete deposit-refund system can now be seen to have a wider field of application than was suggested by the examples just given. The refunding of a deposit if a specific condition is met by the depositor provides an instrument for protecting certain rights that the government may wish to transfer to the buyers of a product. For example, deposits (or similar arrangements) made by producers could protect consumers from producers who fail to honor contracts or warranties because of bankruptcy. The system could also be used to correct for market failure in servicing consumer durables or in the provision of spare parts. It could be applied to help protect people from hazards that could arise from production sites left unattended after shutdown of production. In addition, deposit-refund systems could even be used to protect consumers of public services from unlimited delays or unfulfilled promises of service delivery. . . .

In some cases a deposit-refund system may be the only possible policy solution. For example, it is hardly likely that the authorities could catch a significant number of those people who throw away small hazardous products such as cadmium batteries and thus damage the environment. Therefore, neither a ban nor a tax on improper disposal could be expected to work in this case. In contrast, a deposit on the sale of such batteries and a refund for properly returned batteries could be designed to provide appropriate incentives to protect the environment.

To cite another example, assume that 90 percent of a returnable commodity or a kind of scrap is returned without the support of any government initiative. To introduce a subsidy to increase this figure to, say, 100 percent would rarely be worthwhile, given the "costs" of having to finance the subsidy of those 90 percent already being returned. Thus, to get the remaining 10 percent of the units returned, the actual cost would be the social costs of the total volume of subsidies. This cost may be too high for whatever increased returns are desired, and it has consequently been used as an argument against such subsidies.

A deposit-refund system could accomplish the same thing that a subsidy would, but at a much lower social cost. Those who already return their units will pay and receive the same amount (an appropriate rate of interest may be paid to the depositor, of course), and so no cost other than that of temporary forced saving in the amount of the deposit will hit them. The cost to the people who are not already returning their units will be the inconvenience of using this disposal alternative (in addition to the probably negligible cost of forced saving). But it will not cost the government anything more than the administrative effort. Thus, if administrative costs and inconvenience costs are small enough, it may pay to have recovery rates increased in this way.

In other cases a deposit-refund system may be the only realistic way to introduce economic incentives; hence it may be the only alternative to regulation. Assume, for example, that government would like to consider making those who burn oil and release sulfur dioxide into the atmosphere pay for their sulfur emissions. A charge on all actual emissions would presumably be ruled out because of prohibitive measurement costs. In contrast, an economic incentive system could be introduced consisting of a deposit on the sulfur content of the fuel and a refund on the sulfur recovered. For firms to qualify for refunds at rates far above the market price for sulfur, the authorities may in some cases need to check actual oil purchases and verify the existence of sulfur recovery devices. The control costs would still be small in comparison with the costs of administering a tax on sulfur emissions.

As was pointed out earlier, deposit-refund systems may provide the same economic incentives as taxes or subsidies and at the same time avoid some of the disadvantages of these alternatives. That deposit-refund systems have the same incentive effects is clear from the fact that the deposit becomes a fee if the decision maker's behavior does not qualify for a refund. In certain applications, such systems may provide stronger or more well-focused incentives or involve a smaller amount of policy costs (administrative, enforcement, and information costs) than alternative solutions. For example:

1. If a commodity on which a deposit has been paid is disposed of in a way that does not qualify for a refund, someone else may take care of it and get the refund; this would not happen with regulations or fees on an improper disposal.
2. In a deposit-refund system the owner of a commodity has an incentive to prove that the commodity has not been disposed of in an improper fashion; in alternative systems the owner may have an incentive to hide the fact that it has been disposed of in an improper fashion.
3. In some cases it is simpler and less expensive to administer deposit-refund systems in which one is paid for choosing a certain kind of activity or disposal than systems in which one has to pay for alternative kinds of activity or disposal.
4. It may be simpler in some cases to formulate the conditions under which there is a refund than to state the conditions under which it is forbidden to dispose of the commodity or under which there is a fee for doing so.
5. By paying a deposit or by being told about the refund prospect by the seller as a sales argument, the buyer or user is informed about the conditional refund and thus about (maximum) liability; making similar information available and effective is usually quite costly under alternative systems.
6. The collection costs in deposit-refund systems may in some cases be lower than the corresponding costs under a regulatory system or a system of charges that, to be effective, may require extensive checking operations, prosecution, and so on (see the cadmium battery example mentioned earlier).

Thus policy costs may be lower for deposit-refund systems than for alternative solutions. In addition, budgetary effects of deposit-refund systems may be more attractive to policy makers. Whereas subsidies and regulations with high policy costs create a need for additional government funds, and charges or other allocative taxes may be disliked by administrators when they give rise to an unstable volume of government revenue, deposit-refund systems tend to leave the budget intact to the extent that refunds (paid by the government) approach the volume of deposits (directly or indirectly paid to the government).

Because distributional considerations play a fundamental role in economic policy, such aspects of deposit-refund systems should be discussed in some detail at this point. Let us focus on a deposit-refund system designed to influence disposal behavior in order to reach a given policy goal. In the case in which refunds are set at a level such that a maximum return rate is achieved—say, all beverage containers are returned for a refund by the original buyers—there will be no effects on the net *money income* distribution. In the case in which less than a maximum return rate is achieved, the return rate may differ among income groups; for example, it may be higher for households with a low opportunity cost for time, that is, for low-income households. This means that the deposit-refund system would have a progressive distributional effect for deposits on commodities with unitary income elasticity. In contrast, a product charge or any other policy alternative of an excise tax type—say, in the amount of the expected average negative environmental effect—would be proportional to income under the same circumstances. It would, moreover, definitely raise commodity prices and so, for example, definitely hit households at or below the poverty line. And it would hit those who buy inexpensive versions of the affected commodity harder, relatively speaking, than those who buy luxury versions. Apart from the effects on different income groups, product charges would not differentiate between "bad" and "good" behavior, such as littering and nonlittering, and thus would not be as equitable as a refund would, in the normative sense implied here. Finally, the deposit informs the buyer about different disposal options and their costs at the time of purchase, in contrast to a tax on a nonreturn option or a subsidy on a return option, neither of which it is relevant to observe until the time of disposal. This increase of the information level in the economy could benefit, in particular, households that are uninformed because of poor education.

Liability for Harm Versus Regulation of Safety

STEVEN SHAVELL

Liability in tort and the regulation of safety represent two very different approaches for controlling activities that create risks of harm to others. Tort liability is private in nature and works not by social command but rather indirectly, through the deterrent effect of damage actions that may be brought once harm occurs. Standards, prohibitions, and other forms of safety regulation, in contrast, are public in character and modify behavior in an immediate way through requirements that are imposed before, or at least independently of, the actual occurrence of harm.

As a matter of simple description, it is apparent that liability and safety regulation are employed with an emphasis that varies considerably with the nature of the

Steven Shavell, "Liability for Harm Versus Regulation of Safety," 13 *Journal of Legal Studies* 357. Reprinted by permission of the author and The University of Chicago Law School *Journal of Legal Studies*, copyright © 1984.

activity that is governed. Whether I run to catch a bus and thereby collide with another pedestrian will be influenced more by the possibility of my tort liability than by any prior regulation of my behavior (informal social sanctions and risk to self aside). Similarly, whether I cut down a tree that might fall on my neighbor's roof will be affected more by the prospect of a tort suit than by direct regulation. But other decisions—whether I drive my truck through a tunnel when it is loaded with explosives or mark the fire exits in my store, or whether an electric utility incorporates certain safety features in its nuclear power plant—are apt to be determined substantially, although not entirely, by safety regulation. There are also intermediate cases, of course; consider, for instance, the behavior of ordinary drivers on the road and the effects of tort sanctions and regulation of automobile use.

What has led society to adopt this varying pattern of liability and safety regulation? What is the socially desirable way to employ the two means of alleviating risks? These are the questions to be addressed here, and in answering them I use an instrumentalist, economic method of analysis, whereby the effects of liability rules and direct regulation are compared and then evaluated on a utilitarian basis, given the assumption that individual actors can normally be expected to act in their own interest. In making this evaluation, I have not counted compensation of injured parties as an independent factor on the grounds that first-party insurance (augmented if necessary by a public insurance program) can discharge the compensatory function no matter what the mix of liability and regulation. . . .

Theoretical Determinants of the Relative Desirability of Liability and Safety Regulation

To identify and assess the factors determining the social desirability of liability and regulation, it is necessary to set out a measure of social welfare; and here that measure is assumed to equal the benefits parties derive from engaging in their activities, less the sum of the costs of precautions, the harms done, and the administrative expenses associated with the means of social control. The formal problem is to employ the means of control to maximize the measure of welfare.

We can now examine four determinants that influence the solution to this problem. The first determinant is the possibility of a *difference in knowledge about risky activities* as between private parties and a regulatory authority. This difference could relate to the benefits of activities, the costs of reducing risks, or the probability or severity of the risks.

Where private parties have superior knowledge of these elements, it would be better for them to decide about the control of risks, indicating an advantage of liability rules, other things being equal. Consider, for instance, the situation where private parties possess perfect information about risky activities of which a regulator has poor knowledge. Then to vest in the regulator the power of control would create a great chance of error. If the regulator overestimates the potential for harm, its standard will be too stringent, and the same will be the case if it underestimates the value of the activity or the cost of reducing risk. If the regulator makes the reverse mistakes, moreover, it will announce standards that are lax.

Under liability, however, the outcome would likely be better. This is clear enough under a system of strict liability—whereby parties have to pay damages regardless of their negligence—for then they are motivated to balance the true costs of reducing risks against the expected savings in losses caused. Now assume that the form of liability is the negligence rule—according to which parties are held responsible for harm done only if their care falls short of a prescribed level of "due" care—and suppose further that once harm occurs, the courts could acquire enough information about the underlying event to formulate the appropriate level of due care. Then parties, anticipating this, would be led in principle to exercise due care. The situation is altered for the worse if the courts are unable to acquire sufficient information to determine the best level of due care; but the outcome would still be superior to that achievable under regulation if the information obtained ex post at trial would be better than that which a regulator could acquire and act upon ex ante.

These conclusions are reversed, of course, if the information possessed by a regulator is superior to private parties' and the courts'; converse reasoning then shows that the use of direct regulation would be more attractive than liability.

The question that remains, therefore, is when we can expect significant differences in information between private parties and regulators to exist. And the answer is that private parties should generally enjoy an inherent advantage in knowledge. They, after all, are the ones who are engaging in and deriving benefits from their activities; in consequence, they are in a naturally superior position to estimate these benefits and normally are in at least as good a position to estimate the nature of the risks created and the costs of their reduction. For a regulator to obtain comparable information would often require virtually continuous observation of parties' behavior, and thus would be a practical impossibility. Similarly, the courts—when called upon under a negligence system—should have an advantage, though a less decisive one, over a regulator. One would indeed expect courts to adjust the due care level to take into account the facts presented by litigating parties more easily than a regulator could individualize its prior standards or modify them to reflect changed conditions.

Yet this is not to say that private parties or the courts will necessarily possess information superior to that held by a regulatory authority. In certain contexts information about risk will not be an obvious by-product of engaging in risky activities but rather will require effort to develop or special expertise to evaluate. In these contexts a regulator might obtain information by committing social resources to the task, while private parties would have an insufficient incentive to do this for familiar reasons: A party who generates information will be unable to capture its full value if others can learn of the information without paying for it. For parties to undertake individually to acquire information might result in wasteful duplicative expenditures, and a cooperative venture by parties might be stymied by the usual problems of inducing all to lend their support. Continuing, once a regulator obtains information, it may find the information difficult to communicate to private parties because of its technical nature or because the parties are hard to identify or are too numerous. Thus we can point to contexts where regulators might possess better information than private parties to whom it cannot easily be transmitted, even if the usual expectation would be for these parties to possess the superior information.

The second of the determinants of the relative desirability of liability and regulation is that *private parties might be incapable of paying for the full magnitude of harm done*. Where this is the case, liability would not furnish adequate incentives to control risk, because private parties would treat losses caused that exceed their assets as imposing liabilities only equal to their assets. But under regulation inability to pay for harm done would be irrelevant, assuming that parties would be made to take steps to reduce risk as a precondition for engaging in their activities.

In assessing the importance of this argument favoring regulation over liability, one factor that obviously needs to be taken into account is the size of parties' assets in relation to the probability distribution of the magnitude of harm; the greater the likelihood of harm much larger than assets, the greater the appeal of regulation.

Another factor of relevance concerns liability insurance. Here the first point to make is that a party's motive to purchase liability insurance against damage judgments exceeding his assets will be a diminished one, as the protection will in part be for losses that the party would not otherwise have to bear. A party with assets of $20,000 might not be eager to purchase coverage against a potential liability of $100,000, as four-fifths of the premium would be in payment for the $80,000 amount that he would not bear if he did not buy coverage. Hence, it might be rational for the party not to insure against the $100,000 risk. If this is the case, then the assertion that liability does not create an adequate motive to reduce risk is clearly unrebutted. . . .

Let us turn next to the third of the four general determinants, the chance that *parties would not face the threat of suit for harm done*. Like incapacity to pay for harm, such a possibility results in a dilution of the incentives to reduce risk created by liability, but it is of no import under regulation.

The weight to be attached to this factor depends in part upon the reasons why suit might not be brought. One reason that a defendant can escape tort liability is that the harms he generates are widely dispersed, making it unattractive for any victim individually to initiate legal action. This danger can be offset to a degree if victims are allowed to maintain class actions, whose application has problematic features, however. A second cause of failure to sue is the passage of a long period of time before harm manifests itself. This raises the possibility that by the time suit is contemplated, the evidence necessary for a successful action will be stale or the responsible parties out of business. A third reason for failure to sue is difficulty in attributing harm to the parties who are in fact responsible for producing it. This problem could arise from simple ignorance that a given harm or disease was caused by a human agency (as opposed to being "natural" in origin) or from inability to identify which one or several out of many parties was the cause of harm.

The problems here are aggravated when the potential liability rests on large firms, where complications analogous to those mentioned before exist. Namely, even if the harms can be attributed to an individual firm, the prospect of a successful suit may exert only slight influence on the behavior of corporate decisionmakers. With the passage of time, for example, there might be no clear way of determining which were the responsible employees, or those who were responsible may no longer be with the firm. The actual decisionmakers therefore may be beyond both the threat of suit and the prospect of sanctions internal to the firm.

The last of the determinants is the magnitude of the *administrative costs incurred by private parties and by the public* in using the tort system or direct regulation. Of course, the costs of the tort system must be broadly defined to include the time, effort, and legal expenses borne by private parties in the course of litigation or in coming to settlements, as well as the public expenses of conducting trials, employing judges, empaneling juries, and the like. Similarly, the administrative costs of regulation include the public expense of maintaining the regulatory establishment and the private costs of compliance.

With respect to these costs, there seems to be an underlying advantage in favor of liability, for most of its administrative costs are incurred only if harm occurs. As this will usually be infrequent, administrative costs will be low. Indeed, in the extreme case where the prospect of liability induces parties to take proper care and this happens to remove all possibility of harm, there would be no suits whatever and thus no administrative costs (other than certain fixed costs). Moreover, there are two reasons to believe that even when harm occurs administrative costs should not always be large. First, under a well-functioning negligence rule, defendants should in principle generally have been induced to take due care; injured parties should generally recognize this and thus should not bring suit. Second, suits should usually be capable of being settled cheaply by comparison to the cost of a trial. A final cost advantage of the liability system is that under it resources are naturally focused on controlling the behavior of the subgroup of parties most likely to cause harm; for because they are most likely to cause harm (and presumably most likely to be negligent), they are most likely to be sued.

Under regulation, unlike under liability, administrative costs are incurred whether or not harm occurs; even if the risk of a harm is eliminated by regulation, administrative costs will have been borne in the process. Also, in the absence of special knowledge about parties' categories of risk, there is no tendency for administrative costs to be focused on those most likely to cause harm, again because these costs are incurred before harm occurs. On the other hand, a savings in administrative costs can typically be achieved through the use of probabilistic means of enforcement. But there is a limit to these savings because there is some minimum frequency of verification necessary to insure adherence to regulatory requirements.

Joint Use of Liability and Regulation

Examination of the four determinants has thus shown that two generally favor liability—administrative costs and differential knowledge—and the other two favor regulation—incapacity to pay for harm done and escaping suit. This suggests not only that neither tort liability nor regulation could uniformly dominate the other as a solution to the problem of controlling risks, but also that they should not be viewed as mutually exclusive solutions to it. A complete solution to the problem of the control of risk evidently should involve the joint use of liability and regulation, with the balance between them reflecting the importance of the determinants.

If, then, some combination of liability and regulation is likely to be advantageous, two questions immediately arise: Should a party's adherence to regulation relieve him of liability in the event that harm comes to pass? On the other hand, should a party's failure to satisfy regulatory requirements result necessarily in his liability? Our theory suggests a negative answer to both questions.

As to the first, if compliance with regulation were to protect parties from liability, then none would do more than to meet the regulatory requirements. Yet since these requirements will be based on less than perfect knowledge of parties' situations, there will clearly be some parties who ought to do more than meet the requirements—because they present an above-average risk of doing harm, can take extra precautions more easily than most, or can take precautions not covered by regulation. As liability will induce many of these parties to take beneficial precautions beyond the required ones, its use as a supplement to regulation will be advantageous. At the same time, just because this is true, regulatory requirements need not be as rigorous as if regulation were the sole means of controlling risks.

A similar analysis is appropriate for the second question. If failure to satisfy regulatory requirements necessarily resulted in a finding of negligence, then some parties would be undesirably led to comply with them when they would not otherwise have done so. In particular, there will be some parties (a) who ought not to meet regulatory requirements because they face higher than usual costs of care or because they pose lower risks than normal and (b) who will not have been forced to satisfy regulatory requirements due to flaws in or probabilistic methods of enforcement. By allowing these parties to escape liability in view of their circumstances, the possibility that they would still be led to take the wasteful precautions can be avoided.

Informational Approaches to Regulation
WESLEY A. MAGAT AND W. KIP VISCUSI

Information as a Regulatory Tool

When government regulators are faced with a situation involving a risk, they have four policy options. The first is to do nothing, which is a desirable course of action in situations where the market functions effectively. The second option is to ban the product or activity generating the risk. The Food and Drug Administration, for example, screens new pharmaceutical products to ensure that drugs posing substantial risks are not marketed. Third, the government can undertake an action to directly alter the risk, as in the case of Occupational Safety and Health Administration regulations that control conditions of the workplace, or Environmental Protection

Agency requirements on household pesticides that limit the potency of commercially marketed products.

A fourth option . . . is to adopt a hazard-warning program. From a political standpoint this policy option is often viewed as a compromise between taking no action at all and taking an extreme form of action such as a product ban. As an intermediate policy option it has political appeal that is somewhat independent of its merits.

There are also sound economic reasons why such an intermediate course might be preferred. Informational programs often are potentially more effective than taking no regulatory action. Such programs can potentially remedy the informational inadequacies in markets. A major source of market failure that is often cited in the case of health and safety regulations is that if consumers and workers are not fully informed of the risks they face, market compensation for risk will not produce efficient safety incentives, thus leading to a justification for a government intervention. If these informational problems can be solved directly through informational regulation, more stringent forms of regulation will not be required, so the market outcome will be efficient.

Informational regulation also may be a more effective means of promoting safety than direct technological controls. Most accident situations are the result of two sets of influences: (1) the technological characteristics of the accident context, such as whether a lawn mower has an engine cutoff device, and (2) the role of the potential accident victim. The manner of operation of the lawn mower and the choice of which family member is to operate the mower play a potentially important role. In most instances consumer precautions are required for the amount and nature in which a product is used, as well as for particular precautions and precautionary equipment that should be employed. Once the role of individual action in contributing to risks is recognized, the potential role for regulatory intervention through an informational approach is apparent. Altering behavior may be a more effective regulatory strategy than is the technological approach that dominated the first decade of social regulation in the 1970s.

Partly in recognition of this role of individual behavior and partly because of the increasing cost of technological controls to reduce risk, the emphasis of health, safety, and environmental regulations in the 1980s became increasingly shifted toward informational regulations. In addition there has been increasing pressure to engage individuals in decisions about the risks they face by right-to-know activists.

A further argument for informational remedies, rather than command-and-control regulation, applies to informational problems prevalent in both risk and non-risk-related decisions. Consumer and worker preferences differ, as do the stocks of existing goods that they own, implying that the optimal regulation should differ across them. A product "improvement," such as a lawnmower engine cutoff system which makes the product safer but more costly, is more likely to be desired by consumers who are wealthy, risk averse, and careless than those with limited incomes who are risk prone and carefully operate the product. Similarly, a product requirement, such as the building code requiring adequate caulking around windows, is more likely to be desired by consumers who are wealthy and know little about new homes than those with less income who are well informed about building characteristics.

Although informational regulations have proliferated, they have not been without controversy. Informational regulations generally are not completely effective, and there are some circumstances in which they may not be effective at all. Not all people will read the information provided, and not all recipients of the information will act upon it. Perhaps the main difficulty is that informational regulations do not involve simple technological prescriptions to problems. There is an important human element involved in the processing of the information and in the decisions using the information, and one must take into account both this limitation and the decision-making orientation in which the information will be used. For informational regulations to be effective, they must provide new information that can potentially alter individual decisions. If the information cannot be processed reliably or is viewed as not contributing any new information or perspective on a decision, then the informational program will not be successful.

Notes and Questions

1. BAT standards can take one of two forms. First, the regulator could require the use of the best available technology; such requirements are often referred to as equipment or design standards. Second, the regulator could require the attainment of standards set by reference to what the best available technology can accomplish, but allow each source to choose the actual technology that it intends to use in order to meet the standard; such standards are generally referred to as emission standards in the case of air and effluent standards in the case of water. With few exceptions, the federal environmental laws follow the latter approach. The statutes sometimes specify explicitly (as is the case, for example, in section 111(h) of the Clean Air Act) that the former approach can be followed only when it is infeasible to set a numerical limitation. Which type of BAT standard is preferable? In particular, consider two issues:
   (a) ease of enforcement
   (b) incentive effects

2. Are BAT standards for new sources open to the same criticisms as BAT standards for existing sources? What are the relevant differences? Is it likely to be cheaper to design plants to meet a given standard or to retrofit existing plants? What are possible sources of the differences in compliance costs between new and existing sources? For a particular category of plants, for example, electric utilities, is there likely to be more variation in the costs of meeting a particular standard for new sources or existing sources?

3. Ackerman and Stewart contend that BAT standards discourage technological innovation. Consider a contrary argument: nationally uniform BAT standards create a national market for new technology that is labeled "best" by the regulatory authority; in contrast, under a system of marketable permits, there is likely to be great variation across firms in the chosen levels of emissions, and, consequently, different firms are likely to favor different types of pollution control technology. Assess the strength of this argument. What steps could a regulatory authority take in order to maximize the incentives of BAT standards for technological innovation? From this perspective, are equipment or design standards more desirable than emission or effluent standards? For discussion of the effects of environmental regulation on technological innovation, see Ashford, "Using Regulation to Change the Market for Innovation," 9 *Harvard Environmental Law Review* 419 (1985); Hay, "Innovation under Technology-Based Regulation," Discussion Paper No. 168, Program in Law and Economics, Harvard Law

School (September 1995). For defenses of BAT, see Latin, "Ideal Versus Real Regulatory Efficiency: Implementation of Uniform Standards and 'Fine Tuning' Regulatory Reforms," 37 *Stanford Law Review* 1267 (1985); Shapiro and McGarity, "Not So Paradoxical: The Rationale for Technology-Based Regulation," 1991 *Duke Law Journal* 729.

4. BAT standards are generally set as a function of a unit of output, for example, pounds of sulfur dioxide per million BTU of energy. In contrast, marketable permit schemes and effluent fees are directly linked to units of pollution. Under the former, a firm can emit one unit of pollution for each permit that it holds, whereas under the latter, the firm must pay a set tax for each unit of pollution. What are the implications of this difference between BAT standards on the one hand and marketable permit schemes and effluent fees on the other? For discussion, see Helfland, "Standards Versus Standards: The Effects of Different Pollution Restrictions," 81 *American Economic Review* 622 (1991).

5. In the transition from command-and-control regulation to marketable permit schemes, should there be an initial auction of permits or should the permits be given for free to firms that previously had the right to pollute? The latter approach is generally followed in the acid rain provisions of the 1990 amendments to the Clean Air Act—the most extensive marketable permit scheme under the federal environmental statutes—which create a national market for the trading of permits to emit sulfur dioxide. See 42 U.S.C. § 7651b. Which approach is more desirable? Consider the effects of transaction costs in trading, as well as equity issues between new and existing plants. Which allocation approach is likely to have more political support? This question is considered in more detail in Chapter VIII. Interestingly, the excerpts by Ackerman and Stewart on the one hand and by Baumol and Oates on the other reach different conclusions about the desirability of having the regulatory authority raise substantial revenues through an initial auction. These and other issues concerning the structuring of markets are discussed in T. H. Tietenberg, *Emissions Trading: An Exercise in Reforming Pollution Policy* (Washington, D.C.: Resources for the Future, 1985).

6. How serious is the "hot spot" problem that is likely to arise under marketable permit schemes? How does it depend on the size of the area in which trades are permitted? What competing factors affect the delimitation of this areas? How are such hot spots avoided under command-and-control regulation? As is discussed in Chapter V, the environmental justice movement believes that environmentally undesirable effects disproportionately impact people of color and the poor. What are the environmental justice implications of hot spots? How do hot spots affect the claim that marketable permit schemes are a tool to meet a particular level of environmental quality at least cost? How should one trade off better quality in some areas against worse quality at a hot spot? What particular problems would arise from the implementation of marketable permit schemes under the Clean Air Act, which has as its centerpiece the nationwide achievement of uniform ambient air quality standards (maximum permissible concentrations of pollutants)? The structure of the Clean Air Act is discussed in detail in Chapter IX.

7. In a marketable permit system, should environmental groups be able to purchase permits with the intention of retiring them from circulation, so as to reduce the aggregate amount of pollution? Such purchases are allowed under the acid rain provisions of the Clean Air Act, and the environmental law societies of several law schools have purchased permits at recent auctions. In light of such purchases, is it appropriate for the government to increase the number of permits in circulation? Do these answers depend on how the number of permits in circulation was initially determined? Consider two options:

(a) determination by reference to the level that maximizes social welfare
(b) determination through the political process, without a well-articulated rationale

8. In a marketable permit system, should brokers be permitted to buy permits as an investment and hold them out of circulation for a period of time? Should polluters be allowed to buy permits and hold them for possible future expansion of their operations? Should they be allowed to purchase permits with the objective of increasing the costs of their competitors? Does the legal system have adequate protections with respect to these matters?

9. If, having established a marketable permit system, a regulatory authority decides to dismantle it and return to command-and-control regulation, do holders of permits have a Fifth Amendment takings claim? Does it matter whether the initial allocation was by auction? Does it matter whether the holder bought the permit or got it for free? Are the issues raised by the elimination of a marketable permit scheme conceptually different from those that arise if the regulator makes a BAT standard for existing sources substantially more stringent than the previously applicable standard, thereby making valueless a firm's investment in pollution control technology?

10. The acid rain provisions of the Clean Air Act state:

> An allowance allocated under this subchapter is a limited authorization to emit sulfur dioxide in accordance with the provisions of this subchapter. Such allowance does not constitute a property right. Nothing in this subchapter or in any other provision of law shall be construed to limit the authority of the United States to terminate or limit such authorization. (42 U.S.C. §7651b[f])

Should this provision be dispositive with respect to a possible takings claim?

11. Ackerman and Stewart propose a marketable permit system in units of emissions. How would the hot spot problem be avoided by markets in units of environmental degradation? What are the characteristics of such markets? Consider the following issues:

    (a) the number of markets in which a single source would have to buy a permit
    (b) the problems of coordination raised by the need to purchase permits in several markets
    (c) scientific problems in translating units of emissions into units of environmental degradation
    (d) the role of the regulator

Are the latter problems any less serious in the case of a command-and-control approach to meeting the Clean Air Act's ambient standards?

The impact of emissions on environmental quality is often a product not only of the level of emissions but also, among other factors, of the geographic location, prevailing winds, stack height, and climatic conditions. How does this feature affect the relative desirability of permits in units of emissions versus permits in units of environmental degradation? For discussion of different types of marketable permit schemes, see Montgomery, "Markets in Licenses and Efficient Pollution Control Programs," 5 *Journal of Economic Theory* 395 (1972); Tietenberg, "Transferable Discharge Permits and the Control of Stationary Source Air Pollution: A Survey and Synthesis," 56 *Land Economics* 391 (1980).

As noted in the Baumol and Oates excerpt, the same issues are raised by effluent fees. The question there is whether such fees should be geographically uniform or whether they should be differentiated to reflect that one unit of emissions can have different impacts on environmental degradation as a result of the factors listed above. For further discussion, see Rose-Ackerman: "Effluent Charges: A Critique," 6 *Canadian Journal of Economics* 512 (1973); Tietenberg, "Spatially Differentiated Air Pollutant Emission Charges: An Economic and Legal Analysis," 54 *Land Economics* 265 (1978).

12. Even before the passage of the 1990 amendments, adding the acid rain provisions, the Clean Air Act contained some marketable permit schemes. For example, a source wishing to locate in a nonattainment area (an area that did not meet the ambient standards) had to obtain offsets from existing sources in the area. Studies of the operation of these trading schemes show that they have been vastly underutilized. For discussion, see Hahn and Hester, "Where Did All the Markets Go?: An Analysis of EPA's Emissions Trading Program," 6 *Yale Journal on Regulation* 109 (1989). Recent scholarship suggests that marketable permit schemes may have substantial transaction costs as a result, primarily, of the costs of search and information, bargaining and decision, and monitoring and enforcement. See Stavins, "Transaction Costs and Tradeable Permits," *Journal of Environmental Economics and Management* (forthcoming). What steps can be taken to encourage trades? For prescriptions about how to make marketable permit schemes work, see Hahn and Noll, "Environmental Markets in the Year 2000," 3 *Journal of Risk and Uncertainty* 351 (1990).

13. Under the Clean Air Act, the Environmental Protection Agency (EPA) has also authorized intrafirm trades through the "bubble" program. Under the program, a firm can place an imaginary bubble over its various emission points as long as they are contiguous. The firm can then increase its pollution at certain points, or add additional points, as long as it reduces its pollution elsewhere by an equivalent amount. Thus, for example, a utility can add another smokestack without having to meet the otherwise applicable BAT standards, as long as it sufficiently reduces (or closes down) another smokestack. For discussion, see Richard A. Liroff, *Reforming Air Pollution Regulation: The Toil and Trouble of EPA's Bubble* (Washington, D.C.: Conservation Foundation, 1986). Is the bubble program desirable? As discussed in Chapter VIII, existing sources generally face far less stringent environmental standards than new sources. Under what conditions do bubbles therefore slow down environmental improvements? How do they affect the competitive position of new firms?

14. Marketable permit schemes are sometimes attacked as "licenses to pollute." BAT, however, gives away for free the permission to pollute. Are marketable permit schemes morally more suspect? If so, why? For discussion, see Steven Kelman, *What Price Incentives?: Economists and the Environment* (Boston: Auburn House, 1981).

15. An environmental regulator wants to obtain a total of six units of effluent reduction from two firms that manufacture the same product. Unregulated, these firms would each have effluents of five units. If the regulator imposed BAT standards, which are uniform across categories of similar polluters, it would require each firm to reduce its emissions by three units. Alternatively, it could set up a scheme of marketable permits or impose effluent fees. Each firm's total costs of effluent reduction are set forth in Table VI.1:

Table VI.1.

Units of Effluent Reduction	Cost of Effluent Reduction	
	Firm 1	Firm 2
0	0	0
1	1	2
2	3	6
3	6	12
4	10	20
5	15	30

What is the cost-minimizing allocation of the pollution control burden? How do the aggregate costs under the cost-minimizing allocation compare to the aggregate costs under BAT?

16. Assume, again by reference to Table VI.1, that there is a transition from BAT to a marketable permit system and that the initial permits are given out, for free, to those with rights to emit under BAT (which requires each firm to reduce by three units its unregulated emissions of five units). What trades will take place? At what prices will these trades take place?

17. If effluent fees are imposed, consider the effect of a tax of $2.50 per unit of effluent. The first unit of emission reduction costs $1.00 to Firm 1; thus, Firm 1 will prefer to expend this money rather than pay the tax of $2.50. The second unit of emission reduction costs this firm $2.00, so it will choose to engage in this reduction as well. The third unit costs $3.00, so the firm will prefer to pay the tax. In summary, faced with a tax of $2.50, Firm 1 reduce its emissions two units. Similar logic establishes that this tax leads Firm 2 to reduce its emissions by only one unit. Thus a tax of $2.50 per unit of effluent is too low to achieve the regulator's objective. What is the effect of a tax of $5.50? What tax should the regulator set to achieve an aggregate reduction of six units?

18. If the regulatory authority initially sets the wrong tax (so that either too much or too little pollution is produced), can it correct its mistakes? What are the costs of frequent modifications of the tax rates?

19. How should one determine the relative desirability of marketable permits schemes and effluent fees? The former fix pollution control levels but leave cleanup costs uncertain. The latter fix the incremental (though not total) cleanup costs but leave the quantity of pollution uncertain. Which is more desirable under the following circumstances?
   (a) a sharp increase in the damage of pollution as the level of pollution increases
   (b) a sharp increase in the incremental cost of pollution abatement as the stringency of the controls increase
For further discussion, see Spence and Weitzman, "Regulatory Strategies for Pollution Control," in Ann F. Friedlaender, ed., *Approaches to Controlling Air Pollution* (Cambridge, Mass.: MIT Press, 1978), at 199; Weitzman, "Prices vs. Quantities," 41 *Review of Economic Studies* 477 (1974); Yohe, "Towards a General Comparison of Price Controls and Quantity Controls Under Uncertainty," 45 *Review of Economic Studies* 229 (1978).

Does it matter whether the regulatory authority has accurate information about each firm's costs of pollution control? Which tool is preferable as a replacement for command-and-control regulation if the objective is to meet existing ambient standards? Which tool is preferable given the competition that firms face in global markets? In this connection, does it matter how the initial allocation is performed?

20. Which of the following regulatory tools are more desirable from the perspective of ease of enforcement?
   (a) BAT standards
   (b) marketable permits
   (c) effluent fees
What are the enforcement difficulties raised by each approach?

21. Subsidies are an alternative to effluent fees. If there is no entry of new firms into the market, both approaches ought to have equal properties. Why? What incentives do subsidies create with respect to the entry of new firms into the market? Are such incentives desirable? This issue is discussed in depth in a portion of the Baumol and Oates book (pp. 212–13, 216–28) that is not excerpted.

22. How should one set the deposit amounts in deposit-refund systems? What are the consequences of an amount that is too low? What are the consequences of an amount that is too high?

23. Consider two different types of uses for deposit-refund systems:
    (a) to create incentives for the return of products that can be recycled
    (b) to create incentives for the disposal of waste that ought to be segregated from the remainder of the waste stream
What factors are relevant to determining whether deposit-refund systems are attractive regulatory tools in each of these instances? What is the relevance of the costs of administering the system?

24. With respect to the first category, how should one determine whether to use deposit-refund systems for the following products
    (a) newspapers
    (b) plastic bottles
    (c) cans
If the recycling of these products is economically desirable, why should there be any government involvement? Does a government-sponsored deposit-refund system offer any advantages over entrepreneurship in the private sector? Are the arguments for deposit-refund systems stronger for the second of the categories defined in the previous question?

25. Consider the issues raised by the use of a deposit-refund system for hazardous substances. How could such a system deal with the following issues?
    (a) hazardous substances that are subsequently mixed with other hazardous substances, becoming either more or less hazardous
    (b) hazardous substances that are subsequently mixed with nonhazardous substances
    (c) hazardous substances purchased in other jurisdictions
With respect to the disposal of hazardous substances, compare the relative desirability of deposit-refund systems and liability rules?

26. The Clean Air Act (discussed in Chapter IX) imposes a regulatory approach to the control of air pollution, whereas the Superfund statute (discussed in Chapter X) imposes liability for the release of hazardous substances into the environment, generally the soil, surface water, and groundwater. Can the choice of different policy tools be explained by Shavell's four factors?

27. Under a negligence rule, the court must determine the appropriate standard of care. Shavell calls on courts to choose a standard that would maximize social welfare. Is a negligence rule likely to compare favorably to regulation? Is a negligence rule likely to compare favorably to strict liability?

28. Shavell states that in some cases private parties will not possess information superior to that held by regulatory authorities. In what instances is that likely to be the case?

29. In connection with Shavell's third determinant, consider the following potential roadblocks to bringing suits for environmental harms:
    (a) statutes of limitations
    (b) proof of causation
    (c) free-rider problems among victims
How can the legal system ameliorate these problems? For discussion of the shortcomings of the common law in the environmental area, see Menell, "The Limitations of Legal Institutions for Addressing Environmental Risks," 5 *Journal of Economic Perspectives* 93 (1991).

30. Shavell notes that under a liability system administrative costs are incurred only in the face of harm. He adds: "As this will usually be infrequent, administrative costs will be low." To what extent is this likely to be true? Compare the administrative costs of negligence and strict liability rules. If few cases are, in fact, brought under negligence rules, how will the relevant actors know the standard of care? For further discussion of the choice between regulation and liability, see White and Wittman, "A Comparison of Taxes, Regulation, and Liability Rules Under Imperfect Information," 12 *Journal of Legal Studies* 413 (1983); Wittman, "Prior Regulation versus Post Liability: The Choice Between Input and Output Monitoring," 6 *Journal of Legal Studies* 193 (1977).

31. Consider the relative desirability of regulatory schemes and liability rules from the perspective of a factory that is contemplating the purchase of relatively new and expensive pollution control equipment to reduce the risk of a harm with a long latency period. Which rule is it likely to prefer? What factors are relevant to this choice?

32. In discussing the joint use of liability and regulation, Shavell notes: "If failure to satisfy regulatory requirements necessarily resulted in a finding of negligence, then some parties would be undesirably led to comply with them when they would otherwise not have done so." What penalties should actors face when they fail to meet regulatory standards? Of course, if these penalties are sufficiently severe, the actors will have incentives to meet the regulatory standard regardless of whether the failure to meet this standard is conclusive evidence of negligence. The joint use of liability and regulation in connection with Superfund and the Resource Conservation Act (RCRA) is considered in Chapter X.

33. More generally, does the effect of regulatory standards on behavior depend exclusively on the penalty structure for noncompliance? Does a regulatory standard coupled with a particular penalty for noncompliance have the same effect as a tax of zero for risk below the standard and a tax equal to the penalty for risk above the standard? What view of regulatory standards is implicit in Shavell's discussion?

34. The free-market environmentalist movement contends that markets can take care of environmental problems without government regulation. First, it believes that cooperative organizations interested in environmental protection will be able to solve the free-rider problem, for example, by purchasing conservation easements. Second, it places faith in technological innovations that will make it easier in the future to privatize common resources. Third, it believes that common law courts can play an effective role in protecting private property. See Terry L. Anderson and Donald R. Leal, *Free Market Environmentalism* (San Francisco: Westview, 1991). For criticism of this movement, see Menell, "Institutional Fantasylands: From Scientific Management to Free Market Environmentalism," 15 *Harvard Journal of Law and Public Policy* 489 (1992). What position are free market environmentalists likely to have with respect to marketable permits and effluent fees? Why do they place trust in common law courts but distrust in the political branches of government?

35. For what kinds of risk is a hazard-warning system likely to be a desirable alternative to regulation? Consider the following possibilities:
   (a) disperse air pollution
   (b) highly localized air pollution
   (c) indoor air quality at industrial plants
   (d) disposal of hazardous wastes
   (e) automobiles
   (f) lawn mowers
In these examples, to what extent are affected individuals able to benefit from the additional risks? To what extent do they impose risks on other individuals?

36. Magat and Viscusi give two distinct arguments for hazard-warning systems:
   (a) that some risks are, at least in part, a function of the victims' decisions
   (b) that different individuals have different preferences for risk, or more precisely, different trade-offs between the riskiness and compensation of an occupation or riskiness and price of a product

Does the first argument call for informational approaches as an alternative to regulation or as an adjunct to regulation? With respect to the second approach, to what extent should well-informed individuals be permitted to accept high risks? In this connection, reflect upon the environmental justice discussion in Chapter V.

37. What are the implications for the desirability of informational approaches to regulation of the discrepancy between expert and lay perceptions of risk (discussed in Chapter III)?

38. In recent years, green labeling has begun to play an important role in advertising. Hazard-warning systems are designed to tell consumers about the risk of a product so that the consumer can make an informed decision about whether the positive attributes of the product outweigh its risk. In contrast, green labeling informs consumers about the positive effects on the environment of the *processes* by which the product is manufactured and disposed, so that the consumer can decide whether to value environmental benefits to society at large more than the product's added cost. Should the government promote green labeling? Should it police the accuracy of green labeling claims by private organizations? Is it likely that green labeling will be an effective tool of environmental policy? For discussion, see Grodsky, "Certified Green: The Law and Future of Environmental Labeling," 10 *Yale Journal on Regulation* 147 (1993); Menell, "Eco-Information Policy: A Comparative Institutional Perspective," Working Paper No. 104, John M. Olin Program in Law and Economics, Stanford Law School (April 1993).

39. A different type of informational approach is embodied in the National Environmental Policy Act (NEPA), the first of the modern federal environmental statutes. NEPA requires federal agencies to prepare an environmental impact statement (EIS) before engaging in environmentally destructive activity. The idea behind NEPA was that such information would lead government agencies to make better and environmentally more sensitive decisions. For discussion, see Herz, "Parallel Universes: NEPA Lessons for the New Property," 93 *Columbia Law Review* 1668 (1993); Sax, "The (Unhappy) Truth About NEPA," 26 *Oklahoma Law Review* 239 (1973). Do you believe that the preparation of EISs is likely to lead government agencies to make decisions that are environmentally more protective?

The Political Dimensions of Environmental Law

VII

Federalism and Environmental Regulation

This chapter examines the central arguments for vesting control over environmental regulation in a federalist system at the federal level. This issue is currently one of intense academic and political debate in both the United States and the European Union.

The article by Richard B. Stewart presents several rationales for the centralization of environmental regulation. First, states may decline to adopt protective environmental standards for fear of losing industry to states with laxer standards. Second, centralization may produce economies of scale. Third, environmental interests may have a comparatively greater impact on policy decisions at the national level. Fourth, states will underregulate because, as a result of interjurisdictional spillovers, some of the negative effects of pollution affect other states. Finally, communities may be more willing to make the sacrifices necessary to achieve high levels of environmental quality if they have assurances that other communities are making the same sacrifices.

Stewart then presents rationales for decentralization. First, with respect to some regulatory activities, there may be diseconomies of scale. Second, centralized decision making impairs self-determination. Third, the costs of environmental regulation may be distributed in a regressive manner; this regressive impact may be particularly offensive if it is not the product of local political mechanisms.

The article by Richard L. Revesz notes that two arguments for regulation at the federal level appear most prominently both in the academic literature and in the legislative history of the federal environmental statutes: that states will underregulate because some of the benefits of regulation would accrue to other states (the interstate externality justification); and that, even in the absence of interstate externalities, states would underregulate in an effort to attract industry (the race-to-the-bottom justification). He shows that the race to the bottom is a form of prisoner's dilemma (a concept that is introduced in the Notes and Questions to Chapter I), in which, as a result of their inability to coordinate their actions, two jurisdictions pick suboptimally lax environmental standards even though they would be both better off with more stringent standards.

Revesz then argues that the race-to-the-bottom justification encounters no support in the theoretical literature on interjurisdictional competition. In the absence of interstate externalities, a jurisdiction interested in maximizing the aggregate welfare of their citizen should properly trade off the additional benefits, in terms of jobs and taxes, of attracting new industry against the resulting environmental harms suffered by its citizens. The article shows, moreover, that if a race to the bottom exists, federal regulation will not solve the underlying problem: states will then set suboptimally lax standards in other regulatory areas, such as worker safety, and social welfare will similarly be impaired.

This chapter does not deal with an important set of issues concerning the regulation of environmental quality in a federal system: the extent to which states can impose environmental standards that are more stringent than the federal standards and invoke them to ban or otherwise restrain the entry of out-of-state products that do not meet these standards. This issue arises also in the international context and is discussed in Chapter XI, which deals with international trade.

Pyramids of Sacrifice? Problems of Federalism in Mandating State Implementation of National Environmental Policy

RICHARD B. STEWART

As a nation, we have traditionally favored noncentralized decisions regarding the use and development of the physical environment. This presumption serves utilitarian values because decisionmaking by state and local governments can better reflect geographical variations in preferences for collective goods like environmental quality and similar variations in the costs of providing such goods. Noncentralized decisions also facilitate experimentation with differing governmental policies, and enhance individuals' capacities to satisfy their different tastes in conditions of work and residency by fostering environmental diversity.

Important nonutilitarian values are also served by noncentralized decisionmaking. It encourages self-determination by fragmenting governmental power into local units of a scale conducive to active participation in or vicarious identification with the processes of public choice. This stimulus to individual and collective education and self-development is enriched by the wide range of social, cultural and physical environments which noncentralized decisionmaking encourages. . . .

In our nation, the factors favoring noncentralized decisionmaking have been powerfully reinforced by geography, history, and the structure of our politics. Nonetheless, the presumption in favor of decentralization has in recent years been repeatedly overridden by congressional legislation imposing federal standards and federal measures to control environmental degradation. . . .

The Rationales for Centralization

Four structural factors hinder governmental provision of high environmental quality on a noncentralized basis and help to explain why resort to federal legislation is a necessary or appropriate complement to rising public concern with environmental quality.

The Tragedy of the Commons and National Economies of Scale

The Tragedy of the Commons arises in noncentralized decisionmaking under conditions in which the rational but independent pursuit by each decisionmaker of its own self-interest leads to results that leave all decisionmakers worse off than they would have been had they been able to agree collectively on a different set of policies.

Reprinted by permission of the author, the Yale Law Journal Company, and Fred B. Rothman & Company from *The Yale Law Journal*, Vol. 86, pp. 1196–1272 (1977).

States and local communities whose citizens desire environmental quality are also concerned with employment and economic growth. Given the mobility of industry and commerce, any individual state or community may rationally decline unilaterally to adopt high environmental standards that entail substantial costs for industry and obstacles to economic development for fear that the resulting environmental gains will be more than offset by movement of capital to other areas with lower standards. If each locality reasons in the same way, all will adopt lower standards of environmental quality than they would prefer if there were some binding mechanism that enabled them simultaneously to enact higher standards, thus eliminating the threatened loss of industry or development. The costs and uncertainties of bargaining among many state or local government units render such a compact improbable. If more rigorous environmental standards were imposed nationwide by the federal government, the transaction cost impediments to common agreement would be much less and all localities might well be better off.

The characteristic insistence in federal environmental legislation upon geographically uniform standards and controls strongly suggests that escape from the Tragedy of the Commons by reduction of transactions costs has been an important reason for such legislation. The statutory structure of federal environmental programs also reflects other economies of scale that help explain centralizing tendencies. Collection of data and analysis of environmental problems, standard setting, and (in some instances) selection of control measures involve recurring, technically complex issues; such steps can often be taken far more cheaply once on the national level than repeatedly at the state and local level.

Disparities in Effective Representation

Much of the politics of pollution control involves conflict between environmental groups and industrial and union interests. There are persuasive grounds for believing that, on the whole, environmental groups have a comparatively greater impact on policy decisions at the national level, even disregarding "commons" factors that make state and local governments particularly vulnerable to industry and union pressures. It is therefore not surprising that environmental groups should favor determination of environmental policies by the federal government, and that increased public support for environmental protection should translate into increased legislation and regulation at the federal level.

Industrial firms, developers, unions and others with incentives to avoid environmental controls are typically well-organized economic units with a large stake in particular decisions. The countervailing interest in environmental quality is shared by individuals whose personal stake is small and who face formidable transaction costs in organizing for concerted action. These factors tend to produce more effective and informed representation before legislative and administrative decisionmakers of interests favoring economic development as opposed to those favoring environmental quality. The technical complexity of environmental issues exacerbates this disparity by placing a premium on access to scarce and expensive scientific, economic, and other technical information and analytical skill.

The comparative disadvantage of environmental groups will often be reduced, however, if policy decisions are made at the national level. In order to have effective influence with respect to state and local decisions, environmental interests would be required to organize on a multiple basis, incurring overwhelming transaction costs. Given such barriers, environmental interests can exert far more leverage by organizing into one or a few units at the national level.

Centralized decisionmaking may imply similar scale economies for industrial firms, but these are likely to be of lesser magnitude—particularly if such firms are already national in scope. Moreover, effective representation may be less a function of comparative resources than of attainment of a critical mass of skills, resources, and experience. Industry and development interests can probably deploy these requisites regardless of whether decision is local or national. But a national forum for decision may greatly lessen the barriers to environmental interests' achievement of organizational critical mass, sharply reducing the disparity in effective representation.

In addition to reducing aggregate costs and facilitating achievement of critical mass, environmental advocacy at the national level also affords scale economies in fundraising. Donors may want national coverage in environmental advocacy, or donors in one region (e.g., New York City) may be primarily interested in environmental problems in some other region (e.g., Alaska). These factors, together with the significant start-up costs for effective fundraising, mean that a few national organizations may be able to raise more total resources than numerous local ones.

Finally, environmental groups may enjoy comparatively greater influence at the national level because of the nature and outlook of national decisionmakers. Environmental groups are likely to find more powerful bureaucratic allies in Washington. Both the scale economies of national decisionmaking and fiscal "commons" problems at the state level[1] result in larger, better-funded, and better-staffed health and environmental protection agencies at the federal level. Elected politicians serving in Washington may be somewhat less sensitive to local concerns regarding the costs of environmental protection; in addition, they are subjected to greater cultural and media encouragement to take a "long-run" or "national" perspective favoring environmental concerns.

Centralized decisionmaking does not invariably work to the comparative advantage of environmental groups. For example, intense local environmental concerns may be able to generate a critical mass for effective representation at the local level; if the ultimate decision is made in Washington, the costs of effective representation will rise and the local commitment to influence that decision may wane. Industry may seek preemptive uniform national standards or controls to escape more restrictive state controls. But the factors favoring (from the environmental advocate's view) national decisionmaking seem to predominate. This conclusion is supported by the fact that the most effective champions of environmental interests have been national organizations that concentrate on influencing policy at the federal level.

---

[1] A fiscal "commons" problem arises because each state acting independently fears that unilaterally raising taxes will displace industry and wealthy taxpayers to other states, even though all states might be better off if they could all agree to raise taxes together.

Spillovers

Even if the "commons" problem were eliminated, decentralized environmental decisionmaking would remain flawed because spillover impacts of decisions in one
jurisdiction on well-being in other jurisdictions generate conflicts and welfare losses
not easily remedied under a decentralized regime.

The most obvious form of spillover is physical pollution. Prevailing winds or
river flows may transport pollution generated in one state to another and visit
damage there. These spillovers are in many instances pervasive and far-reaching.
For example, a significant percentage of sulfate pollution in the eastern states is
attributable to emissions originating hundreds of miles westward. Spillovers can
also be psychic and economic. Environmental degradation in pristine areas often
impose substantial welfare losses on individuals in other states who value the
option of visiting such areas or who take ideological satisfaction in their preservation.[2] A state that encourages economic development at the expense of environmental quality may inflict economic loss (in the form of industrial migration or
decreased economic growth) on other states that prefer a higher level of environmental quality.

Bargaining among states to minimize the losses occasioned by such spillovers is
costly (particularly given the complexity and wide dispersal of many forms of environmental degradation), and may do little to improve the lot of states in a weak position (such as those in a downwind or downstream location). These states are likely
to favor federal intervention to eliminate the more damaging forms of spillover. If
spillover losses are sufficiently significant and multidirectional then all states may
gain (to a greater or lesser degree) from centralized determination of environmental
policies. Predictably, therefore, the need to deal with spillover problems and interstate conflicts was advanced as a justification for federal legislation concerning air
and water pollution.

Moral Ideals and the Politics of Sacrifice

The groundswell of public concern with environmental quality that arose in the late
1960s had undeniable aspects of a moral crusade with powerful emotional, even
religious, undercurrents. This development cannot be fully explained by utilitarian
models that explain individual behavior in terms of calculated preference satisfaction. On the contrary, it partially reflects the sacrifice of preference-satisfaction in
order to fulfill duties to others, or to transform existing preference structures in the
direction of lessened dependence upon consumption of material goods and greater
harmony with the natural environment. For example, the Clean Air Act assumes an
obligation to protect the health of susceptible individuals, such as children or those
already suffering from disease, through geographically uniform standards that
ensure every individual a minimum healthy environment even though such unifor-

---

[2]Because states with high environmental quality are precluded from taxing the benefits of such "long-distance" satisfactions and are also unable to recoup fully the gains of those who visit the state to enjoy its environmental amenities,
the resulting external economies will create a market imperfection that will lead the state to adopt a lower level of environmental quality than would be desirable from the viewpoint of society as a whole.

mity is economically inefficient. The preservation of pristine areas may be understood in part as reflecting a special obligation to future generations (despite the fact that they may, in economic terms, be wealthier than we) to prevent the potentially irreversible loss of important categories of human experience. These measures to preserve natural environments, together with programs to protect endangered species, could also be viewed as an assumption of duties to nature. Alternatively, they could be understood as a deliberate renunciation of maximum economic progress in order to affirm a different view of the ends of human life to which the society should aspire.

National mechanisms for determining environmental policies facilitate, to a greater degree than their state and local counterparts, the achievement of commitments entailing material sacrifice; the moral content of rising environmental concern thus helps explain the increasing resort to centralized decision. Communities no less than individuals may be far more willing to undertake sacrifices for a common ideal if there are effective assurances that others are making sacrifices too. National policies can provide such assurances and also facilitate appeals to sublimate parochial interests in an embracing national crusade. It is accordingly not surprising that federal pollution control legislation in the early 1970s was conceived and advertised as a national war against degradation and disease.

Centralization also makes less apparent the sacrifices involved in public expenditures to promote environmental quality. The relation between one's tax payment into the large and complex federal fisc and any particular federal expenditure is obscure; the correlation between a state or local bond issue for sewage treatment facilities and personal financial sacrifice is more direct and immediate. The ambitious municipal waste treatment programs adopted in federal legislation would probably have been rejected in many states and localities.

As noted above, the federal health and environmental protection bureaucracies are generally larger and more professional than their state and local counterparts. Once a substantial program of environmental protection is launched, these federal bureaucracies' very size, professional orientation, and remoteness also makes them comparatively less sensitive to public discontent when the economic and social costs of such programs become apparent, particularly if these costs fall disproportionately on a few regions. For analogous reasons, public protests, especially if localized, will have less impact on federal judges and legislators than on their state and local counterparts.

Thus a variety of "ratchet" factors make it less likely that federal (as opposed to state or local) environmental programs initially undertaken in part out of moral concern will be abandoned or compromised because of the sacrifices entailed. Under centralized decisionmaking these sacrifices may be less visible (because of fiscal mechanisms) or more palatable (because widely shared). Or the sacrifices may be discounted because federal officials are simply less sensitive to short-term swings in public attitudes. These features of national decisionmaking would be welcomed by those who embrace a genuine moral commitment to environmental protection but fear their inability to maintain that commitment in the face of subsequent privations. Delegation of environmental programs to the federal government can accordingly be viewed as a self-binding mechanism—an insurance policy against *akrasia*. . . .

The Antithetical Rationales: Local Resistance to National Environmental Policies

. . . Having catalogued the reasons favoring centralized determination of environmental policies, we are now in a position to understand more precisely the corresponding grounds of state and local resistance to such policies. For the virtues of federal dictation are matched by corresponding vices.

Diseconomies of Scale

While centralized decisionmaking may be necessary in order to overcome the commons problem and deal with spillovers, it also often generates burdens that are or will appear to be unjustified in particular localities. Federal environmental programs typically place heavy reliance on nationally uniform standards or controls. These uniformities, which reflect both political and administrative constraints in federal decisionmaking, impose economic and social costs on certain areas that are unnecessary or excessive in relation to the benefits obtained. For example, the uniform federal emission limitations on new automobiles impose high costs on rural areas while yielding few compensating benefits. . . .

Even when the federal government utilizes nonuniform measures to deal with spillover problems, the resulting distribution of benefits and burdens may plausibly be regarded as unjust by some of the states affected. For example, requirements of nondegradation in federal air pollution policy limit development in "clean" states for the benefit of citizens in "dirty" states without providing compensation for the sacrifices (in the form of development forgone) thereby imposed on "clean" states. Local inequities (perceived or real) are almost inevitable when Congress attempts to deal with interstate conflicts on a "wholesale" basis through statutes applicable to all states and dealing with a broad class of problems, in contrast to the resolution of such disputes at the "retail" level by courts adjudicating a particular controversy between two states.

The Impairment of Self-Determination

Environmental interests may well enjoy relatively more influence if environmental decisions are shifted from the state and local to the national level. But this shift is accomplished at the expense of local political self-determination. Decisions about environmental quality have far-reaching implications for economic activity, transportation patterns, land use, and other matters of profound concern to local citizens. Federal dictation of environmental policies depreciates the opportunity for and value of participation in local decisions on such matters. The impairment of local self-determination is considerably aggravated when . . . local fiscal resources and governmental powers are conscripted by federal agencies. Nor is it clear that this loss of self-determination always purchases a net gain in social welfare. Even if unorganized interests (such as environmentalists) are underrepresented at the local level, they may well be . . . overrepresented at the national level.

National Ideals as "Pyramids of Sacrifice"

Moral crusades enjoy little credit with the nonbelievers who are taxed to underwrite such ventures. Motorists facing drastic curtailment of mobility, the poor with increased utility bills, and the unemployed in rural areas closed to development may understandably view the sacrifices they are called upon to make as excessive. Resistance and resentment may be heightened by the fact that many environmental programs distribute the costs of controls in a regressive pattern while providing disproportionate benefits for the educated and wealthy, who can better afford to indulge an acquired taste for environmental quality than the poor, who have more pressing needs and fewer resources with which to satisfy them. These circumstances may foster, and in part justify, a cynical attitude towards the moral justifications advanced by upper-middle class advocates for environmental programs which benefit that class disproportionately. The impairment of local political mechanisms of self-determination and official accountability involved in federally dictated environmental programs affords further grounds for resentment.

It is not too fine a conceit to mark a parallel between the local impact of national environmental policies and Peter Berger's assessment of the social and moral costs of development in third-world nations. In his book *Pyramids of Sacrifice*, Berger decries the insensitive willingness of governmental elites to impose severe sacrifices on the populace, repressing opposition to such sacrifices on the grounds that they are necessary to "development" but will not be undertaken voluntarily, and that once development has occurred the society will look back upon the sacrifices as justified. Aspects of national environmental policy might similarly be viewed as the insensitive imposition of sacrifices on local communities, viewed as unjustified by those that bear them (in particular the poor communities), for the sake of a national elite's vision of a better society. Why should Washington force San Francisco to have cleaner air than it apparently wants?

Rehabilitating Interstate Competition: Rethinking the "Race-to-the-Bottom" Rationale for Federal Environmental Regulation

RICHARD L. REVESZ

Perhaps the most widely accepted justification for environmental regulation at the federal level is that it prevents states from competing for industry by offering pollution control standards that are too lax. This competition is said to produce a "race to

Reprinted from 67 *New York University Law Review* 1210 (1992).

the bottom"—that is, a race from the desirable levels of environmental quality that states would pursue if they did not face competition for industry to the increasingly undesirable levels that they choose in the face of such competition. . . .

This Article challenges the accepted wisdom on the race to the bottom. It argues that, contrary to prevailing assumptions, competition among states for industry should not be expected to lead to a race that decreases social welfare; indeed, as in other areas, such competition can be expected to produce an efficient allocation of industrial activity among the states. It shows, moreover, that federal regulation aimed at dealing with the asserted race to the bottom, far from correcting evils of interstate competition, is likely to produce results that are undesirable.

This challenge to the validity of race-to-the-bottom arguments should lead to serious questioning of the federal environmental statutes. While there are other rationales for regulation at the federal level, they rest upon different empirical foundations and justify different forms of federal intervention than does the race-to-the-bottom rationale. Most importantly, the other prominent market-failure argument for federal environmental regulation is that, in the absence of such regulation, interstate externalities will lead states to underregulate because some of the benefits will accrue to other states. But interstate externalities explain only isolated parts of the federal environmental statutes, with a good portion of the remainder being justified on race-to-the-bottom grounds. Alternatively, one might justify federal regulation on public-choice grounds by arguing that state political processes systematically undervalue the benefits of environmental protection or overvalue the corresponding costs, whereas at the federal level the calculus is more accurate. But this rationale rests upon an empirical claim about failures in the political process rather than failures in the market for industrial location. Thus, at the very least, a different predicate would have to be constructed to defend the federal statutes. . . .

The Race to the Bottom over Environmental Regulation

Because commentators have not paid sufficient attention to the characteristics a race to the bottom over environmental regulation would have, I start by defining the elements of the race. . . .

First, consider an "island" jurisdiction—a single jurisdiction surrounded by ocean, which is unaffected by what occurs beyond its borders. This island jurisdiction has a number of firms engaged in industrial activity that produces air pollution. The citizens of the jurisdiction suffer adverse health effects as a result of the pollution.

In the absence of regulation, the firms will choose the level of pollution that maximizes their profits and, as is the case generally with externalities, will ignore the social costs produced by their activities—the costs borne by the citizens who must breathe air of poor quality. The firms will be able to produce their goods more cheaply and will pollute more than if they were forced to bear these social costs.

Traditional economic theory holds that the socially optimal level of pollution reduction is the level that maximizes the benefits that accrue from such reduction to the individuals who breathe the polluted air, minus the costs of pollution control. To achieve this optimal reduction, a regulator must force polluters to internalize the costs that they impose on breathers. For the purposes of this discussion, it does not matter whether the regulator achieves this goal through command-and-control regulation, Pigouvian taxes, marketable permit schemes, or other strategies. Finally, for comparative purposes, assume that in this island jurisdiction the level of pollution reduction chosen by the regulator does not affect entry into or exit from the market. Thus, the number of polluters in the jurisdiction will be independent of the actions of the regulator.

Second, consider instead a "competitive" jurisdiction. This jurisdiction is affected by the actions taken in other jurisdictions, and, in turn, its own actions have effects beyond its borders. I have in mind a state within a federal system.

In order to focus the discussion on the competition among states to attract industry, assume for now that there are no interjurisdictional pollution externalities. Assume further, for ease of exposition, that the total number of firms across jurisdictions remains fixed—that although firms can move from one jurisdiction to another, there is no entry into or exit from the national market. Within the national market then, other factors being equal, firms will try to reduce the costs of pollution control by moving to the jurisdiction that imposes the least stringent requirements. Industrial migration will occur whenever the reduction in the expected costs of complying with the environmental standards is lower than the transaction costs involved in moving.

As in the island situation, competitive jurisdictions will want to set a pollution reduction level that takes account of the benefits to its citizens of such reduction and of the costs to polluters in the jurisdiction of complying with this level. There will be, however, an additional factor to consider: the location of a firm can lead to the creation of jobs, and thus to increases in wages and taxes—important benefits for a state. As a result of this additional factor, competitive jurisdictions will consider the potential benefits, in terms of inflows of industrial activity, of setting standards that are less stringent than those of other jurisdictions, and, conversely, the potential costs, in terms of outflows of industrial activity, of setting more stringent standards.

With this background in mind, I present the structure of the race-to-the-bottom argument. Remember, however, that I am not positing that a competitive jurisdiction will in fact engage in a race to the bottom. I am, instead, merely explaining the theoretical structure of race-to-the-bottom claims.

The simplest example of the race to the bottom is one in which there are two identical jurisdictions. Assume that State 1 initially sets its level of pollution reduction at the level that would be optimal if it were an island. State 2 then considers whether setting its standard at the same level is as desirable as setting it at a less stringent level. Depending on the benefits of pollution reduction, costs on polluters, and benefits from the migration of industry, the less stringent standard may be preferable, and industrial migration from State 1 to State 2 will ensue.

To recover some of its loss of jobs and tax revenues, State 1 then considers relaxing its standard, and so on. This process of adjustment and readjustment continues until an equilibrium is reached, in which neither state has an incentive to change its standard further.

At the conclusion of this race, both states will end up with equally lax standards, and they will not experience any inflow or outflow of industry. Each of these competitive states will thus have the same level of industrial activity that it would have had as an island jurisdiction. Social welfare in these states, however, will be less than it would be in identical island jurisdictions, because, as a result of the race to the bottom, the states will have adopted suboptimally lax standards.

The race to the bottom is the result of non-cooperative action on the part of the states. If they could enter into an enforceable agreement to adopt the optimally stringent standard, they could maximize social welfare without the need for federal regulation.

An alternative to an agreement among the states is pressure by the states for federal regulation. Federal regulation is justified under the race-to-the-bottom theory because it can eliminate the undesirable effects of the race. If the federal regulation sets the standard at the level that the states would find optimal if they were islands, the states will be precluded from competing for industry by offering less stringent standards. They will end up with optimal, rather than suboptimally lax, standards, and they will not suffer the resulting loss in social welfare. In short, both states will be better off as a result of the federal regulation. The problem can thus be described in principal-agent terms, in which the principals, the states, empower an agent, the federal government, to achieve their goal of obtaining protective environmental standards.

The race to the bottom is a form of the prisoner's dilemma. . . . First, each individual has a dominant strategy—a course of action that he follows regardless of what the other individual does. . . . Second, if each individual uses his dominant strategy, the final outcome is Pareto-inferior in that both would have been better off with another outcome. . . . Third, even if the individuals can communicate beforehand and agree to avoid the Pareto-inferior outcome, unless they can somehow enter into a binding agreement, they will ultimately defect and follow their dominant strategies.

To see the applicability of the prisoner's dilemma, consider a simple race-to-the-bottom example in which each of two competitive jurisdictions has only two choices: it can either set the optimally stringent standard that it would choose if it were an island, or it can set a suboptimally lax standard. In the presence of a race to the bottom, if one jurisdiction sets the optimally stringent standard, the other will set the lax standard and will benefit from industrial migration; in contrast, if one jurisdiction sets a suboptimally lax standard, the other will do so as well to avoid the outflow of industry. Thus, where the jurisdictions must choose between only two environmental standards, the dominant strategy for each is to pick the suboptimally lax standard, even though they would both be better off picking the stringent standard. . . .

Finally, it is important to stress that the existence of interstate competition for industry is not sufficient, by itself, to produce a race to the bottom or, consequently, to justify federal regulation. Obviously, a race to the bottom requires not just the existence of a "race," but also that the race be "to the bottom." This latter element requires, first, that a competitive jurisdiction adopt a less stringent pollution control standard than an otherwise identical island jurisdiction would have adopted. Second, it requires that the less stringent standards that emerge from the competitive process be socially undesirable. . . .

The Uncertain Theoretical Foundation of Race-to-the-Bottom Arguments

. . . Until now, the legal literature has done little more than assert in a conclusory fashion that interstate competition to attract industry will reduce a jurisdiction's social welfare. It has not shown why, when an island jurisdiction is placed in a competitive situation, it becomes a participant in a race to the bottom. This Part demonstrates that there is no support in the theoretical literature on interjurisdictional competition for the claim that, without federal intervention, there will be a race to the bottom over environmental standards.

An Initial Hurdle

Race-to-the-bottom advocates must clear an initial hurdle: for the competition among states to attract industry to be a race to the bottom, interstate competition must be socially undesirable. But interstate competition can be seen as competition among producers of a good—the right to locate within the jurisdiction. These producers compete to attract potential consumers of that good—firms interested in locating in the jurisdiction. Even though states might not have the legal authority to prevent firms from locating within their borders, such firms must comply with the fiscal and regulatory regime of the state; the resulting costs to the firms can be analogized to the sale price of a traditional good.

If one believes that competition among sellers of widgets is socially desirable, why is competition among sellers of location rights socially undesirable? If federal regulation mandating a supra-competitive price for widgets is socially undesirable, why is federal regulation mandating a supra-competitive price for location rights socially desirable? . . .

The Theoretical Literature

Rather than undertake the daunting task of describing the numerous economic models that shed some light on the race-to-the-bottom question, I center my account of the development and current state of the literature on a discussion of three principal works. . . .

In his influential article published in 1956, Charles Tiebout argues that a decentralized governmental structure, with multiple jurisdictions competing for residents, produces a Pareto-optimal outcome.

Tiebout argues that individuals will sort themselves into the jurisdictions offering the mix of taxes and public services that they prefer and that these jurisdictions will be of the optimal size. He concludes that it is therefore preferable to provide public services at the local level, rather than at the federal level and, more pertinently, that interjurisdictional competition is desirable.

Two of Tiebout's assumptions are problematic for my purposes. His . . . assumption that individuals do not consider the employment opportunities of their prospective communities and that no productive activities take place in those communities assumes away the issue that is central to the race-to-the-bottom argument: the effects on the environment of efforts by jurisdictions to attract firms in order to provide jobs for their residents. Second, much of the legal literature has dismissed as unrealistic the assumption of perfect mobility by individuals. There may, indeed, be substantial transaction costs in exiting one jurisdiction and moving to another, particularly in a world in which individuals have jobs and do not live solely on dividend income.

In his article published in 1975, William Fischel extends the Tiebout analysis to deal with the problem of industrial location. In his model, firms are owned by outsiders to the jurisdiction, do not employ any of the jurisdiction's residents, and all of their production is for export from the jurisdiction. The environmental externalities produced by these firms are uniformly distributed over all the residents, and there are no interjurisdictional spillovers. . . .

Fischel contemplates that jurisdictions are able to exclude firms through zoning decisions. They will refuse to permit a firm to locate unless it makes a direct cash payment to the zoning board, to be divided equally among the residents. The minimum amount that a jurisdiction would demand for the deterioration of its environmental quality is equal to the environmental harm caused by the firm. Any amount over this minimum constitutes a "profit," which competition among jurisdictions will drive to zero. . . .

The Fischel model suggests that competition among jurisdictions leads to economically desirable results, rather than to a race to the bottom. Two of his assumptions, however, are troubling. First, under his model, firms do not hire residents of the jurisdiction in which they locate. Thus, like Tiebout, he assumes away one of the cornerstones of race-to-the-bottom arguments: that jurisdictions will relax their environmental standards to suboptimal levels in order to provide jobs for their residents. Second, also like Tiebout, Fischel assumes that individuals are perfectly mobile.

These two shortcomings are addressed by Wallace Oates and Robert Schwab in an article published in 1988. In their model,

> jurisdictions compete for a mobile stock of capital by lowering taxes and relaxing environmental standards that would otherwise deflect capital elsewhere. In return for an increased capital stock, residents receive higher incomes in the form of higher wages. The community must, however, weigh the benefits of higher wages against the cost of foregone tax revenues and lower environmental quality.

Oates and Schwab envision jurisdictions that are large enough to allow individuals to live and work in the same jurisdiction. Moreover, they assume that there are no interjurisdictional externalities: pollution generated in one jurisdiction does not spill over into another.

Each jurisdiction produces the same single good, which is sold in a national market. The production of the good requires capital and labor and produces waste emissions. The instrument of environmental policy is command-and-control regulation: each jurisdiction sets the total amount of allowable emissions. In addition, each jurisdiction raises revenues by levying a tax on each unit of capital. Capital is perfectly mobile across jurisdictions and seeks to maximize its after-tax earnings.

Unlike capital, however, labor is perfectly immobile.[1] Each individual in the community, who is identical in both tastes and productive capacity, puts in a fixed

---

[1] . . . In a companion, unpublished manuscript, Oates and Schwab argue that their conclusion that competition among states produces efficient outcomes holds even if individuals are mobile. . . .

If individuals are mobile, they will sort out, as in the Tiebout model, by reference to their preferences for environmental protection. Individuals who are willing to trade off a great deal in wages for better environmental quality will move to jurisdictions that impose stringent controls on industry; individuals who attach less importance to environmental quality will go to dirtier areas. . . .

period of work each week, and everyone is employed. Additional capital raises the productivity of workers and, therefore, their wages.

Oates and Schwab describe the role of an individual resident of a jurisdiction as follows:

> First, he is a consumer, seeking in the usual way to maximize utility over a bundle of goods and services that includes a local public good, environmental quality. And, second, he supplies labor for productive purposes in return for his income. From the latter perspective, residents have a clear incentive to encourage the entry of more capital as a means to increase their wages. But this jurisdiction must compete against other jurisdictions. To attract capital, the community must reduce taxes on capital (which lowers income and, therefore, indirectly lowers utility) and/or relax environmental standards (which lowers utility directly). These are the tradeoffs inherent in interjurisdictional competition.

Each jurisdiction makes two policy decisions: it sets a tax rate on capital and an environmental standard. Oates and Schwab show that competitive jurisdictions will set a tax rate on capital of zero. For positive tax rates, the revenues are less than the loss in wages that results from the move of capital to other jurisdictions; subsidies would cost the jurisdiction more than the increase in wages that additional capital would generate.

In turn, competitive jurisdictions will set an environmental standard that is defined by equating the willingness to pay for an additional unit of environmental quality with the corresponding change in wages. Pollution beyond this level generates an increment to wage income that is less than the value of the damage to residents from the increased pollution; in contrast, less pollution creates a loss in wage income greater than the corresponding decrease in pollution damages.

Oates and Schwab show that these choices of tax rates and environmental standards are socially optimal. . . .With respect to the environmental standard, competitive jurisdictions equate the marginal private cost of improving environmental quality (measured in terms of forgone consumption) with the marginal private benefit. For tax rates of zero, the marginal private cost is, as noted above, the decrease in wage income produced by the marginal unit of environmental protection. This decrease is also the marginal social cost, since it represents society's forgone consumption. Oates and Schwab conclude that "competition among jurisdictions is thus conducive to efficient outcomes." Thus, there is no race to the bottom. . . .

In their model, Oates and Schwab assume that capital does not require the provision of public services, such as roads, and police and fire protection. If it does, the optimal tax rate on capital, rather than zero, is the rate that exactly covers the cost of these services. It follows from the preceding discussion that, for this rate of taxation, jurisdictions would set environmental standards at the optimal level. . . .

Lessons from the Theoretical Literature

The conclusions that emerge from this review of the theoretical literature point strongly against race-to-the-bottom claims. Tiebout, Fischel, and Oates and Schwab all conclude, in situations progressively more analogous to the problem of this Article, that interstate competition is not inconsistent with the maximization of social welfare. There are no formal models supporting the proposition that competition among states creates a prisoner's dilemma in which states, contrary to their interests, compete for industry by offering progressively laxer standards. . . .

One should not overstate the nature of my claim against race-to-the-bottom justifications for environmental regulation. The fact that there are no models consistent with race-to-the-bottom claims does not rule out the possibility that further research will yield such models. Modeling, by necessity, involves making strong sets of assumptions. The Oates and Schwab work, which has studied the problem in the most systematic way, is no exception. A theoretical literature evolves as assumptions are relaxed, often one at a time. It is certainly conceivable that the next generation of theoretical work will provide support for race-to-the bottom arguments. But the fact remains that race-to-the-bottom arguments in the environmental area have been made for the last two decades with essentially no theoretical foundation.

The Implications of the Environmental Race to the Bottom

Having shown . . . that the race-to-the-bottom hypothesis, though influential, lacks a sound theoretical basis, this Part shows that even if there *were* such a race in the environmental arena, federal regulation would not necessarily be an appropriate response. The analysis centers on two important consequences of race-to-the-bottom arguments in favor of federal environmental regulation. First, if the premises underlying the race to the bottom hold, federal environmental regulation will have undesirable effects on other state regulatory or fiscal interests; the supposed benefits of federal environmental regulation should therefore be balanced against these undesirable effects. Second, logic compels the conclusion that arguments in favor of federal environmental regulation are a frontal challenge to federalism, because the problems that they seek to correct can be addressed only by exclusive federal regulatory and fiscal powers. Both these consequences, which have been unexplored in the literature, ought to cast even more doubt on the validity of race-to-the-bottom arguments for federal environmental regulation.

Race-to-the-bottom arguments appear to assume, at least implicitly, that jurisdictions compete over only one variable—in this case, environmental quality. So, jurisdictions that would choose to have stringent environmental quality standards in the absence of interstate competition adopt less stringent standards as a result of such competition, and, consequently, suffer a reduction in social welfare.

Consider, instead, the problem in a context in which states compete over two variables—for example, environmental protection and worker safety. Assume that, in the absence of federal regulation, State 1 chooses a low level of environmental protection and a high level of worker safety. State 2 does the opposite: it chooses a high level of environmental protection and a low level of worker safety protection. Both states are in a competitive equilibrium: industry is not migrating from one to the other.

Suppose that federal regulation then imposes on both states a high level of environmental protection. The federal scheme does not add to the costs imposed upon industry in State 2, but it does in State 1. Thus, the federal regulation will upset the competitive equilibrium, and unless State 1 responds, industry will migrate from State 1 to State 2. The logical response of State 1 is to adopt less stringent worker-safety standards. This response will mitigate the magnitude of the industrial migration that would otherwise occur.

Thus, federal environmental standards can have adverse effects on other state programs. Such secondary effects must be considered in evaluating the desirability of federal environmental regulation. Most importantly, the presence of such effects suggests that federal regulation will not be able to eliminate the negative effects of interstate competition. Recall that the central tenet of race-to-the-bottom claims is that competition will lead to the reduction of social welfare; the assertion that states enact suboptimally lax environmental standards is simply a consequence of this more basic problem. In the face of federal environmental regulation, however, states will continue to compete for industry by adjusting the incentive structure of other state programs. Federal regulation thus will not solve the prisoner's dilemma. . . .

One might respond to these arguments by saying that worker safety should also be the subject of federal regulation. But states would then compete over minimum wage laws, fair labor standards, and so on. It is difficult to imagine a federal system in which all the regulatory requirements that impose costs on industry are mandated at the federal level.

Suppose, however, that this were the case. States impose burdens on industry not only through regulation but also through taxes, which fund a variety of state programs and functions. So, if all regulatory programs are federalized, states still will be able to compete through their fiscal powers. Consider, now, an example in which State 1 and State 2, as island states, would impose both stringent regulatory standards and high corporate taxes. When placed in a competitive situation, State 1 chooses stringent regulatory standards and low corporate taxes, whereas State 2 does the opposite. If the federal government then requires stringent regulatory standards, State 2 will respond by lowering its taxes, and by, say, decreasing the size of its income maintenance programs. This reduction is a direct by-product of the federal regulatory scheme.

Thus, even if all regulatory functions are federalized, federal regulation will continue to have an adverse effect on other issues of state concern—in this example, social welfare programs. Moreover, such a scheme will not eliminate the reduction in social welfare that results from competition among the states.

The next logical step, of course, is to suggest preemption of state taxes, because otherwise the supposedly evil effects of interstate competition will persist. The race-to-the-bottom rationale for federal environmental regulation is, therefore, radically underinclusive. It seeks to solve a problem that can be addressed only by wholly eliminating state autonomy. The prisoner's dilemma will not be solved through federal environmental regulation alone, as the race-to-the-bottom argument posits. States will simply respond by competing over another variable. Thus, the only logical answer is to eliminate the possibility of any competition altogether. In essence, then, the race-to-the-bottom argument is an argument against federalism. . . .

Conclusion

This Article should not be read as a definitive refutation of race-to-the-bottom arguments in the environmental area. It is intended, instead, to question the underpinnings of such arguments and to suggest that the forces of interstate competition, far from being conclusively undesirable, are at least presumptively beneficial. If this project proves successful, it will be followed, without a doubt, by studies attempting to define specific circumstances in which federal regulation could improve upon the results of interstate competition.

Notes and Questions

1. Is the "tragedy of the commons" justification presented by Stewart consistent with Hardin's treatment of this concept? Is it, instead, a race-to-the-bottom argument? Is mobility of industry and commerce (a central element of the race-to-the bottom argument) a necessary condition for the existence of a tragedy of the commons in a federal system?

2. Stewart states: "The characteristic insistence in federal environmental legislation upon geographically uniform standards and controls strongly suggests that escape from the Tragedy of the Commons by reduction of transactions costs has been an important reason for such legislation." Why is the uniformity of federal standards necessary for addressing the tragedy of the commons? Might nonuniform standards be an alternative? This issue is considered further in Chapter IX.

3. With respect to which of the following regulatory activities is the economies of scale argument most persuasive?
    (a) scientific studies of the adverse effects of pollution
    (b) the setting of ambient standards
    (c) the setting of emission standards
    (d) the enforcement of emission standards
What types of information might be relevant to allocating, by means of emission standards, the pollution control burden necessary to meet ambient standards? Is the federal government likely to have better information than the states with respect to these matters?

4. Under most of the federal environmental statutes, regulations with national scope can be challenged only in the United States Court of Appeals for the District of Columbia Circuit. In light of Stewart's argument about disparities in effective representation, are these exclusive venue provisions desirable? In the 1970s, environmental groups were primarily engaged in litigation, challenging regulations that they considered too lax (and, by the second half of the decade, challenging EPA's failure to promulgate regulations pursuant to the schedules prescribed by Congress). In the 1990s, environmental groups rely more heavily on nonlitigation strategies, and more of their activities are conducted at the state and local level. From these trends, can one draw any conclusions about the relative effectiveness of environmental representation at the federal level as opposed to the state and local levels?

5. Stewart maintains that "[a] state that encourages economic development at the expense of environmental quality may inflict economic loss (in the form of industrial migration or decreased economic growth) on other states that prefer a higher level of environmental quality." A state that seeks to attract industry in this manner is essentially lowering the price for industrial location. Is its decision different from the decision of a manufacturer to lower the price of its product, where this price change has undesirable economic effects on a competitor?
    Consider the following distinction between technological and pecuniary externalities:

[N]ot all relationships that appear to involve externalities will produce resource mis-allocation. There is a category of pseudo-externalities, the *pecuniary externalities*, in which one individual's activity level affect the financial circumstances of another, but which need not produce a misallocation of resources in a world of pure competition. . . . Pecuniary externalities result from a change in the prices of some inputs or outputs in the economy. An increase in the number of shoes demanded raises the price of leather and hence affects the welfare of the purchasers of handbags. But unlike a true externality (. . . a *technological externality*), it does not generate a *shift* in the handbag production function. (William J. Baumol and Wallace E. Oates, *The Theory of Environmental Policy* [New York: Cambridge University Press, 2d ed., 1988], at 29)

Is the externality described by Stewart technological or pecuniary? Does it affect social welfare?

6. In discussing the "politics of sacrifice," Stewart maintains that "[c]ommunities no less than individuals may be far more willing to undertake sacrifices for a common ideal if there are effective assurances that others are making sacrifices too." Is this phenomenon simply a product of the tragedy of the commons or of the free-rider problem? Does it have an independent basis?

7. Assess the strength of Revesz' criticisms of race-to-the-bottom arguments for environmental regulation. An implication of the Oates and Schwab article, on which Revesz relies, is that if states set a positive tax rate on mobile capital (that exceeds the cost of providing public services), they will set suboptimally lax environmental standards, but if they give mobile capital a net subsidy, they will set suboptimally stringent environmental standards. Which outcome is more likely? The divisions in the economic literature on this question are reviewed in Mieszkowski and Zodrow, "Taxation and the Tiebout Model: The Differential Effects of Head Taxes, Taxes on Land Rents, and Property Taxes," 27 *Journal of Economic Literature* 1098 (1989). For related discussion, see Rose-Ackerman, "Environmental Policy and Federal Structure: A Comparison of the United States and Germany," 47 *Vanderbilt Law Review* 1587 (1994).

8. Revesz also claims that even if there were a race to the bottom, its undesirable effects would not be cured by federal regulation. Is that claim persuasive?

9. In a recent article, Stewart argues that there are ways in which a race to the bottom might occur:

A . . . possibility is that A is uncertain about the exact value that B's political system places on environmental protection and how B's government will respond to A's choice of standard, and vice-versa. If each nation's choice of standards depends on those chosen by other nations, if each is uncertain as to what choices others will make, and if each is unsure how others will respond to its choices, it is possible that each nation might indeed adopt lower standards. ("Environmental Regulation and International Competitiveness," 102 *Yale Law Journal* 2039, 2059 [1993])

Might this type of uncertainty also lead countries to adopt suboptimally stringent standards? If so, under what circumstances?

10. Under the Clean Air Act, does compliance with the National Ambient Air Quality Standards (NAAQS) necessarily reduce interstate externalities? Does compliance with emission standards have this effect? Might federal ambient standards exacerbate interstate spillovers? In connection with the latter question, consider ways in which states might attempt to meet the ambient standards without reducing their emissions? The issues discussed in this

question and in the two that follow are explored in Revesz, "Federalism and Interstate Environmental Externalities," 144 *University of Pennsylvania Law Review* (forthcoming 1996).

11. Sections 110(a)(2)(D) and 126 of the Clean Air Act are the most comprehensive provisions aimed directly at constraining interstate pollution spillovers. Even though Congress adopted these provisions in 1977, indicating in the legislative history that the prior scheme had been largely ineffective at addressing the problem of spillovers, no court has ever constrained upwind pollution. What would be a desirable set of rules for determining when to enjoin upwind pollution? More specifically, consider the following scenarios, defined by reference to whether an upwind and downwind state meets the NAAQS:

(a) The downwind state would violate the NAAQS even absent the upwind pollution
(b) The downwind state violates the NAAQS only as a result of the upwind pollution
(c) The downwind state meets the NAAQS even despite the upwind pollution, but this pollution reduces the downwind state's margin for industrial growth.
(d) The same as scenario (c), except that the upwind pollution causes the downwind state to violate a state ambient standard that is more stringent than the corresponding federal standard.

In general, should the relative stringency of emissions regulations in the two states matter? In connection with the latter question, should it matter whether the downwind state has a substantial industrial base that will be affected by these standards, or whether, instead, the bulk of the burdens will be borne by the upwind state?

12. If a state could stop polluted air at its borders, as it can stop garbage, and if there were no federal regulation, the permissibility of restrictions on out-of-state air pollution would be governed by the dormant commerce clause. Can the understanding of federalism implicit in the dormant commerce clause inform, as a matter of policy though not of constitutional law, the standards to be used under sections 110(a)(2)(D) and 126 of the Clean Air Act? In connection with the dormant commerce clause, the Supreme Court has repeatedly held that a state cannot ban the disposal of out-of-state garbage and that it cannot impose discriminatory conditions such as higher disposal fees on out-of-state garbage. How could this standard be translated to the air pollution context under the hypothetical scenario presented in this question?

13. Consider the rationales for federal regulation in the following components of the Clean Air Act (the statutory scheme is discussed in Chapter IX):

(a) NAAQS, which impose minimum levels of ambient environmental quality nationwide
(b) prevention of significant deterioration (PSD), which imposes more stringent ambient standards in areas that have air quality that is better than that prescribed by the NAAQS
(c) new source performance standards (NSPS), which impose nationally uniform emission standards on certain categories of new sources, such as electric utilities
(d) emission standards for automobiles
(e) acid rain provisions, which are designed to counteract the adverse effects from the long-distance transport of certain pollutants

14. In the case of emission standards for automobiles, the federal standards create both a floor and a ceiling (they preempt both less stringent and more stringent state requirements). In the case of the other components of the Clean Air Act that are discussed in the preceding question, the federal standards set only a floor (they preempt only less stringent state requirements). Is this distinction justifiable?

15. Consider the rationales for federal regulation in the following components of the Superfund statute (the statutory scheme is discussed in Chapter X):

    (a) the liability scheme, which, in the event of the release of hazardous substances from a site, imposes liability on generators and transporters, as well as the site owner and certain former owners

    (b) the cleanup standards, which specify the extent to which the contamination at hazardous waste sites must be remedied

    (c) the tax scheme, which raises revenues primarily from chemical and petroleum companies to finance cleanups at sites at which liable parties are either insolvent or cannot be found

For which of these components are the claims for federal regulation most persuasive? Least persuasive?

16. The European Union (then known as the European Economic Community) was established by the Treaty of Rome in 1957. This treaty did not contain any specific rules dealing with environmental protection. Environmental regulation between 1957 and 1986 was based principally on Article 100, which authorizes the issuance of directives "for the approximation of such provisions laid down by law, regulation or administrative action in Member States as directly affect the establishment or functioning of the common market." A paradigmatic measure authorized by this provision would be the harmonization of product standards, so that a particular product could be sold freely in all the member countries.

This provision was also invoked, for example, to justify the issuance of both ambient standards and emission standards for air pollutants. How do such standards "directly affect the establishment or functioning of the common market?" One answer might be that the equalization of the costs of production is necessary for manufacturers from all member countries to be on an equal footing. Do either uniform ambient standards or uniform emission standards further this goal? If, with respect to a particular pollutant, the Union has adopted only an ambient standard, what is the effect of such a standard on countries that, prior to its adoption, had different levels of air quality? What is the effect of uniform emission standards on countries that have different investments in infrastructure, education, health care, wages, and productivity, as well as different taxing structures?

17. The Treaty of Rome was amended by the Single European Act in 1986 and by the Treaty on European Union (the Maastricht Treaty) in 1992. Now, Articles 100a and 130r through 130t explicitly deal with environmental protection. Most importantly, Article 130r(1) provides:

> Community policy on the environment shall contribute to pursuit of the following objectives:
>
> • preserving, protecting, and improving the quality of the environment;
>
> • protecting human health;
>
> • prudent and rational utilization of natural resources;
>
> • promoting measures at international level to deal with regional or worldwide environmental problems.

Moreover, Article 130t states, consistently with the prevailing practice in the United States: "The protective measures adopted . . . shall not prevent any Member State from maintaining or introducing more stringent protective measures. Such measures must be compatible with this Treaty." Is there tension between this provision and the pre-1986 rationale for environmental regulation? If so, what might explain the evolution?

18. The Maastricht Treaty adopted a "subsidiarity" principle, which provides in Article 3b:

> In areas which do not fall within its exclusive competence, the Community shall take action, in accordance with the principle of subsidiarity, only if and in so far as the objectives of the proposed action cannot be sufficiently achieved by the Member States and can therefore, by reason of the scale or effects of the proposed action, be better achieved by the Community.

(A similar principle had applied exclusively to environmental regulation under the Single European Act.) What inquiry should be undertaken to determine whether a proposed directive is consistent with the subsidiarity principle? Does Article 3b consititutionalize the policy debate that is ongoing in the United States concerning which level of government should have responsibility for environmental regulation? Are there differences?

For discussion of the application of the subsidiarity principle to environmental law, see Brinkhorst, "Subsidiarity and European Environmental Policy," in *Proceedings of the Jacques Delors Colloquium* (Maastricht: European Institute of Public Administration, 1991); Wils, "Subsidiarity and EC Environmental Policy: Taking People's Concerns Seriously," 6 *Journal of Environmental Law* 85 (1994). Wils argues that one basis for justifying regulation at the European Union level is the presence of "psychic spillovers." He states (p. 89):

> Many people in Denmark or the Netherlands are affected by the hunting and trapping of birds in France, even if these birds are not migratory, and irrespective of any touristic interest. Similarly, a large number of EU citizens in various Member States are appalled by bullfighting in Spain and would derive satisfaction from a ban.

Are Wils's "psychic spillovers" coextensive with existence values? How large must these spillovers be to justify communitywide regulation? How would an appropriate inquiry about the magnitude of this spillover be made? Is subsidiarity likely to ever be a constraint under Wils's theory? If so, under what circumstances?

VIII

Environmental Law and Public Choice

So far, the book has generally taken a public interest view of environmental regulation. Under this view, governmental action is designed primarily to mitigate the social welfare losses that arise from the presence of externalities. In contrast, under a public choice view, environmental regulation is seen as the response to the pressure of powerful groups that seek to further their individual interests, generally at the expense of aggregate social welfare.

It is difficult to explain environmental regulation in public choice terms if such regulation is viewed simply as imposing costs—in terms of pollution control requirements—on industrial firms in order to provide benefits—in the form of better health—to the mass of citizens. Public choice theory would predict that the industrial firms would be able to organize more effectively than the mass of citizens: as a result of their smaller numbers and larger individual stakes in the legislative outcome, their free-rider problems would be less serious. Thus, they would be able to block the enactment of environmental legislation.

This account, however, is too naive. Often, environmental regulation protects firms from worse outcomes or imposes on them differential costs, creating some relative gainers and some relative losers. Consistent with public choice theory, such regulation can be explained, at least in some instances, as the response to the pressures of the relative gainers. Four of the five readings in this chapter provide such accounts. The fifth challenges the basic premises of the public choice explanation for environmental regulation.

The article by E. Donald Elliott, Bruce A. Ackerman, and John C. Millian seeks to explain the enactment of the federal Clean Air Act. Their argument proceeds in two steps. First, by the mid-1960s, state environmental regulation had begun to pose a serious threat to powerful groups, particularly the automobile and coal-mining industries. In response, these groups lobbied successfully for federal legislation as a means of either preempting state standards or diminishing the impetus for further state legislation. Second, in 1970, President Richard Nixon, who was preparing to seek reelection in 1972, and Senator Edmund Muskie, his likely opponent, competed to appear more pro-environmental than the other. The authors argue that, as a result of these two phenomena, stringent environmental legislation was adopted despite the weakness of environmental groups. Their account challenges two widely accepted tenets: (1) that federal regulation was a response to inaction on the part of the states, and (2) that industry opposed such regulation.

The excerpt from the book by Bruce A. Ackerman and William T. Hassler describes the forces that led to the 1977 amendments to section 111 of the Clean Air Act, which requires new sources to comply with New Source Performance Standards (NSPS), set by reference to the emission reductions that can be achieved through the use of the best available technology (BAT). Under the regime established in 1970, coal-fired power plants could meet these standards in either of two ways: by burning untreated low-sulfur coal or by "scrubbing" high-sulfur coal. The result of these standards was to increase the relative market share of low-sulfur coal, which is primarily produced in the West, at the expense of Eastern high-sulfur coal. Ackerman and Hassler recount how the high-sulfur coal lobby fought to regain its lost market. Ultimately, aided by the intervention of some environmental groups, it prevailed in securing an amendment to section 111 requiring, in addition to an emission limitation, a percentage reduction over the emissions that would result from the burning of untreated coal. According to Ackerman and Hassler, this standard, though nominally more protective of the environment, in fact increased air pollution in parts of the country.

The article by B. Peter Pashigian examines another aspect of the 1977 amendments to the Clean Air Act: the adoption of the Prevention of Significant Deterioration (PSD) program. The Clean Air Act of 1970 had adopted uniform National Ambient Air Quality Standards (NAAQS), which specified the maximum permissible concentrations of air pollutants that had to be met throughout the country. The statute did not indicate whether areas that had better air quality could degrade up to the level of the NAAQS or whether, instead, they would be subject to more stringent ambient standards. Pashigian shows that support for the PSD program, which prescribed more stringent standards in clean air areas, came disproportionately from areas with comparatively dirty air, where growth was constrained by the NAAQS and which were therefore not going to be affected by the PSD policy. In contrast, opposition came primarily from areas with comparatively clean air, where the PSD program was going to have the effect of constraining economic growth.

The article by Michael T. Maloney and Robert E. McCormick explains that environmental regulation may enhance the profits of existing firms. In particular, regulation typically imposes more stringent controls on new firms, thereby restricting entry and potentially aiding the competitive position of existing firms. The authors provide empirical support for their theory by providing an example in which environmental regulation produced an otherwise unexplained rise in the stock prices of firms in the affected industry.

The article by Daniel A. Farber challenges public choice explanations for environmental regulation. He maintains that such regulation cannot be explained on the basis of rent seeking by either industry or consumers of environmental amenities. Instead, Farber argues, environmental legislation is explained by strong public demand, as well as by the exploitation of this demand by politicians. Relying on the civic republican tradition, he sees the 1970 Earth Day as a "republican moment," defined by widespread public participation fueled by moral discourse. He notes, moreover, that the public interest about environmental regulation has persisted throughout the subsequent decades. With respect to the role of politicians in satisfying the public demand for environmental protection, Farber claims that legislators find it politically advantageous to seek credit on the basis of their progressive environmental record.

Toward a Theory of Statutory Evolution:
The Federalization of Environmental Law

E. DONALD ELLIOTT, BRUCE A. ACKERMAN, AND JOHN C. MILLIAN

Collective Action and Prisoners' Dilemma

The Problem of Collective Action

Modern theories of voluntary organization, derived from Mancur Olson, imply that national environmental groups will be difficult, if not impossible, to organize. Large numbers of citizens, each with only a small stake in clean air, will, if they are rational in the narrow economic sense, decline to invest their time or money in the cause of cleaning up the environment in the hope that they will be able to "free-ride" on the efforts of others. Since everyone will be inclined to "let George do it," it won't get done at all. The paradox, of course, is that everyone ends up worse off than they would have been if they had been able to organize their actions for their collective benefit.

It is a small step from Olson's theory of voluntary organizations to the political corollary that the interest of citizens in a clean environment will be systematically underrepresented in any lawmaking process in which interest group politics plays a significant role. Individual citizens who wish to breathe clean air are a classic example of a large, disorganized population seeking a collective good which will benefit each individual by only a small amount. The costs of environmental regulation, on the other hand, tend to fall heavily on a relatively small number of companies, which are already reasonably well-organized and thus presumably less subject to free-rider problems. According to most popular theories of political influence, well-organized industries would be systematically overrepresented and diffuse environmentalists systematically underrepresented in formulating policy. How, then, is one to explain the passage of strong environmental legislation in the late 1960s and early 1970s. . . .

Politicians' Dilemma

The answer, or at least a more complete answer, can be discovered by considering the problems of environmental organizing and passing environmental legislation as analogous to the game of Prisoners' Dilemma.

Prisoners' Dilemma gets its name from a story about two prisoners who are separately interrogated about a crime. The two were the only witnesses, so if they both refuse to testify, the worst that can happen to them is a one-year conviction for illegal possession of firearms. However, a clever prosecutor approaches each prisoner

Reprinted by permission of the authors and Oxford University Press from 1 *Journal of Law, Economics, and Organization* 313 (1985).

and offers him a proposition: "If you confess and testify against your partner, he'll get life but you'll go free; the only hitch is that if you both confess, you'll both get a sentence of six years for armed robbery. I should tell you that I'm offering the same deal to your partner."

Assuming that the game is played only a single time, and assuming further that the prisoners are rational and motivated only by self-interest, they will both confess—and get six years in jail, rather than keep quiet and get off with only a year. The paradox, of course, is that by pursuing their individual self-interest, the prisoners behave in a way that is contrary to their shared collective interest in shorter sentences. If they could only organize their actions for their common benefit, they would both be better off. . . .

This . . . perspective helps to explain the evolution of environmental law during the late 1960s and early 1970s. Not that the evolving institutional structure technically complied with all the conditions for the game of Prisoners' Dilemma—in contrast to the standard game, our story involves many relevant players, no single subset of which could have coordinated their strategies in a way that guaranteed them an optimal result. Nonetheless, like the prisoners, many of the key actors responded to institutional threats of terrible outcomes by rationally choosing strategies that were very far from first-best from their point of view. We shall, then, use the term *Politicians' Dilemma* to describe situations which are analogous to the game of Prisoners' Dilemma in that the structure of incentives facing the players creates a strong incentive for them to pursue a less than ideal outcome in order to avoid an even less desirable result. . . .

Politicians' Dilemma and Environmental Statutes

The first significant federal statutes regulating air pollution, the Motor Vehicle Pollution Control Act of 1965 and the Air Quality Act of 1967 were not passed because of the political power of environmentalists at the national level but because two well-organized industrial groups, the automobile industry and the soft coal industry, were threatened with a state of affairs even worse from their perspective than federal air pollution legislation—namely, inconsistent and progressively more stringent environmental laws at the state and local level. As a consequence of the structure of our federal lawmaking system, environmentalists were able to organize industry to do their bidding for them. Thus, the first federal legislation regulating air pollution was passed not because environmentalists solved their own organizational problems on the national level but because environmentalists exploited the organizational difficulties of their industrial adversaries at the state and local level.

The auto industry and the soft coal industry undoubtedly would have preferred no government regulation of air pollution rather than federal legislation. When faced with the threat of inconsistent and increasingly rigorous state laws, however, they resolved their Politicians' Dilemma by using their superior organizational capacities in Washington to preempt or control the environmentalists' legislative victories at the state level.

It does not matter to our argument whether environmentalists and industrialists were consciously pursuing the strategy we outline. Like the characters in the Prisoners' Dilemma, they may have been simply reacting rationally to the strategic implications of their situation. What we have found, in short, is empirical support for a paradox previously elaborated by theorists of federal systems.

We will analyze the dynamic by first showing why a federal system gives environmentalists important strategic advantages at the state level, then showing that the strategic situation at the national level was vastly preferable from the polluters' point of view, and finally showing how the strategic interests of polluters coincided with the pursuit of political self-interest by reelection-maximizers and presidential aspirants in the Congress. . . .

The Period of Political Cost-Externalization

The existence of states aids environmentalists in three ways. First, and most obvious, the existence of the states makes it possible for environmentalists to seek piecemeal solutions to their organizational difficulties. Not that the effort to transcend their free-ride problems will be easy—even in smaller states, thousands of people will have to be convinced to take seriously the signals activists are beaming in their direction, and states such as California and New York present the problem of organizing a population the size of Canada's. Nonetheless, even here, the demands on a variety of resources—from political savvy to hard cash—do not compare with the challenges involved in achieving organizational credibility in a nation of a quarter of a billion. . . .

Second, federalism opens up the possibility of a distinctive credit-claiming strategy for aspiring politicians on the state level, which we call *cost-externalization*. Quite simply, dividing the nation into fifty geographic zones makes it almost inevitable that some pollution problems will be generated by out-of-staters. Since midwestern auto workers don't vote on whether California should ban the internal combustion engine to control smog and Appalachian coalminers don't vote on whether New York should ban coal to control sulfur oxides from power plant smoke stacks, these issues promise politicians on the state level the equivalent of a free lunch—"tough" legislation allows them to garner public credit for bringing a benefit to *their* constituents at somebody else's expense.

Finally, as scattered environmental victories begin to appear, this evidence of success will feed efforts in other states. Activists will be prompted to continue the fight, rather than seek out other issues; the media and the public will gradually begin to take greater notice and express increased interest. A bandwagon effect becomes possible: victories in one state may promote the marshalling of the resources necessary for victory in another. Indeed, legislation in one state can stimulate other states to adopt even more stringent laws. . . .

Preemptive Federalization

The Motor Vehicle Air Pollution Control Act of 1965. The first statute which gave the federal government regulatory power over air pollution was the Motor Vehicle Air Pollution Control Act of 1965. The roots of this federal legislation run deep into the California of the 1950s and 1960s. . . . Suffice it to say that through a combination of a cost-externalization strategy by California politicians, auto industry ineptitude, and local environmental organizing, state air pollution legislation had begun to pose a serious threat to the automobile industry by the middle 1960s. California had already adopted a regulatory program requiring the installation of emission controls on all new cars sold in the state, an auto emissions bill was pending in the Pennsylvania state legislature, and New York was considering an emission standards bill even more stringent than California's. . . .

Unlike most other industries, the automobile industry has strong reasons to prefer national legislation over state and local regulation of air pollution. Most manufacturing industries would rather have state and local governments set air pollution standards, because the political and economic costs of controlling their pollution are concentrated at the local level. It is a rare politician who is immune to the charge that a proposal will harm a local, job-creating industry. In addition, some manufacturing industries may be able to play one state off against another by threatening to move their factories out of states which set stringent air pollution standards (thereby creating a true Prisoners' Dilemma from the standpoint of the states).

The automobile industry is in a very different strategic position, however, because it is geographically concentrated and its product, not its factories, is the main source of its pollution. Local politicians can set strict antipollution standards for motor vehicles without fear of being accused of putting their constituents out of work. It is true that pollution controls tend to increase the price of new cars, but the connection between government action and particular price increases is only dimly perceived by voters. And unlike other industries, Detroit could not credibly threaten to stop selling cars in California or other states which established stringent pollution standards. Moreover, differing or inconsistent air pollution standards set at the state and local level were perceived as a serious threat to Detroit's assembly lines. Finally, the companies feared a kind of political domino effect, in which one state legislature after another would set more and more stringent emission standards without regard to the costs or technical difficulties involved.

Ideally the auto companies would have preferred to remain free of any substantial government regulation of pollution, but if they were going to be regulated, federal legislation was preferable to state legislation—particularly if federal standards were set based on technical presentations to an administrative agency rather than through symbolic appeals to cost-externalizing politicians.

During the early 1960s, the automobile industry successfully opposed federal emission standards for motor vehicles. In mid-1965, however, the industry abruptly reversed its position on the advice of Washington lawyer Lloyd Cutler: provided that the federal standards would be set by an administrative agency, and provided that they would preempt any state standards more stringent than California's, the industry would support federal legislation. As a result, Senator Muskie's pending bill to have the federal government set emission standards for motor vehicles was amended to provide that standards would be set by HEW,[1] rather than in the legislation itself, and legislative history was written to leave no doubt that more stringent state laws were preempted. . . .

[1]The Department of Health, Education, and Welfare, which had jurisdiction over environmental regulation before the establishment of the Environmental Protection Agency (EPA)—ED.

The Air Quality Act of 1967. The Air Quality Act of 1967 was the first federal statute to give the federal government a significant role in regulating air pollution from stationary sources such as factories and powerplants. Under the 1967 act, the federal government was to promulgate criteria based on the latest scientific evidence concerning the adverse effects of air pollution. Each state was then to develop its own air pollution control plan based on the federal criteria. If any state failed to adopt a satisfactory plan, the federal government could promulgate one for it.

The story behind the Air Quality Act of 1967 is complicated, but here too the threat of state and local legislation provided the impetus for a crucial industry to acquiesce in federal legislation in the hope that it might dampen local legislative initiatives. Like the automobile industry, the high-sulphur, soft (bituminous) coal industry is geographically concentrated, and its product, not its factories, constitutes the primary source of its air pollution. Soft coal provided a logical target for local politicians anxious to place the blame for pollution on out-of-state sources.

During the mid-1960s, the soft coal industry faced increasingly strict air pollution regulations in the Northeast, which eventually threatened it with the loss of a major market. In 1965, Mayor John Lindsay of New York proposed—and despite strong opposition mounted by the coal industry, the city council eventually passed—a program to ban the use of coal as a heating fuel and to greatly restrict the sulfur content of coals used for other purposes. In 1966, New York, New Jersey, Pennsylvania, and Connecticut announced joint plans to combat air pollution. In March 1967, the threat of strict state legislation which would eliminate markets for high-sulfur coals in most major metropolitan areas increased when the federal HEW released an advisory criteria document reviewing the scientific literature on the health risks of sulfur dioxide, a pollutant which is formed when soft coal is burned.

Its unsuccessful campaign against pollution control legislation in New York City had taught the coal industry that it was virtually impotent in local political arenas. . . . Ideally, the soft coal industry, like the automobile industry, probably would have preferred that there be no government regulation of the pollution produced by its product. However, if there was going to be regulation, federal legislation offered distinct advantages to the coal industry over runaway state and local lawmaking. While the 1967 federal Air Quality Act did not forbid states from setting air pollution standards more stringent than those recommended by HEW, as a practical, political matter, the air quality criteria which HEW established based on the latest scientific evidence would tend to restrain state legislation. Advisory committees within the federal bureaucracy promised to be a far more hospitable forum for the coal industry than the politics of state and local legislatures. . . .

The Air Quality Act of 1967, like the Motor Vehicle Air Pollution Control Act of 1965, passed not because environmentalists were a well-organized pressure group at the federal level, but because their efforts, and the actions of local politicians, created a Politicians' Dilemma for a well-organized industry. Faced with the even less desirable alternative of a significant loss of markets through state and local legislation, the soft coal industry strongly supported passage of [the] bill.

Aspirational Lawmaking

A structurally similar process also accounts for some of the surprisingly stringent provisions of the Clean Air Amendments of 1970. In particular, the requirement that automobile manufacturers reduce their pollution by 90 percent within five years, and the stipulation that EPA ignore economic and technological feasibility, did not result from the success of environmentalists at organizing a strong lobbying presence of their own in Washington. Here too a Politicians' Dilemma was at work. The "prisoners" in this case were politicians, primarily Senator Edmund Muskie and President Richard Nixon. By strategically threatening these political entrepreneurs with the loss of political capital which they had previously worked to build, environmentalists were able to organize them to pass a statute more stringent than the politicians really wanted. In an ideal world both Nixon and Muskie probably would have preferred a compromise statute less likely to alienate either industry or environmentalists, but as in Prisoners' Dilemma, they were confronted with a situation in which they both had to choose the least-worst situation politically.

The structural feature which creates the Politicians' Dilemma is the fragmentation of the lawmaking system between Congress and the Executive, between House and Senate, between legislative committee and legislative committee. This division of lawmaking authority creates a situation in which various politicians can credibly claim credit for any particular law. David Mayhew has argued persuasively that most members of Congress can be thought of as "reelection maximizers." For this breed of legislator, the costs of sponsoring broad legislation such as the Clean Air Act will generally outweigh the benefits. A smaller number of legislators, however, aspire to run for higher office. These aspirants may analyze the political costs and benefits involved in environmental lawmaking in terms that are very different from the simple reelection maximizer.

If he hopes one day to gain the presidency, the aspirant must, somehow or other, gain public recognition as a serious political leader throughout the United States. And to this end, it will not suffice to sprinkle the home district with dams, post offices, and similar goodies, or to help constituents with their Social Security checks, or to return to the district for weekend orgies of baby-kissing and speechifying. To make a national impact, the aspirant must project an image as a statesman seriously concerned with the good of all Americans. And from this perspective, it *may* make sense to invest heavily in environmental lawmaking. . . .

Throughout the 1960s, Senator Muskie carefully invested his time and legislative effort in the environment, long before the issue achieved great public attention. As the primary drafter of the federal air pollution statutes of 1965 and 1967, as well as several water pollution statutes, Muskie stood to gain from the rapid rise in importance which the voting public attached to environmental issues in the early 1970s. However, because of the separation of the lawmaking function into multiple bodies and the difficulty which the voting public has in monitoring all the lawmaking activities in Washington, Muskie was vulnerable to see "his" issue stolen by other politicians, particularly the one in the White House. In addition, because most voters do not bother to follow the details of what goes on in Washington that closely,

Muskie was also vulnerable to charges from the embryonic environmental move-ment that he was really "Mr. Dirty," not "Mr. Clean."

The divisions of lawmaking authority, coupled with the difficulty of credibly communicating with the voters about the political significance of legislative activi-ties, created a situation in which Nixon and Muskie were caught in a Politicians' Dilemma. The result was the passage of the Clean Air Act of 1970 in a form which was more stringent than either of them would have preferred.

The Clean Air Act of 1970. The Clean Air Amendments of 1970 is a complex statute. There is no denying that a number of strands came together to contribute to its passage. One factor was the realization that the Air Quality Act of 1967 had failed to achieve its goal of cleaning up the air. In addition, by 1970 there had been an enormous increase in popular concern about the environment, fueled in part by the attention which the issue was receiving in the press and on television. Finally, 1970 was different from 1967 in that a "loose coalition" of environmentalists was just beginning to organize on the national level, although environmentalists were still nowhere near a match for even a single auto company's lobbyists, either in terms of numbers or funding.

In this political environment, it would not have been surprising for Congress to pass additional air pollution legislation of an incremental sort—perhaps an increase of funding here, or a realignment of federal and state authority there. In fact, on December 10, 1969, the leading proponent of federal air pollution legislation, Sena-tor Muskie, introduced just such a bill, the Air Quality Improvement Act. . . .

One writer has used the term "policy escalation" to refer to the process by which Muskie's weak, original proposal was transformed into a more extreme final statute; another has called it "speculative augmentation." Whatever one calls it, what happened was essentially as follows: on February 10, 1970, two months after Muskie had introduced his Air Quality Improvement bill, President Nixon trans-mitted his own air pollution proposals to Congress. Nixon's proposals called for major structural changes in existing federal air pollution statutes, including national standards for extremely hazardous air pollutants and a requirement that states develop abatement plans to meet mandatory federal air quality standards within one year.

The next significant event occurred in May 1970, when a Ralph Nader task force published a report harshly criticizing Muskie as being soft on industry. The flavor of the report is summed up by a sentence displayed prominently on its dust jacket: "Sen. Muskie's sub-committee on pollution and the federal laws for which it was responsible have resulted in a 'business-as-usual' license to pollute for countless companies across the country." The Nader report went on to claim that Muskie should be "stripped of his title as 'Mr. Pollution Control'" and to demand that he resign his chairmanship of the air and water pollution subcommittee. Muskie was clearly stung by Nader's public criticism.

In August, Muskie's subcommittee reported out a revised air quality bill which essentially followed the outlines of Nixon's proposal but was tougher at every turn than what the president had proposed: where Nixon's proposal would have allowed states one year to develop their implementation plans, Muskie's bill allowed only nine months; where the administration proposed that the auto companies be given until 1980 to achieve a 90 percent reduction in emissions, Muskie's subcommittee cut the deadline to 1975; where Nixon proposed nationwide federal air quality standards, Muskie's subcommittee added the requirement for an additional "margin of safety" and the protection of especially sensitive groups; where Nixon had proposed that we do what "we can do within the limits of existing technology," Muskie deleted technological or economic feasibility as a constraint. . . .

Muskie wrote a "tough" pollution statute in 1970, one which ran a serious risk of alienating industry, only when he was threatened with an outcome which was even worse from his perspective—the loss of his reputation with the public as a crusader to clean up the environment; Nixon went along, reluctantly, because the adverse political consequences of vetoing the bill were perceived as greater than those of signing it. . . .

It is important to recognize that the surprisingly strong environmental legislation in 1970 did not result from superior organization by environmentalists. Indeed, it is possible to speculate that if environmentalists had been more tightly organized as a conventional pressure group in 1970, as they later became, the Clean Air Act amendments might have been less, rather than more, stringent. Had there been a well-organized environmental lobby in 1970, Muskie could have deflected Nader's charges by giving in to its demands. And it is quite likely that this lobby would have settled for far less than the Great Leap Forward achieved by the Clean Air Act. In 1970, however, no group yet existed with whom to bargain. In these circumstances, Muskie had no way of knowing how much would be enough. He did about all that he could have done to prove that he was more "pro-environmental" than Nixon: he proposed a bill which was essentially Nixon's, only more so on every point.

Clean Coal/Dirty Air

BRUCE A. ACKERMAN AND WILLIAM T. HASSLER

Our study . . . focuses upon a crucial substantive policy issue: the future of the coal-burning power plant. At present, these plants contribute 48 percent of all electric power produced in the United States. This share will grow over the next half century. With oil scarce, nuclear risky, solar embryonic, and hydro limited, the nation's rich and cheap coal reserves call for exploitation. At the same time, coal burning generates environmental burdens. Coal-fired power plants are major sources of several pollutants; they are currently the single most important source of sulfur oxides. As a result, the control of new coal burners has gradually emerged

Reprinted by permission from *Clean Coal/Dirty Air* (New Haven, Conn.: Yale University Press, 1981).

as one of the most pressing questions confronting the Environmental Protection Agency (EPA), leading it, in June 1979, to revise the "new source performance standards" (NSPS) it had previously imposed on sulfur emissions from new coal-burning power plants. . . . Our story begins with the way Congress set about to control the environment in 1970. . . .

From Statute to Policy

. . . Despite Congress's aggressive stand on NSPS, it remained for the EPA to translate the statute into practical policy. Its task was to establish a standard that, in the words of the statute, "reflects the degree of emission limitation achievable through the application of the best system of emission reduction which (taking into account the cost of achieving such reduction) the Administrator determines has been adequately demonstrated." As we shall show, however, setting such a standard would have been a tricky business at best—for in its desire to avoid the perils of expertise, Congress had adopted a simplistic formula that was exceedingly difficult to apply to the problem posed by coal burning. . . .

Technology in a Vacuum

By its very terms, two aspects of the statute invited, if they did not require, agency sophistication in setting a standard for new coal burners. First, there was the instruction that the administrator take cost "into account" in making his decision. . . . Second, a proposed system had to be "adequately demonstrated" before it became the basis for a clean-up requirement. . . .

To understand the bureaucracy's view of its own choices, we must consider the state of technology in the early 1970s. At that time, there existed two methods for reducing emissions from coal-fired plants: a relatively old-fashioned technique known as physical coal cleaning, or "coal washing," and an embryonic technology, flue gas desulfurization, commonly known as "scrubbing."

Physical cleaning removes sulfur from coal before the coal is burned. The simplest method involves equipment not much more advanced than a wire screen and garden hose: freshly mined coal is crushed, passed through a screen, and wetted, so that heavy sulfur-bearing fragments can settle out. These relatively simple processes can remove most of the sulfur-bearing particles, called pyrites, that are physically mixed with the coal as it comes from the mine. They cannot, however, remove sulfur that is chemically bonded to the coal. Nonetheless, the gains achieved by primitive washing techniques can be substantial, varying from 20 percent to 40 percent. Within these limits, physical coal cleaning is a cheap and reliable technology. Moreover, it was not obvious that potential gains from coal washing had been exhausted. By grinding coal into a fine slurry, it would be possible to wash more of the pyritic sulfur than could be reached by the primitive crush-and-hose methods practiced in the early 1970s.

Nonetheless, the EPA disregarded such humdrum possibilities and concerned itself exclusively with more symbolically satisfying technologies—devices that, attached to a smokestack and paid for directly by the polluter, promised to cleanse the smoke produced by the boilers below. Among the technologies existing in 1971, the scrubber was the only one that performed this symbolic function. Its ability to perform as an actual control technology, though, was more problematic.

A scrubber does not rely on physical processes such as crushing and washing, but on the maintenance of a large-scale chemical reaction. It is a 70-foot test tube which on a typical day may consume 400 tons of limestone and thousands of gallons of water to remove over 200 tons of SO_2. As exhaust gases flow up a power plant smokestack, they are exposed to a lime or limestone solution that is sprayed in their path. Sulfur dioxide in the gas reacts with the spray and goes into solution, from which it is later removed, dewatered, and extruded in the form of sludge. Maintaining the proper conditions for this reaction requires continuous supervision. For example, the coal burned may contain elements such as chlorine that interrupt the desired reaction; or a variation in the amount of SO_2 in the flue gas may adversely affect the process. Even when the reaction is proceeding apace, the machines must operate in a very harsh environment. Unreacted SO_2 and water may combine to form sulfuric acid which corrodes the inside of the scrubber and smokestack. Reacted SO_2 may form compounds that travel beyond the scrubber and clog smokestacks, pipes, and pumps. While these problems were not difficult to solve under carefully controlled conditions, the early scrubbers were prone to frequent breakdown—operating less than half the time. When the administrator promulgated his new standard of performance in 1971, only three scrubbers were operating in the United States. The oldest, built in 1968, would be abandoned by the end of the year.

In short, the EPA squarely confronted the problem posed by any statute requiring that a system be "adequately demonstrated"—how to trade off certainty and economy against incentives for further technological development. Moreover, creative ways of mediating the tension were available; perhaps a standard based on improved washing should have been required in the intermediate term with reliance on scrubbers projected for a decade or more hence. Rather than give the statutory formula a creative interpretation, however, the language was used as an excuse for thought. The possibility of advanced coal washing was totally excluded and official documents narrowly focused their attention on the question of whether the scrubber was "available" in some engineering sense, divorced from other development opportunities. Such an approach justified the bureaucracy's spending its limited resources on engineering projections to determine if scrubbers could be made operational in the near future. On the basis of these projections, the administrator found that the scrubber's ability to eliminate about 70 percent of a coal burner's sulfur oxides had been "adequately demonstrated," and proceeded to the task of translating this engineering judgment into regulatory policy.

Squaring the Circle

At this point the agency was forced to confront the dilemmas imposed by its own impoverished reading of the statute. The embarrassing fact is that the agency's interpretation made it conceptually impossible to move from its engineering judgment about the scrubber's availability to a policy judgment defining the number of pounds of SO_2 a plant could emit for each MBTU of energy it produced. To see why, consider that a power plant's emissions are not exclusively determined by its treatment technology but are also a function of the amount of sulfur in the coal that the plant burns. The agency's engineering judgment about scrubbing could readily be translated into an emission limitation only if *all* of America's coal contained the same sulfur content. Only then could 70 percent be multiplied by a constant to yield a single nationwide limit on new plant discharges. But, alas, America's coal reserves range in sulfur content from 1 to more than 10 pounds. To make matters even more difficult for the administrator, these coals are distributed unevenly throughout the coal-producing regions. Roughly half of the nation's reserves lie west of the Mississippi in the Northern Great Plains and Mountain regions and consist largely of low-sulfur coal. Eastern reserves, primarily from the Midwest and the Appalachians, contain much larger proportions of higher sulfur coal. . . .

The agency's predicament was intensified by a final statutory artifact. Although the statute directed the administrator to look at the "best system" in defining applicable effluent standards, it did not authorize him to force polluters to install scrubbers if their power plants could meet the effluent limit in some other way. Thus, whatever ceiling the administrator might set, polluters might find it cheaper simply to burn low-sulfur coal than to install scrubbers. The threat of a massive shift away from high-sulfur coal would, in turn, generate powerful political pressures from eastern producers. The conceptual inadequacy of the engineering approach—its lack of explicit concern with the variability of polluting inputs—masked a potentially explosive political problem: how to parry the predictable counterattack by eastern coal?

Surely not by surrendering without a fight. After all, the whole point of environmental regulation is to force producers to bear the social costs of their enterprise. High-sulfur coal had previously gained an unfair competitive advantage over low-sulfur coal precisely because the extra harm it caused the environment had not been reflected in coal prices. A program of controlling sulfur oxides merely removed that advantage. . . .

Despite the conceptual impossibility and political unwisdom of moving immediately from its engineering judgment to regulatory standards, the agency tried to square the circle by treating the problem of coal variability as if it were a minor detail. The agency simply failed to recognize that its findings about scrubbing were compatible with an NSPS ceiling ranging all the way from 3 pounds per MBTU to 0.3 pounds per MBTU—depending on whether ten-pound eastern or one-pound western coal was being scrubbed at 70 percent efficiency. Apparently there was no effort, however rudimentary, to estimate the costs and benefits generated by a range of different possible emission ceilings. Instead the agency finessed its conceptual and political problems by announcing a number and making a few casual remarks in its support. The numerical ceiling was set at 1.2 pounds of SO_2 per MBTU; in support, the EPA merely stated that this ceiling would permit eastern power plants to comply by scrubbing 70 percent of the SO_2 out of the average eastern coal—which was said to contain roughly 4 pounds of SO_2 per MBTU ($[1.0 - .70] \times 4$ pounds = 1.2 pounds). At the same time, the agency recognized that utilities might respond to this ceiling the natural way, by burning 1.2-pound coal. It failed, however, to estimate the impact such decisions would have on the eastern coal industry, let alone whether the benefits of the 1.2 standard outweighed its expected costs. Instead, it blandly proclaimed that burning low-sulfur coal would also satisfy the new standard. . . .

The House Proposals of 1976

. . . When amendments were initially proposed by the subcommittee, they contained . . . a barely perceptible alteration of Section 111[1]. . . . Turning to the House committee report, however, one enters a new world of meaning. It proclaims that scrubbing or some other add-on technology is required of all new coal burners. . . . This clear statement would bear fruit only if the EPA and the courts focused their attention upon the committee report, glancing at the statute itself only when the report's language was ambiguous. Unfortunately, however, this approach to statutory interpretation is hardly unprecedented. . . .

House staffers began to organize political support for the invisible amendment of Section 111.

Their search for politically potent allies for the "new" NSPS led in a surprising— but readily explicable—direction. Although universal scrubbing had only a problematic relation to clean air goals, there *was* an interest group that valued it for its own sake—the producers of high-sulfur coal in the eastern United States. For many years this group had fought and lost the good fight against "unnecessarily" stringent pollution standards. As far as they were concerned, the invisible amendment of Section 111 offered a new strategy that promised to further their basic interests more effectively. Once the eastern utilities were forced to install scrubbers, it would be possible for them to meet the 1.2 NSPS while continuing to use cheap high-sulfur coal. Only if utilities were allowed to substitute low-sulfur coal for scrubbers would a shift away from high-sulfur products be conceivable. Thus it made sense for the dirty coal producers to abandon their campaign to weaken pollution standards and take up the cudgels for the costliest possible clean air solution—universal scrubbing.

As a consequence, when House staffers made contact with eastern coal interests, they met with a sympathetic response. From the point of view of the United Mine Workers Union, the scrubbing issue was particularly straightforward. Because its membership is concentrated in the East, it had no difficulty coming out publicly for universal scrubbing. Politics were more complicated for the National Coal Association. Since western owners were naturally interested in maximizing sales of low-sulfur coal in the East, they would not take kindly to the national lobby endorsing a requirement that would freeze them out of a potentially rich market. Consequently, the National Coal Association refused to take a position on scrubbing, leaving it to eastern members to use their considerable political muscle in ways we shall describe.

The stage had been set for a bizarre coalition between clean air and dirty coal forces. . . . The consummation of this political marriage was evident as early as the

---

[1]The NSPS section—ED.

1976 House committee report, whose language was repeated in the House report the following year. Besides announcing that the "new" Section 111 had invisibly imposed a universal scrubbing requirement, the report justified the innovation by finding six flaws in the "old" NSPS:

1. The standards give a competitive advantage to those States with cheaper low-sulfur coal and create a disadvantage for Mid-western and Eastern states where predominantly higher sulfur coals are available.
2. These standards do not provide for maximum practicable emission reduction using locally available fuels, and therefore do not maximize potential for long-term growth.
3. These standards do not help to expand the energy resources (that is, higher sulfur coal) that could be burned in compliance with emission limits as intended.
4. These standards aggravate compliance problems for existing coal-burning stationary sources which cannot retrofit and which must compete with larger, new sources for low-sulfur coal.
5. These standards increase the risk of early plant shutdowns by existing plants (for the reasons stated above), with greater risk of unemployment.
6. These standards operate as a disincentive to the improvement of technology for new sources, since untreated fuels could be burned instead of using such new, more effective technology.

Although this pronunciamento was written by House staffers with strong environmentalist reputations, its authors might easily have come from a major coal company. While point 6 attempts an inadequate invocation of technology-forcing, the text employs the standard rhetoric of the eastern coal lobby—the need to eliminate unemployment by using "locally available fuels" and to defeat the energy crisis by burning high-sulfur coal. . . .

Political Convenience and Legal Ambiguity

While the House report established the foundation for a clean air–dirty coal alliance, it remained for the environmental lobby in Washington, D.C. to accept the political invitation. . . .

Instead of campaigning for a congressionally mandated reduction in the 1.2 standard, . . . public interest lawyers embraced the dirty coal rhetoric. For example, Joseph Brecher, on behalf of the Sierra Club, condemned the 1971 NSPS decision because "eastern high-sulfur coal, which is now available, is having a hard time getting a market because of the comparative cheapness of bringing in western low-sulfur coal." Richard Ayres, for the Natural Resources Defense Council, bemoaned the fact "that the Clean Air Act became a factor influencing the competition in the marketing for coal, encouraging practices such as shipping mountain state coal to Illinois and Louisiana, with attendant use of oil powering the diesel engines used to transport train after train of coal." These passages represent a remarkable rhetorical turn. Typically, environmentalists do not protest when a government initiative forces industry to discard "dirty" inputs and substitute "clean" ones. In so doing, government is simply eliminating market distortions and requiring the prices of high-polluting products to reflect their true social cost. Rather than condemn the advantage gained by clean coal as artificial, environmentalists characteristically applaud when the "true" costs of dirty coal have finally been revealed. No matter. With such rhetorical assistance, the peculiar coalition between the friends of clean air and dirty coal would be a powerful political force.

Yet rhetoric alone cannot overcome the fundamental conflicts concealed by a marriage of convenience. While both groups wanted scrubbers, they wanted them for different reasons. Environmentalists saw scrubbing as a technique for cutting new plant emissions *below* 1.2 pounds to even lower levels. For example, if a plant using one pound western coal were required to scrub at 90 percent efficiency, only 0.1 pound of SO_2 would be discharged. For eastern coal interests, however, scrubbing was desirable only as long as new plants could keep discharging at the old 1.2 level. If the administrator used scrubbing as a reason for lowering the emission standard dramatically below 1.2, high-sulfur coal producers would be frozen out of the new plant market once again. . . .

Low-Visibility Politics

. . . The invisible revision of Section 111 passed through the House untouched; its lack of salience was emphasized by the failure of the Senate committee to include a comparable revision in its legislative proposals. . . .

[T]he Senate bill . . . proposed *no* change in Section 111, and the Senate report contained nothing resembling the House committee's low-visibility pronunciamento on behalf of dirty coal. Moreover, the scrubbing lobby was content to evade explicit senatorial consideration . . . leaving it to the conference committee to reach a "compromise" in which the House's legislative history would once again expose the hidden meaning of Section 111.

Midnight Lawmaking

The Senate's approval of the Clean Air Act . . . was only a preliminary to the main event—where conferees from both the House and Senate committees would hammer out a final compromise after two years of disagreement.
. . . [T]he House achieved a formal victory. . . . Henceforth the statute would require the administrator to regulate power plants differently from other dischargers. Although the administrator is simply to tell everybody else to reach specified emission limits, an acceptable power plant standard also requires: "the achievement of a percentage reduction in the emissions from such category of sources from the emissions which would have resulted from the use of fuels which are not subject to treatment prior to combustion."

Given this provision, the administrator must not only require a power plant to discharge no more than X pounds of sulfur oxide per MBTU, but also to reduce the sulfur in the coal by Y percent. By setting the percentage reduction requirement at a level only scrubbers can achieve, 85 percent for example, the administrator could effectively force all coal burners to install scrubbers. But the statute falls far short of mandating such a high percentage. Indeed, it does not even require the administrator to establish the same percentage for all coal burners. Although each plant must be told "a percentage reduction" to achieve, another subsection, unchanged from 1970, expressly authorizes the administrator "to distinguish among classes, types, and sizes within categories of new sources for the purposes of establishing such [NSPS] standards." Hence, the administrator has the authority to tell users of low-sulfur coal to reduce their sulfur content by Y_1 percent and require high-sulfur burners to eliminate Y_2 percent. . . .

Indeed, when viewed within the framework of the section as a whole, the new provision does not even bar the administrator from establishing a reduction of zero percent for low-sulfur coal burners. . . .

Rather than try to bring the increasingly complex statutory language under control, the draftsmen turned to their first love: making legislative history. Once again, the House staff gained a victory within the conference report that it failed to achieve on the surface of the statute: "The Senate concurs in the House provision with minor amendments. The agreement . . . preclude[s] use of untreated low sulfur coal alone as a means of compliance." Yet midnight legislative history is a game any number can play. Aides for Senator Dominici were quick to add their own opinion in the next paragraph: "The conferees agreed that the Administrator may, in his discretion, set a range of pollutant reduction that reflects varying fuel characteristics." Given this threat, the House staffers counterattacked with another tack-on: "Any departure from the uniform national percentage reduction requirement, however, must be accompanied by a finding that such a departure does not undermine the basic purposes of the House provision and other provisions of the act, such as maximizing the use of locally available fuels.". . .

The incoherent quality of the legislative history was apparent to the amendment's supporters almost immediately. After spending all of August 3 compiling the conference report that had been completed at 2:20 that morning, the staffers thought it wise to spend the next day formulating a "Clarifying Statement" while both Houses were voting their approval of the statutory language. Among these clarifications is one that reflects the continuing effort to wrestle with the scrubbing confusion. This time it is said that

> while the conferees agree that the Administrator may set the percentage reduction requirement as a percentage range, the conferees expect the Administrator to be exceedingly cautious if he should elect to do so. Such range would be allowed only to reflect varying fuel characteristics and must be based on a carefully and completely documented finding . . . that [this] does not undermine the basic purpose[s] . . . of the House Report. . . .

In short, the draftsmen brewed a mix of statute and legislative history worthy of the occasion. Instead of integrating Section 111 into the basic structure of the act, their task was to avoid a potential conference impasse by writing a document whose legal meaning was hopelessly confused. The new Section 111 is easier to understand as an exercise in small group dynamics than as a serious effort to guide the bureaucratic management of a multibillion-dollar problem.

Environmental Regulation: Whose Self-Interests Are Being Protected?

B. PETER PASHIGIAN

Most economists rely on the established theory of externalities to explain the reason for environmental regulation. Self-interest or rent-seeking is seldom mentioned when the subject is pollution control. The infatuation of economists with the theory of externalities to the exclusion of the self-interest hypothesis is not difficult to understand. The theory of externality is well known and firmly embedded in the economists' arsenal of economic analysis. By and large, the profession accepts the notion that social welfare and not private interest is the guiding principle for environmental legislation and regulation. This easy acceptance of the externalities hypothesis is nevertheless disquieting because the literature on the self-interest hypothesis . . . is considerable, and the evidence from regulated industries suggests private interests have benefitted from regulation. Yet, when it comes to explaining environmental legislation, a substantial number of economists believe the regulatory program has been largely untouched by the special interest considerations and has been successful in improving social welfare. Indifference to the self-interest hypothesis is not the sole province of economists, but is shared by many political scientists. . . .

This paper seeks to elevate the role of self-interest in explaining the regulation of stationary sources. An important part of environmental legislation becomes comprehensible under the self-interest hypothesis and not under hypotheses based on a taste or political preference for environmental protection. The paper focuses on the policy of prevention of significant deterioration (PSD), a policy that is largely unfamiliar to all but specialists in environmental policy. This policy prohibits areas with air quality superior to the *minimum* national standards from permitting a significant deterioration of local air quality by placing limits on economic development. Environmental policy also requires all areas to meet the minimum national standards. While areas with unduly low air quality are ordered to improve air quality and to attain the minimum standards, other areas with air quality superior to the minimum standards must prevent a significant deterioration of air quality.

Why these implicit multiple standards became an integral part of environmental policy is the central focus of this paper. The major premise advanced and tested is that PSD policy was developed to attenuate the locational competition between developed and less developed regions and between urban and rural areas. The votes cast in the House on PSD policy are examined and show opposition to PSD policy comes from the South, the West, and rural locations, areas with higher growth rates and with generally superior air quality. PSD policy is opposed in these areas because it places limits on growth. The strongest supporters of PSD policy are northern urban areas, many of whom have lower air quality and are not directly affected by

Reprinted by permission of the author and publisher from 23 *Economic Inquiry* 551 (No. 4, October 1985).

PSD policy. It is argued that federal PSD policy raised the cost of factor mobility and thereby allowed northern locations with lower air quality to improve local air quality without as large a loss of factors to areas with superior air quality.

Other hypotheses could and have been advanced to explain this North-South/West voting dichotomy. The regional differences in voting position might be due to differences in political representation or philosophy or to tastes for environmental regulation. A considerable part of the modeling effort is designed to separate the locational competition hypothesis from a taste or cultural-political heritage interpretation. This is done by first comparing the votes on PSD policy with votes on auto emission policy, another important environmental issue. A comparative vote analysis shows greater regional differences for votes cast on PSD policy than on auto emission policy. This finding is interpreted to mean that the votes on the PSD issue do not simply reflect regional differences in the taste for environmental protection. Otherwise, the regional patterns of votes on the two issues would be similar. Second, a comparison of the votes on environmental policies with the votes cast on a portfolio of economic issues once again shows greater regional differences in the votes cast on environmental issues than on the economic issues. If the votes cast on environmental issues reflect differences between regions in liberal-conservative political philosophy, these differences are simply too small to explain the regional differences in the votes on environmental issues. The weight of the evidence indicates that PSD policy cannot be explained solely by regional differences in political philosophy or tastes for environmental regulation. . . .

Statistical Model of Legislative Voting

The statistical analysis will be limited to House votes because there are larger differences in the economic and demographic characteristics between Congressional districts than between states. A preliminary examination of Senate votes indicated the regional pattern of Senate votes did not differ substantially from House votes. The statistical analysis will . . . explain 1905 votes cast in 1976 and 1977 on *five* Nondegradation Amendments and 1607 votes cast on *four* Automobile Emission Amendments. . . . The independent variables describe the economic, demographic, industry and party characteristics of the district and thereby determine the net benefits or losses from a more stringent PSD policy. . . .

Results of Vote Analysis

The percent of the votes on the five Nondegradation Amendments in support of a more stringent PSD policy is shown by Census region in column 1 of Table VIII.1. The region's percentage relative to New England's percentage is shown in column 2. Political support for PSD policy does vary across regions. The strongest support for a more stringent PSD policy has come from the New England, Middle Atlantic, Pacific, and East North Central regions, while the regions mounting the strongest opposition to PSD policy are the East South Central, West South Central and South

Table VIII.1. House Votes on Nondegradation and Auto Emission Amendments by Census Region

| Region | Percent of Votes in Favor of More Stringent Policy | | | |
| | Nondegradation Amendments | | Auto Emission Amendments | |
	%	Region % Relative to New England %	%	Region % Relative to New England %
UnitedStates	46.1	.59	36.8	.54
New England	78.8	1.00	67.7	1.00
Middle Atlantic	72.0	.91	57.8	.85
East North Central	50.7	.64	23.0	.34
West North Central	40.4	.51	30.1	.44
South Atlantic	29.9	.38	32.5	.48
East South Central	10.0	.13	12.5	.18
West South Central	11.5	.15	9.2	.14
Mount	38.0	.48	27.1	.40
Pacific	58.3	.74	60.2	.89

Atlantic. The results in Table VIII.1 . . . suggest that House members from areas with relatively dirty air were staunch defenders of the relatively clean air in *other* parts of the country. Congressmen from New England and other northern states demonstrated more concern about degrading air quality in Oklahoma and South Carolina than did the Congressmen from these states. . . .

The votes cast on the Nondegradation Amendments may be compared with the votes cast on the Automobile Emission Amendments where regional conflicts are expected to be less pronounced. Columns 3 and 4 of Table VIII.1 show the percent of votes supporting a more stringent automobile emission policy and the region's percentage relative to New England's percentage. Among the faster growing regions of the country, the South Atlantic, East South Central and Pacific regions deviate more from New England on PSD policy than on automobile emission policy. . . . Among the faster growing regions, only the West North Central and the Mountain regions do not deviate more from New England on the PSD issue than on the auto-mobile emission issue. The second interesting finding in column 3 is the opposition to tighter auto emission controls by the East North Central region. This region devi-ates substantially more from New England on the automobile emission policy than on PSD policy. This is not surprising since the automobile and related industries are located throughout the East North Central region. This casual review of the raw votes suggests that voting position on the two environmental issues changes as the self-interest of the region changes.

A more systematic logit analysis is needed to isolate the effects of region from the effects of other independent variables. . . . After taking account of the economic, demographic, party and industry effects, the Mountain and the Pacific regions sup-ported the Automobile Emission Amendments to the *same* degree as did the New England region. The strongest opposition came from the East North Central, West

South Central and the East South Central regions. There is a significant realignment of the regional coefficients for the Nondegradation Amendments. For PSD policy, the Middle Atlantic and the East North Central regions are closer to New England in their support for PSD policy. . . . There is greater regional variation in the support for Nondegradation Amendments than for the Automobile Emissions Amendments. Support for PSD policy is more concentrated in the northern regions of the country. The younger and less developed regions of the country offered more active opposition to the Nondegradation Amendments than to the Automobile Emission Amendments. This changing pattern of the regional coefficients lends some support to the locational competition hypothesis. Locational competition considerations help explain why the southern and western regions deviated more from New England on PSD policy than they did on auto emission policy.

Other results also suggest that the pattern of regional coefficients on the two environmental issues reflects the effects of self-interest and not tastes. As noted above, the regional coefficients of the Pacific and the Mountain regions on the auto emission issue are close to the regional coefficient for New England. California has a state auto emission policy that is more stringent than the federal policy. A more stringent federal auto emission policy would place California at less of a competitive disadvantage and would explain why the California delegation supported a more stringent federal auto emission policy. The support for a more stringent auto emission policy by the Mountain region is less easily explained even though several cities in this region have severe auto emission problems, for example, Denver. Still, the voting position of the Mountain region appears to be a departure from the implications of the locational competition hypothesis.

Still other results indicate voting position changes from one environmental issue to another as the perceived net benefits change. This is well illustrated by the votes cast in districts with a large share of district employment connected with the automobile industry (new car). Not surprisingly, these districts strongly opposed the Automobile Emission Amendments and supported or were neutral on the Nondegradation Amendments. Support for a less stringent auto emission policy after a neutral stance on PSD policy indicates voting position changes as the interests of the district change. Another result is consistent with the self-interest hypothesis. Rural and less densely populated districts opposed a tightening of auto emission standards. Since average air quality in rural areas is superior to that in urban areas, the reluctance of rural residents to support more stringent emission controls and to pay higher automobile prices so that the lower air quality in central cities can be improved is understandable. These different results all suggest that more is at work here than simply regional differences in the taste for environmental protection. . . .

Conclusions

The self-interest hypothesis gives a more complete understanding of why northern-urban constituencies are for, and southern, western and rural constituencies against, PSD policy. Other hypotheses that relate the votes on PSD policy to party or political philosophy or to tastes for environmental protection are lacking in one or another dimension and are unable to explain the interesting and subtle variations in votes across environmental issues or between environmental and economic issues. These findings suggest that self-interest has played a prominent role even in environmental legislation, a prime example of social legislation.

There is a temptation to misinterpret the locational competition hypothesis as simply a contrived opportunity to impose costs on less developed regions by the developed regions of the country. Such an interpretation is misguided since it ignores the important role played by northern residents demanding improved environmental conditions. . . .

The locational competition hypothesis, while unquestionably useful in understanding PSD policy, has its limitations as well. It has difficulty explaining the relatively strong support for PSD policy by House members from the Mountain region. . . . While the competition among locations was a major determinant of the vote on PSD policy, it was not the only determinant. The basic point is that the self-interest hypothesis need not take a back seat to hypotheses based on other motives.

A Positive Theory of Environmental Quality Regulation

MICHAEL T. MALONEY AND ROBERT E. McCORMICK

The purpose of this paper is to demonstrate that environmental quality regulation may enhance producer wealth while it simultaneously reduces an externality problem by restricting access to common property. Regulation not only corrects a resource misallocation, but it creates a scarcity rent as well. In the recent history of environmental quality, the common access problem has been addressed by federal and state agencies through a standards-based approach, rather than through the enforcement of tradable property rights. As a consequence, rents from the right to use these assets have accrued to producers.

Considerable attention has been paid in welfare theory to the conflict between producers of externalities and the victims. However, the interests of environmentalists and producers may coincide against the welfare of consumers, because both may profit from output reductions. In the context of the economic theory of regulation, environmental regulation, that is, regulation of a competitive industry in a negative externality setting, carries with it the implication that producers and the victims of pollution may find it in their self-interest to form a like-minded

Michael T. Maloney and Robert E. McCormick, "A Positive Theory of Environmental Quality Regulation" 25 *Journal of Law and Economics* 99 (1982) (Various pages, edited). Reprinted by permission of the authors and The University of Chicago Law School *Journal of Law and Economics*.

coalition to lobby for input restrictions and/or output reductions. The difference in this case of regulation and the standard producer-consumer conflict is clear—there is extra political support from environmentalists on the side of the producers.

The maze of environmental quality regulation is overwhelming and bears little resemblance to the efficiency criteria proposed in the economics literature. Many regulatory techniques—such as technology-specific regulations, differential standards for old versus new firms, uniform percentage reductions across pollution sources, and the inalienability of pollution permits—are hard to understand on the surface. Some of the confusion is the result of only focusing on the externality effects of air and water pollution and ignoring the effects of price changes on the regulated industry. It is the purpose of this paper to resolve part of this confusion by focusing attention on the cartel-like gains and losses that may accompany improved environmental quality. . . .

Input Restrictions and Profits

There are several mechanisms by which the state might improve environmental quality. Effluent fees can be imposed; limitations on the production of final output can be enforced; standards or restrictions on the output of pollutants can be put in place; or technology-specific production functions can be imposed on the producers within an industry. Although economists usually focus on taxes and emission fees, regulators typically restrict pollution emissions or impose technological requirements. . . .

[T]axation will always be detrimental to producer interests as the tax amounts to a charge for a resource which was previously free to the industry, and, moreover, a tax is an assignment of environmental quality rights to all society. Juxtaposed to taxes, direct-output regulation can be equivalent to cartelization, with its well-known benefits (to producers), and amounts to giving the environmental assets to the producing firms. To be successful, regulation must not only restrict output, but must restrict entry as well, and the output restriction must not be so severe that average cost is raised above price.

In actual practice, environmental quality regulation usually dictates a certain input use or mandates a specific technology that firms would otherwise not have chosen. Like a shadow-price increase, regulation induces the firm to substitute out of the environmental input into other factors of production. The basic purpose of this section is to derive the conditions under which these technology-imposed regulations, combined with entry limitations, will lead to enhanced industry profitability.

Entry limitations must be enforced if the entire industry is to experience increased profits. This is frequently achieved by imposing differential pollution-control requirements on new firms—the environmental quality analogue to grandfather clauses. For example, the 1970 Clean Air Act and its amendments imposed standards on existing pollution sources as a function of the ambient air quality, while new firms had to meet the strictest standards regardless of local air quality. Moreover, the ambient air standards have been the tightest in the cleanest air regions, further restricting the entry of rivals. Entry restrictions seem to pervade every aspect of this regulatory process. . . .

[The authors then derive a set of sufficient conditions, concerning the cost functions of firms, under which regulation leads to the increase of industry profits—ED.]

Keep in mind that these are sufficient conditions for potential profit enhancement. When these conditions exist, there is always a regulation which can aid the existing firms. . . . Whether the regulatory authorities will choose such a regulation is, of course, an empirical issue.

Intraindustry Transfers Via Regulation

It does not necessarily follow that, even if some firms suffer a decline in value as a result of regulation, all firms will be worse off. . . .

Depending on the age and location of capital, managerial expertise, and many other factors, regulation via standards will not normally impose costs uniformly across firms. If the most efficient firms in the industry can comply most cheaply with the law, then . . . market price will increase more than costs for some firms. The rents of marginal firms will decline and some will exit as they face higher costs. The potential for intraindustry transfers presents an amendment to the economic theory of regulation— there is also a demand for regulation to redistribute wealth among competing firms in the industry. Consequently, there may be cases where all firms in an industry do not benefit from a regulation, even if sponsored by industry. Furthermore, some firms may gain, although the industry majority opposes a law. Even when entry is not restricted, cost-imposing regulations can increase the profits of inframarginal firms. . . .

Empirical Evidence

[T]he major deductive consequence and distinguishing difference of our theory is profit enhancement. In this part of the paper we . . . test . . . the basis hypothesis. . . . The test method employs financial market analysis to assay the extent of profit enhancement. . . .

[W]e consider the June 1973 Supreme Court decision ordering the Environmental Protection Agency to "prevent significant deterioration" of air quality in the clean regions of the United States. The "PSD" ruling was a court reinterpretation of the EPA mandate under the 1970 Clean Air Act (CAA). While the 1970 CAA is a law with broad scope, the PSD ruling affected only a few types of pollution, primarily nonferrous metal smelting and electric generation. The PSD ruling did not affect existing plants, but made new plant construction more difficult. We tested the value-enhancing potential of this regulation by examining the stock-market rates of return for a portfolio of copper, lead, and zinc smelters.

Stock Market Event Analysis

One convenient technique for assaying the effects of regulation is financial market analysis. . . . The efficient markets theory of modern finance holds that the price of any security incorporates at each instant in time all current information and expected future returns discounted to the present at a rate of interest which includes the relative risk of the security. The use of financial data must center on event analysis. At some point in time an event occurs which signals to the capital market a change in either future returns or risk. The expected effect of this event on future cash flows is discounted into the stock-price adjustment at that time. It is possible that the capital market may receive the information slowly or that the researchers may not be able exactly to identify the point in time when the information was available, so a "window" of time preceding the event is examined to see if the returns behaved abnormally. . . .

The PSD Ruling

In the initial formulation of the 1970 Clean Air Act, Congress did not provide for regulation of air quality in areas which met the National Ambient Air Quality Standards (NAAQS) as established by the Environmental Protection Agency. In 1972 the EPA was in the process of evaluating and approving state implementation plans as authorized by the 1970 act. These plans detailed the circumstances under which new pollution sources would be granted permits. Up to this point, the EPA had not imposed any requirement on the states to control new sources of pollution that posed no threat to ambient standards. This feature would have meant that there were no entry restrictions for firms in areas which had relatively clean air, so long as they met the technological standards imposed on all new firms in the industry.

In May of 1972 the Sierra Club and three other environmental groups brought suit against the EPA to block "significant deterioration" of air quality in those regions of the country where the ambient standards were being met. On June 12, 1972, the District Court for the District of Columbia issued a preliminary injunction requiring the EPA to disapprove state plans which did not prevent deterioration. On November 2, 1972, the U.S. Court of Appeals, D.C. Circuit, affirmed the lower court's decision. In January 1973, the Supreme Court agreed to hear the case and, on June 11, 1973, they affirmed the appellate court's decision upholding the "prevention of significant deterioration." Following this ruling, in late 1973, the EPA proposed rules to incorporate PSD requirements into each state plan. In December 1974, the final regulations were promulgated which dictated that each state include a PSD requirement.

We interpret this court ruling as a strong, binding, entry-limiting device. Firms which could otherwise have constructed facilities would now either be barred from entering or forced to meet higher standards. . . . [T]he EPA argued that it could only regulate sulphur oxides and particulates for PSD because there were currently no analytical procedures available to assess the impact of the other . . . pollutants by individual sources on ambient air quality. Sulphur oxide and particulate pollution sources are heavily concentrated in two industries, nonferrous ore smelting and electricity production, and hence the PSD ruling should have had a direct effect on copper, lead, and zinc smelters, in that future entry was significantly impaired. Such a restriction would raise the value for the existing firms in the industry.

For a . . . test of the commonality of interest between environmentalists and producers, we performed an event analysis on a portfolio of nonferrous metal smelters. We regressed the smelting-portfolio return on the market return using monthly data from 1965 through June 1973—the Supreme Court decision. We chose an event window of twelve months which covers the appellate court's ruling. . . .

We find support for the claim that environmental regulation is beneficial to some of the existing firms in the industry. . . . [T]he PSD ruling was associated with a smelting-portfolio return 44 percent higher than predicted by the market model.

After obtaining these results, we investigated the possibility that events other than the PSD ruling might have accounted for the positive earnings of the smelting portfolio. Two other events are contemporaneous with the PSD court case. In early 1973 copper prices increased. The *Wall Street Journal* carried stories on 15 February and 5 March 1973 in which it was speculated that increasing demand for copper was expected to boost prices. Reasons cited included production problems in Chile, political friction between Rhodesia and Zambia, and the "monetary crisis." Of course, if the copper market was responding to future output restrictions because of the PSD ruling, then the price increases may have also been due to the future effects of reduced capacity.

Second, in March of 1973 the U.S. Treasury Department announced that it was investigating the possibility that lead was being sold in the United States by Canada and Australia at illegally low prices. In October 1973, the Treasury concluded that Canada and Australia had in fact "dumped" lead in the United States at less than fair market value. These events could have led the financial market to anticipate higher lead prices through increased tariffs on foreign lead. . . .

Even taking into account the effect of metal price changes, there is a significant change in the value of the smelting portfolio contemporaneous with the Supreme Court PSD ruling.

Conclusions

Environmental quality regulation is a relative newcomer to government restrictions on the right to contract. We have attempted to explain some of the regulatory peculiarities by incorporating the effects of regulation on the affected firms. Specifically, environmental use restrictions create valuable rents, some of which may accrue to the regulated firms. In effect, many of the existing laws and institutions can be explained as devices for distributing the rents created by regulation. . . .

[T]here is a strong interplay between producers, consumers, and the victims of pollution. Each party has some influence on the outcome, and one-sided analysis neglects many important aspects of regulation. Environmental quality regulation is complicated, but many of the observed perplexities are consistent with a rent-seeking, self-interest theory of government.

Politics and Procedure in Environmental Law

DANIEL A. FARBER

As Mancur Olson pointed out, the dynamics of collective action are actually quite adverse to the passage of statutes such as these, which involve the production of public goods. . . . The Olson paradigm appears to have a straightforward implication for environmental legislation: there should not be any. For example, air pollution legislation benefits millions of people by providing them with clean air; it also imposes heavy costs on concentrated groups of firms. The theory predicts that the firms will organize much more effectively than the individuals, and will thereby block the legislation. We would also expect to find little regulation of other forms of pollution. . . . Yet the reality is quite different. . . .

In this article, I seek not to "debunk" interest-group theory, but to explain how the political system manages to overcome the inherent advantages of special interests. Synthesizing prior scholarship on the subject, I identify a symbiotic relationship between legislators and environmental groups. Briefly, the passage of environmental laws is attributable to strong public demand, coupled with exploitation of that demand by ideological and credit-seeking politicians. . . .

The Sources of Environmental Legislation

Special Interests and Environmental Law

The simplest explanation for the passage of environmental laws is that they actually reflect the influence of special interests rather than the preferences of the general public. The special interest involved may be variously depicted as either a producer interest or a consumer interest.

Environmental statutes as industry rent-seeking. The producer theory is that environmental statutes are actually blinds for rent-seeking by the regulated industries themselves. Thus, as Jon Macey puts it, "environmental protection statutes, which appear to be classic public interest statutes designed to ensure the optimal production of a public good (i.e., a clean environment), often contain features consistent only with the protection of special interests." Taking advantage of the public desire for environmental protection, firms procure statutes that are actually designed to profit the purportedly regulated industries.

Environmental statutes clearly do impose heavy costs on firms. The special-interest story, then, involves differential costs between different segments of industry, which are thought to give the favored segments an advantage. There are two problems with this theory.

Reprinted by permission of the author and Oxford University Press from 8 *Journal of Law, Economics, and Organization* 59 (1992).

First, the broad scope of environmental legislation is at odds with the theory. The number of firms involved seems too high to be plausible. Environmental legislation such as the Clean Air Act involves virtually every industry. Even if only large firms benefit, passage of the legislation would still require organizing the efforts of many politicians and of firms in multiple industries. Apart from the transaction costs involved in organizing a lobbying effort on this scale, such a widespread effort would create a high risk of public disclosure. There would be obvious rewards to any media firm that uncovered the industry conspiracy behind major pending legislation. Even if these organizational problems are not severe enough to make broad-scale legislation impossible, it clearly could be easier for the key firms in particular industries to organize than to organize a single economy-wide effort. (Note that there may be conflicts between industries, since efforts to increase rents for suppliers will usually disadvantage firms that purchase their products.) Hence, if firms were providing the primary impetus for environmental legislation, we would expect to find that most environmental regulation originated in industry-specific legislation. This pattern was characteristic of earlier economic regulation, but has not been typical of environmental laws. . . .

This does not mean, of course, that industry groups play no role in shaping environmental legislation. To the extent that alternate regulatory methods can achieve similar environmental results, environmental groups may be relatively indifferent, so the final choice may be heavily influenced by industry lobbying. On occasion, environmental groups and industry also may form a coalition, to obtain legislation that for varying reasons is beneficial to both. Thus, domestic car producers may support safety standards that discriminate against foreign producers, or large textile or chemical firms may support environmental standards that discriminate against their smaller competitors. While the original "push" for this regulation may not have been to particularly hurt foreign or small competitors, that may have been the price that had to be paid to have the support of the affected industry. Moreover, industry opposition may often result in legislative compromises between economic and environmental goals. Industry may even promote environmental legislation on its own, as a way of heading off an environmental backlash. Thus, in various ways, industry may be an active participant in the passage of environmental legislations. Nevertheless, industry does not seem to provide the primary impetus for environmental legislation. . . .

Rent-seeking by consumers. An alternative theory views environmental legislation as rent-seeking by "consumers" of environmental quality. In this view, environmental statutes are primarily designed to serve the interest of upper-middle-class backpackers, who have an unusually high demand for environmental amenities.

One flaw in this theory is that some crucial environmental legislation passed before groups such as the Sierra Club had become politically formidable. For example, the National Environmental Protection Act and the Wilderness Act were passed in the 1960s, when the Sierra Club was a small California-based club of 35,000 members. Although it became an active litigant in the 1970s, not until 1980 did the Sierra Club engage in any political endorsement, and then on only a limited scale. . . .

More fundamentally, this theory underestimates the prevalance of environmentalist attitudes. In 1989, 80 percent of the population agreed that "[p]rotecting the environment is so important that regulations and standards cannot be too high, and continuing environmental improvements must be made regardless of cost." It is doubtful that voters would really support environmental regulation "regardless of cost," but their willingness to endorse this statement does show that they place a high value on the environment. Other studies of public opinion characterize environmentalism as a "consensual" value in American society. Indeed, environmentalist attitudes are now well-nigh omnipresent in American society. . . .

In either its producer or consumer form, the special-interest theory falls short as an explanation of environmental legislation. This is not to say, of course, that particular features of specific environmental statutes never reflect the influence of producer groups or elite groups of nature lovers; it would be surprising to find that these groups had zero political influence. Thus, environmental legislation will tend to represent a balance between industry and environmental interests. But taken as a whole, the existence of modern environmental legislation cannot be explained by the special-interest model. That model has a variety of weaknesses, and it is not surprising to find that it fails here. But its failure leaves a major puzzle. If environmental statutes do in significant part reflect a broad public demand, how does that demand get translated into legislation?

Mass Politics and Political Entrepreneurs

Public opinion and environmental legislation. Classical public-choice theory tends to cut any link between public opinion and legislation, giving center stage to special-interest groups, with more widely dispersed preferences having only a peripheral role. . . .

An alternative tradition of political thought, that of civic republicanism, is represented in a recent article by James Pope:

> Our history has from the outset been characterized by periodic outbursts of democratic participation and ideological politics. And if history is any indicator, the legal system's response to these "republican moments" may be far more important than its attitude toward interest group politics. The most important transformations in our political order . . . were brought on by republican moments.

As Pope defines these republican moments, their major features are (a) widespread public participation, taking the form of social movements and voluntary associations; and (b) utilizing a moral discourse appealing to concepts of the common good. For present purposes, it is the first of these characteristics that is most significant.

Putting aside the normative aspects of this theory, it adds a significant temporal dimension to the analysis. The implication is that politics alternates between normal periods, in which public attention to an issue is weak, and extraordinary periods, in which the issue has high salience for the public. In those extraordinary periods when broad segments of the public are intensely involved with an issue, legislators find themselves in the spotlight, and their positions shift closer to those of the public at large. During republican moments, voters acquire information about legislative positions, but they also acquire information about the state of the world that may lead to a change in their own expressed preferences. These periods are likely to be attended by new legislative initiatives responding to this public demand, which is less likely than legislation passed in other periods to be responsive to the demands of conventional interest groups.

The original 1970 Earth Day looks very much like a "republican moment." An estimated 20 million Americans participated in a variety of public events that day. More than 2,000 colleges, 10,000 high schools and elementary schools, and 2,000 communities took part. Some 20 years later, the reverberations were still being felt, as millions of people took part in a celebration of the 20th anniversary of Earth Day. It is little wonder that environmental law has been used as a paradigm by republicans such as Sagoff.

The observation that environmental legislation has been driven by broad public opinion is not necessarily tied to republicanism or any normative political theory; it is ultimately an empirical assertion. Positive theory confirms that in periods of heightened public attention, legislative "shirking" will diminish, and legislative outcomes will be pushed in the direction preferred by the median voter.

This view of the popular origins of environmental legislation is confirmed by an important historical study. In their investigation of the passage of the 1970 Clean Air Act, EAM[1] found that organized environmental groups did not play a major role, nor did the activities of traditional industry interest groups. Instead, the primary motivating force was public pressure for environmental protection.

The term "republican moment" is perhaps misleading, to the extent that it suggests that very short periods of high-pitched public interest alternate with periods of nearly total public apathy. Rather, there is a continuum. Earth Day of 1970 represented a peak, but there have been lesser peaks of public pressure sparked by events such as Love Canal or Three Mile Island. In between these peaks, public attention is lower, but not nonexistent.

Legislative motivation. Popular enthusiasm cannot by itself produce federal legislation; the Constitution provides no mechanism for direct democracy. If popular enthusiasm is to be translated into legislation, then legislators have to be actively involved. What is their motivation?

The EAM study suggests that one powerful motivation is the desire to earn a public reputation by taking credit for major reform legislation. This incentive is especially strong in the Senate, where a significant number of members have presidential aspirations. For example, one driving force behind the Clean Air Act was Senator Muskie's desire to establish himself as "Mr. Environment." Once the legislative leadership defined the "environmentally sound" position, other legislators had a strong incentive to "get on board" or risk the wrath of an aroused public; the leadership proposal became the minimum measure of environmentally sound policy.

Symbolic Versus Substantive Legislation

"Lash yourself to the mast" and "strike when the iron is cold." The combination of republican moments and legislative credit-seeking provides a convincing explanation for the passage of environmental legislation. As the reference to republican

---

[1]The article by Elliott et al., which appears earlier in this chapter—ED.

"moments" itself suggests, however, waves of popular mobilization are usually intense but short. The rational strategy for legislators is, in the language of Roger Noll and James Krier, "to 'lash themselves to the mast' while waiting out the temporary siren calls for immediate overreaction"; the politicians then delegate to an agency on terms that will allow the "agency to 'strike when the iron is cold,' after the issue has lost its political salience." Thus, as Noll and Krier say, we would expect to find a temporal instability, in which apparently bold legislative measures result in little actual implementation. . . .

If the only operative forces were popular mood and legislative credit taking, symbolic legislation would be ubiquitous. Yet, as pointed out earlier, it is hard to see how this situation could be sustained in the long run unless voters are not just ignorant or even irrational but also outright fools, incapable of learning even after sustained experience. . . .

Second, in general, environmental laws seem to have greater long-term efficacy than this model suggests. They produce years of regulatory effort, often prompted by active litigation. Participants in the legislative process are quite aware that only implementation, rather than mere passage of a law, produces environmental improvements.

Third, the programs contain far more substantive detail than can be easily accommodated by this model. While some of the legislation is symbolic, much of it seems to mean business. For instance, the 1990 amendments of the Clean Air Act demonstrate a serious commitment to implementation. . . . More generally . . . environmental statutes typically contain procedural mechanisms intended to preserve the original legislative deal. Usually, these mechanisms provide for hearing rights and judicial review at the behest of environmental groups, who can thereby keep the legislative effort from dissipating in the administrative process. In short, environmental law has much more durability and "bite" than seems plausible under the symbolic legislation model.

Incentives for nonsymbolic legislation. Why do legislatures enact real as opposed to symbolic legislation? Part of the motivation for environmental legislation may be ideological. For example, Earth Day was the brainchild of Senator Gaylord Nelson, who had previously demostrated his environmental allegiance as a governor of Wisconsin. Ideological legislators seek to screen legislation for substance as well as symbolism, since they care about policy outcomes.

This explanation assumes, of course, that legislators' ideology has some causal influence on their behavior. There is a wealth of empirical evidence supporting this proposition. There are also two strong theoretical explanations. Ideological behavior may represent "shirking" by legislators—that is, the use of their offices to serve their own goals rather than those of constituents or interest groups. There is at least some empirical support for this view of legislator ideology. An alternative view is that ideology serves constituents by decreasing monitoring costs. By voting for a legislator with a strong ideological commitment on issues of concern to them, constituents can be assured that the legislator will implement those interests. Hence, constituents need not invest so much in monitoring the actual performance of legislators. . . .

Credit-seeking legislators also have good reason to seek effective legislation. Typically, they will not be able to establish their positions as leaders on particular issues overnight. Rather, only a series of legislative initiatives can convince the media of their leadership role. Because of this delay factor, however, information will become available on the implementation of the legislator's earlier proposals. If these earlier proposals were purely symbolic, the media will dismiss the legislator as a lightweight. To be an effective credit seeker, it is important to avoid the appearance of pure credit seeking—to show that one is not "all hat and no cattle" or "all sizzle and no steak." Yet it is difficult to develop an image as a substantive player without actually delivering some substance.

Notes and Questions

1. Elliott, Ackerman, and Millian use the term "Politicians' Dilemma" to describe situations in which "the structure of incentives facing the players creates a strong incentive for them to pursue a less than ideal outcome in order to avoid an even less desirable result." Is the Politicians' Dilemma that Elliott and his colleagues ascribe to the automobile and soft-coal industries analogous to the prisoner's dilemma, which is explained in the notes and questions to Chapter I? Who are the actors in this game? What is the structure of their incentives? Is the Politicians' Dilemma that the authors ascribe to Nixon and Muskie analogous to the prisoners' dilemma? Is the Politicians' Dilemma label useful?

2. Is the preference of the automobile industry for federal regulation likely to be contingent on the federal standards being (1) less stringent than *all* state standards or (2) less stringent than at least some state standards? Why is the connection between more stringent standards and higher car prices "only dimly perceived by voters"? Under the 1990 amendments to the Clean Air Act, states can opt for either the federal standards or the more stringent California standards. See 42 U.S.C. § 7507. For some states, the impact on car prices of the California standards had an important effect on this choice. Had voters become more sophisticated by 1990?

3. If the purpose of the Air Quality Act of 1967 was to prevent more stringent regulation at the state level, why did the statute not preempt more stringent state standards, as the Motor Vehicle Air Pollution Control Act of 1965 had done? Was there not a risk that the coal industry would get the worst possible outcome: regulation at the federal level coupled with more stringent regulation at the state level? Is the promulgation of federal air quality criteria, which are merely compilations of scientific evidence on the adverse health effects of air pollutants and do not contain enforceable standards, sufficient to make such a result unlikely, as these authors maintain?

4. Elliott, Ackerman, and Millian focus on the cost-externalization permitted by a federal system. They give the example of a ban by New York on the use of high-sulfur coal in power plants affecting coal miners in Appalachia. But there also is benefit-externalization: some of the benefits of lower power plant emissions will accrue to residents of other states. What accounts, then, for state regulation? Are Elliott and his co-authors likely to be advocates of "race-to-the-bottom" arguments for vesting responsibility for environmental regulation at the federal level? Are they likely to believe that federal regulation is desirable because environmental interests are systematically underrepresented at the state level? Such arguments are discussed in detail in Chapter VII.

5. Were the plights of the automobile and soft-coal industries comparable? Do they both have the same interest in uniform regulation?

6. Is the Politicians' Dilemma faced by Nixon and Muskie a distinctive feature of the American political system? Might similar competition for credit-claiming exist in parliamentary governments? Unlike the actors in the prisoners dilemma, Nixon and Muskie were repeat players: the Clean Air Act was not the only piece on legislation about which they had a mutual interest. Game theory suggests that repeat players are more likely to be able to coordinate their actions. Why might such coordination not have been possible for Nixon and Muskie?

7. Ackerman and Hassler's argument can be best understood by reference to a stylized example. Assume that all low-sulfur coal has 1.2 pounds of SO_2 per MBTU and is concentrated along the West Coast, and that all high-sulfur coal has 8 pounds of SO_2 and is concentrated along the East Coast. Consider the impact on the markets for low-sulfur and high-sulfur coal of the following policies: (1) the pre-1971 regime, under which power plants emissions are essentially unregulated; (2) the regime under the Clean Air Act Amendments of 1970, which led to the adoption of regulations in 1971 requiring new power plants to meet a standard of 1.2 pounds per MBTU; (3) the regime under the administrator's proposed regulations published in 1978 pursuant to the Clean Air Act Amendments of 1977, which limited emissions to 1.2 pounds per MBTU and required an 85 percent reduction from the emissions that would have resulted from the use of untreated fuel; (4) the regime under the administrator's final regulations, adopted in 1979, which imposed (a) a 1.2 pounds per MBTU limitation and (b) a 70 percent reduction if the resulting emissions are no more than 0.6 pound per MBTU, and a 90 percent reduction otherwise. Table VIII.2 shows the effects of the various regulatory

Table VIII.2.

Pre-1970 Regime

Low-sulfur coal 1.2 pounds per MBTU no scrubbing	*High-sulfur coal* 8 pounds per MBTU no scrubbing

1971 Regulations

Low-sulfur coal 1.2 pounds per MBTU no scrubbing	*High-sulfur coal* 1.2 pounds per MBTU 85 percent scrubbing

1978 Proposed Regulations

Low-sulfur coal 0.18 pounds per MBTU 85 percent scrubbing	*High-sulfur coal* 1.2 pounds per MBTU 85 percent scrubbing

1979 Final Regulations

Low-sulfur coal 0.36 pounds per MBTU 70 percent scrubbing	*High-sulfur coal* 0.8 pounds per MBTU 90 percent scrubbing

measures on coal markets. The various rectangles are proportional to the market shares of the two different types of coal.

Before 1971, the relative markets for low-sulfur and high-sulfur coal were determined by the relative prices of the two types of coal and by transportation costs. On the line dividing the two regions, the total costs (purchase price plus transportation) of both types of coal are equal. To the west of this line, it is cheaper to use low-sulfur coal; conversely, to the east of this line it is cheaper to use high-sulfur coal. The 1971 regulations increase the cost of using high-sulfur coal by effectively requiring scrubbing for this coal to meet the 1.2 pounds per MBTU standard, but leave unchanged the cost of using low-sulfur coal. As a result, the line separating the two regions moves to the east, and high-sulfur coal producers lose part of their market.

In contrast, the 1978 proposed regulations increase the cost of using low-sulfur coal by effectively requiring scrubbing of such coal in order to meet the percentage reduction requirement, but leave unchanged the cost of using high-sulfur coal, which already had to be scrubbed as a result of the 1971 regulations. Thus, they expand the market for high-sulfur coal. In fact, assuming that the costs of scrubbing a given percentage of sulfur dioxide are the same of low-sulfur and high-sulfur coal, the 1978 proposed regulations return the two markets to their pre-1971 division by imposing the same pollution control costs on both types of coal. (The final regulations, by imposing a differential scrubbing requirement, once again move the dividing line to the east, but not as far as the 1971 approach.) What regulatory scheme should the administrator have promulgated pursuant to the 1977 amendments? Was he constrained in his ability to adopt differential reduction requirements?

8. By making new source standards more stringent, the 1979 regulations exacerbated the old-plant effect, that is, they created incentives for highly polluting existing plants to remain in business rather than be replaced by new, cleaner plants. Consider an example in which the annual operating costs of an existing plant, including pollution costs, is $100, and in which the plant emits 10 pounds of sulfur dioxide per MBTU. The annual operating costs for a new plant with the same production capacity, including pollution control costs and capital costs, are $90 if it must meet a standard of 2 pounds per MBTU and $110 if it must meet a standard of 1 pound per MBTU. Thus, if the regulator chooses the less stringent new source standard, the old plant is replaced and the resulting emissions are 2 pounds per MBTU (as compared to the 10 pounds per MBTU emitted by the old plant). In contrast, if the regulator chooses the more stringent new source standard, the old plant continues in operation and the resulting emissions are 10 pounds per MBTU. Thus, the more stringent regulation leads to the worse environmental outcome. For further discussion of the old-plant effect, see Crandall, *The Political Economy of Clean Air: Practical Constraints on White House Review,* at 205, 212–14, in V. Kerry Smith, ed., *Environmental Policy Under Reagan's Executive Order: The Role of Benefit-Cost Analysis* (Chapel Hill, N.C.: University of North Carolina Press, 1984). How should regulators attempt to avoid the old-plant effect? What information do they need? What is a desirable relationship between the standards imposed on new and existing plants, respectively?

9. The 1977 amendments had different environmental effects throughout the country. Table VIII.3 summarizes, for the 1971 and 1979 regulations, respectively, the effects illustrated in Table VIII.2.

Table VIII.3.

Region	1971 Regulations	1979 Regulations
West	Low-sulfur, unscrubbed	Low-sulfur, scrubbed
Midwest	Low-sulfur, unscrubbed	High-sulfur, scrubbed
East	High-sulfur, scrubbed	High-sulfur, scrubbed

The 1977 amendments produced several potentially negative environmental effects. The old-plant effect is discussed above. In addition, scrubbers are likely to malfunction or not be used at full capacity; while scrubbed high-sulfur coal produces the same emissions as unscrubbed low-sulfur coal, if the scrubber malfunctions the resulting emissions for high-sulfur coal will be 8 rather than 1.2 pounds per MBTU. Moreover, scrubbers produce sludge, creating a solid waste problem. The first two of these negative effects are felt not only in the Midwest, but also in the East. Indeed, as sulfur dioxide travels in the atmosphere, it gets transformed into sulfates, which in turn cause acid rain. Because prevailing winds blow from west to east, the adverse consequences of forced scrubbing in the Midwest exacerbate the acid rain problem in the East. See Bruce A. Ackerman and William T. Hassler, *Clean Coal/Dirty Air*, at 20, 67–68, 70–72 (New Haven, Conn.: Yale University Press, 1981).

What were the environmental effects in the West? Which environmental groups are more likely to have supported the amendments? You should consider that sulfates impair visibility and that the reduced visibility in the Grand Canyon and other national parks in the West was an important environmental concern at the time of the passage of the 1977 amendments. Were there better ways to improve the air quality in the West? Was there a better national solution?

10. Ackerman and Hassler urge courts to give little weight to the legislative history of the New Source Performance Standards (NSPS) (id. at 108–9). They note: "There can . . . be no doubt that the clean air–dirty coal coalition would have vastly preferred a single statutory phrase commanding full scrubbing to a ream of legislative history in praise of 'locally available coal.' The reason they settled for less is that they feared they would lose if they went for broke and forced a statutory showdown on NSPS" (id. at 108).

The strategy of the lobbyists, however, must be analyzed against the prevailing techniques of statutory construction. It was not worth their while to take the risks of placing the protection of high-sulfur coal in the statute (and perhaps fail in this attempt) because they were likely to achieve the same result through the use of legislative history. If the prevailing technique of statutory construction had ignored legislative histories, the high-sulfur lobby would have had nothing to lose by pushing for statutory language and might well have gotten it. The impact of rules of statutory construction cannot be evaluated statically. Rather, one must consider the problem dynamically: if the ground rules are different, the lobbying activities will probably be different as well. In light of this complication, are Ackerman and Hassler justified in urging courts to depart from the then-prevailing technique of statutory construction?

11. The NSPS percentage reduction requirement for power plants, which had been added in the 1977 amendments, was repealed in the 1990 amendments to the Clean Air Act. Can this repeal be explained in public choice terms?

12. In 1972 a district judge held that the Clean Air Act of 1970 embodied a nondegradation policy. This conclusion was upheld by the Supreme Court, on a 4–4 vote. In 1974, EPA issued nondegradation regulations, which were similar in structure to the current Prevention of Significant Deterioration (PSD) program. The regulations were upheld by the District of Columbia Circuit in 1976. In 1977, before the congressional passage of the amendments to the Clean Air Act, the Supreme Court granted certiorari to determine whether the statute authorized a nondegradation policy. Following the passage of the amendments, the Court vacated its grant of certiorari. How does this sequence of events affect the strength of Pashigian's argument?

13. Pashigian concludes that "[t]here is temptation to misinterpret the locational hypothesis as simply a contrived opportunity to impose costs on less developed regions by the developed regions of the country." Do you agree that such an interpretation is "misguided?" For another empirical study of congressional self-interest in voting on environmental legisla-

tion, see Crandall, *Controlling Industrial Pollution: The Economics and Politics of Clean Air* (Washington, D.C.: Brookings Institution, 1983).

14. Given the findings by Ackerman and Hassler and by Pashigian, might one hypothesize that the regional losers at the time of a statutory enactment organize and are able to become winners at the time of the subsequent enactments? Why might winners not be able to protect their gains more effectively?

15. Maloney and McCormick suggest that environmental regulation gives existing firms an advantage over entrants into the market. But new entrants are also likely to have existing plants. In fact, the most successful firms might be disproportionately interested in expanding their production. How does this possibility affect the strength of the authors' theoretical argument? Are the authors likely to think that the old-plant effect is a significant problem?

For a discussion about why the House is likely to be more protective of existing industry than the Senate, see McCubbins, Noll and Weingast, "Structure and Process, Politics and Policy: Administrative Arrangements and the Political Control of Agencies," 75 *Virginia Law Review* 431, 459–63 (1989). For a discussion about why both environmentalists and industry are likely to prefer more stringent standards for new plants than for existing plants, see Hahn, "The Political Economy of Environmental Regulation: Towards a Unifying Framework," 65 *Public Choice* 21, 27–28 (1990).

16. In light of Maloney and McCormick's argument, how would you expect the following issues to be resolved?

  (a) whether a new plant can avoid regulation by "bubbling," that is by decreasing by an equal amount the emissions of an existing source owned by the same firm
  (b) whether the initial allocation of permits in a marketable permit scheme is done by auction or by assigning the permits without charge to existing firms
  (c) whether technology-based standards for new plants should be progressively strengthened
  (d) whether standards should be replaced by effluent fees

With respect to these issues, first, since 1981 EPA has been favorably disposed toward bubbles. Second, under the 1990 amendments to the Clean Air Act, marketable permits are assigned to the firms that currently have the right to pollute. Third, after setting technology-based standards for new plants, EPA has typically not subsequently strengthened these standards. Fourth, effluent fees are not in use under any pollution control statute. Are any of these results anomalous? How would you explain any anomalies? An influential article argues that regulated firms will prefer standards to taxes because the former serve as barriers to new entry while the latter do not (and give rise to an additional cost). See Buchanan and Tullock, "Polluters' Profits and Political Response: Direct Control Versus Taxes," 65 *American Economic Review* 139 (1975).

17. Maloney and McCormick chose to study the effect on stock prices of the Supreme Court's decision on PSD. Might the Supreme Court's decision have been seen instead as casting doubt on the legality of the PSD program? How would such a perception have affected the authors' arguments? Recall that the court was evenly divided and Justice Powell did not participate. How might this fact have affected the litigation strategy, following EPA's promulgation of PSD regulations in 1974, of opponents of PSD?

18. Maloney and McCormick's argument would suggest that trade associations, representing existing firms, would support the legality of regulations imposing costs on new entrants. Such trade associations, however, are frequent challengers of such regulations rather than intervenors defending their legality. Can these practices be reconciled with the authors' argument?

19. Farber suggests that the 1970 environmental legislation might be the product of a "republican moment." What explains later legislation? Can the notion of a permanent republican moment be sustained? For studies showing that the ideology of legislators, rather than their self-interest, better explains congressional votes on the Superfund program, see Hird, "Congressional Voting on Superfund: Self-Interest or Ideology," 77 *Public Choice* 333 (1993); Hird, "Superfund Expenditures and Cleanup Priorities: Distributive Politics or the Public Interest?," 9 *Journal of Policy Analysis and Management* 455 (1990).

20. Farber also suggests that environmental legislation is also the result of credit seeking by politicians, primarily by senators aspiring to the presidency. As Elliott, Ackerman, and Millian point out, this incentive might explain the actions of Senator Muskie in 1970. Subsequently, however, with the possible exception of Vice President Gore, no strong contender for the presidency was closely identified with environmental protection. Can a useful theory be constructed by extrapolating from the events of 1970?

21. Farber states that "if firms were providing the primary impetus for environmental legislation, we would expect to find that most environmental regulation originated in industry-specific legislation." Such a pattern, Farber claims, "has not been typical of environmental laws." Environmental statutes, however, commonly delegate to EPA the authority to impose industry-specific restrictions. Consider, for example, the major federal emission limitations under the Clean Air Act. NSPS standards under section 111 and standards for toxic pollutants under section 112 are both to be set by EPA for categories of new sources (e.g., cement plants). Similarly, under sections 301 and 306 of the Clean Water Act, effluent limitations for existing and new sources, respectively, are also set by reference to categories of sources. How does this feature of the legal regime affect the strength of Farber's arguments?

22. Farber relies significantly on Elliott, Ackerman, and Millian. Do both articles present compatible stories? Are they works of the same genre? In what ways does the structure of their arguments differ?

23. Which of the readings in this chapter provides a better explanation for environmental regulation? What strengths and weaknesses does each have? How much of environmental regulation do the readings explain?

Case
Studies

IX

Control of Air Pollution

The basic structure of the Clean Air Act dates to 1970. The statute was subsequently amended extensively in 1977 and 1990. Some background concerning the interrelationships among certain of the statute's provisions is necessary to critically evaluate the readings that follow.

The National Ambient Air Quality Standards (NAAQS), which are set at the federal level by EPA, are the centerpiece of the regulatory scheme enacted in 1970 (the genesis of which is discussed in the article by Elliott, Ackerman, and Millian excerpted in Chapter VIII). The NAAQS prescribe nationally uniform maximum concentrations for "criteria" pollutants (all pollutants other than hazardous air pollutants). These ambient standards are aggregate measures of environmental quality, which do not directly constrain the activities of any polluter. Instead, the enforceable limitations necessary to achieve the ambient standards are imposed by means of emission standards. EPA sets new source performance standards (NSPS), which apply to new stationary sources, as well as standards for moving sources—principally automobiles. In contrast, for existing stationary sources, these standards are imposed by the states through State Implementation Plans (SIPs). The Clean Air Act contemplated that all areas in the country would meet the NAAQS by the late 1970s.

The 1970 statute leaves unanswered two important questions that were addressed by the 1977 amendments. First, what are the requirements of regions that have air quality that is better than the NAAQS? Second, what consequences attach to the failure to meet the NAAQS?

As to the first, the amendments adopted a complex Prevention of Significant Deterioration (PSD) program, which has both an ambient standard and an emission standard component. PSD areas cannot exceed an ambient standard defined by a baseline plus an increment. The baseline is the ambient concentration at the time, following the enactment of the program, that the first major source applies for a permit. The increment is the allowable degradation for the area. The PSD program defines Class I areas, primarily consisting of national parks, in which the allowable increment is very small. All other areas were initially classified as Class II, where moderate degradation is allowed. Subject to various procedural safeguards, the states have the authority to reclassify particular areas to Class III, where there can be considerable degradation. In no event may the ambient air quality levels in PSD areas exceed the NAAQS.

The emission standard component of the PSD program consists of a requirement that new major sources in PSD areas comply with an emission limitation set by reference to the best available control technology (BACT). BACT standards must be at least as stringent as any applicable NSPS standard.

With respect to areas with air quality worse than the NAAQS, the 1977 amendments established a nonattainment program. An alternative ambient standard was imposed on such areas: they would have to make "reasonable further progress" toward the attainment of the NAAQS. The statute contemplated that attainment would occur by the mid-1980s. Moreover, new sources seeking to locate in nonattainment areas have to meet an emission standard defined by reference to the "lowest achievable emission rate" (LAER), which, like BACT, had to be at least as stringent as NSPS.

Despite these provisions, progress toward attainment of the NAAQS was slow, and in the late 1980s, large areas of the country were still far from achieving it. The 1990 amendments created distinctions based on the magnitude of the nonattainment problem and extended the deadlines further. In the case of carbon monoxide, areas of extreme nonattainment will not need to meet the NAAQS until 2010.

The readings in this chapter deal with some of the central questions raised by the structure of the Clean Air Act. The article by James Krier questions whether it is sensible for the NAAQS to be uniform nationwide. He argues that the regulation of air quality should seek to minimize the costs of pollution plus the costs of avoiding the costs of pollution. Uniform standards are undesirable from the perspective of this cost minimization criterion because the costs and the benefits of pollution control vary throughout the country.

The article by George Eads, who was a member of the Council of Economic Advisers during the Carter administration, criticizes judicial decisions prohibiting EPA from considering compliance costs in setting the NAAQS. He argues that supporters of this approach draw a distinction between "goal setting," which they interpret to encompass the setting of the NAAQS, and "goal implementing," maintaining that the consideration of costs might be appropriate at the latter stage but not at the former. Eads states that this distinction is inconsistent with the Clean Air Act's structure because the statutory scheme contemplates no leeway concerning the implementation of the NAAQS. He shows, however, that in practice, the NAAQS are treated as goals, with Congress extending the deadlines through the nonattainment provisions and EPA taking a flexible attitude with respect to violations. More fundamentally, Eads argues that the model of setting the NAAQS without considering costs is unworkable because of the absence of thresholds—levels below which pollution does not cause adverse effects. The most protective standard (in fact, the only fully protective standard) is one that bans pollution altogether, but such a standard is prohibitively expensive. Eads maintains that it is therefore impractical to set standards without regard to costs. He shows that EPA takes costs into account, at least indirectly, even though it is forced to pretend that it is complying with the legal prohibition.

The article by Craig Oren analyzes the PSD program (the genesis of which is discussed in the article by Pashigian excerpted in Chapter VIII). Oren suggests that the program should be seen as having both "control-compelling" components, intended to force sources of pollution to reduce their emissions, and "site-shifting" components, designed to prevent or discourage even well-controlled sources from siting in areas in which they would have an undesirable impact on ambient air quality. Oren uses this distinction to evaluate the legitimacy of the program's stated goals. In particular, he takes issue with the legislative history's justification of PSD as necessary for the protection against harms that may occur at levels below the NAAQS. He argues that if the NAAQS are uniformly too lax, the obvious solution is to strengthen them; thus, PSD must be explained as a reaction to standards that are too lenient for clean areas only. He rejects the cost differential between cleaning up dirty areas and preventing clean areas from getting dirty as a justification for the program, noting that not only the costs but also the benefits of more stringent standards might be higher in dirty areas. He suggests that a better justification might be "hysteresis"—valuing more highly the preservation of an existing resource than the acquisition of a new resource.

The Irrational National Air Quality Standards: Macro- and Micro-Mistakes

JAMES E. KRIER

Uniform Standards: Macro-Mistake

I begin with the assertion that uniform national ambient standards represent a fundamental error in our approach to air quality control. The Clean Air Amendments require the EPA to promulgate two such standards for each pollutant: a primary standard, designed to protect public health, and a secondary standard to safeguard "public welfare"—aesthetic values, plant and animal life, materials and so forth. Each state must submit an implementation plan to achieve the primary standards within three years of approval of the plan by the EPA, unless certain specified exceptions can be granted; secondary standards, which are generally more stringent than the primary, must be achieved within a reasonable time. Both kinds of standards are nationally uniform. . . .

A Test for Resource Allocation

. . . At a minimum, wise resource allocation entails recognition of two propositions: *First*, that while use of a resource in certain ways may result in costs, avoidance of those uses (or the costs of those uses) also results in costs; *second*, that society is using a resource most efficiently (though not necessarily most fairly, an important point to which we shall return) when the total of these costs is as low as possible. Applying these propositions to the air resource, we should first recognize that using the air as a pollutant receptacle entails costs such as discomfort, ill health, crop damage, materials damage, and scenic blight. But we should also realize that avoiding these pollution costs requires expenditures of time, effort, and money. At least from the standpoint of efficiency, therefore, our goal for air quality—for allocating the air to polluting and nonpolluting uses—appears to be quite clear: The air should be cleaned (or polluted, depending upon one's perspective) to the level that minimizes the sum of (a) the costs of pollution, plus (b) the costs of avoiding the costs of pollution.

Applying the Test to Uniform Air Quality Standards

Let us turn to applying the resource allocation test to the federal requirement of uniform national air quality standards. Do such standards represent an efficient allocation of the air resource?

I think it clear that at one level of analysis the answer to this question must be no. To justify uniform standards as efficient in cost-minimization terms one would have to assume that the costs of a given level of pollution and a given level of control are the same across the nation. This assumption, however, is manifestly not valid. For example, aesthetic costs and materials losses will be functions of the

Reprinted by permission from 22 *UCLA Law Review* 323 (1974).

varying resource endowments, degrees of development, and human attitudes that exist in different regions. Even health costs—which were of the greatest concern to Congress in passing the 1970 legislation—vary from place to place. Since such costs represent the aggregate of individual health effects, and since population varies significantly by region, so too will total health costs. If one believes that per capita and not aggregate health costs should be the relevant factor, efficiency considerations would still suggest some variation in air quality levels. This is because the costs of pollution control will also vary, depending upon population, density and nature of development, and meteorological and topographical conditions in any particular region. In short, since the costs of pollution and the costs of control vary across the country, it is difficult to see how a uniform standard can begin to take the varying costs into account. The standard that minimizes total costs for a region in Iowa is hardly likely to do so for all the regions of California or New York or Colorado as well. To require adherence to the same stringent standard everywhere will in many areas result in the imposition of control costs which are much larger than the pollution costs avoided.

At this level of analysis, then, uniform standards appear to labor under a strong presumption of inefficiency. At a somewhat higher level, however, there are several arguments that might be made to rationalize uniformity. To my mind, only the last of these carries any weight at all, and not much at that.

The first argument is that uniform standards are not inefficient because the law permits each state to set more stringent standards (for the whole state or any region within it). Thus, states where the uniform standards are less strict than dictated by efficient resource allocation can solve the problem by tightening up. The problem with this argument is that those states where the standards are too strict are not permitted, under the law, to relax them. . . .

There is a suspicious asymmetry lurking in the tighter-but-not-looser approach of the federal law. The approach recognizes that states have a legitimate interest in setting tighter standards to, say, protect health at the cost of growth; but it denies that states have an equally legitimate interest in setting looser standards to, say, promote growth at the cost of health. One rationale for this approach might be that a relaxed standard in one state could adversely affect a neighboring state by increasing the latter's pollution levels, while a strict standard in one state could have no such adverse impact on its neighbors. The difficulty is that the second premise of this response is not so clearly correct. A strict standard in one state could interfere with the interest of a neighbor—for example, by injuring the neighbor's economic well-being. This result is likely where a particular state has features attractive to industry (a harbor, a transportation hub) which are not shared by a neighboring state. Industrial location in the first state might well yield benefits to the neighbor such as growth within its borders of firms manufacturing supplies for industries in the first state, as well as increased employment for its citizens (who might work for the local firms or cross state lines to work for the industries in the first state). Stringent air quality standards in the first state could cause enterprises to look elsewhere for the features they want, and the neighboring state—not suitable for location by these industries—could suffer losses in well-being as a result.

Another rationale for the tighter-but-not-looser approach brings us to the second argument in favor of uniform standards. This is the assertion that "protection of health is a national priority." Note first that this statement hardly justifies uniform *secondary* standards, which are not based on a concern for public health. But beyond this, the assertion would not appear to support uniform primary standards either. During congressional debate on the Clean Air Amendments, representatives of the Government often tried to justify *uniform* standards with arguments that only justified *federal* standards. The "national priority" argument is one example: Certainly the health of the nation's citizens is of sufficient importance that the federal government can legitimately undertake to be its guardian, but this does not mean that in doing so it should ignore costs. Yet uniform standards do just that, for they do not take account of the fact that health costs and health-cost avoidance costs vary from region to region. The objective of Congress was to "protect the health of persons regardless of where such persons reside." If this objective implies uniform standards, then it seems unacceptable in efficiency terms. . . .

As a further example of trying to support uniform standards with arguments that only support federal ones, consider the Nixon administration's position that a chief advantage of "uniform nationwide air quality standards" is that they provide "an opportunity to take into account factors that transcend the boundaries of any single state. States cannot be expected to evaluate the total environmental impact of air pollutants, or take it into account in standard setting." This line of reasoning might suggest that air quality standards should be set by the federal government because it is in the best position to assess the interdependent needs of the entire nation, and also perhaps best able to avoid parochialism, delay, and pressure from special interests. But the same reasoning would apply if the federal government set standards varying from basin to basin. The logic, in short, supports federal standards but *not* uniform ones.

This observation introduces the final argument offered to rationalize the efficiency of uniform standards. It is to my mind the argument with the most substance, although just how much so turns out to be an empirical question. There is some evidence indicating that the argument ultimately amounts to little.

In essence, the argument concedes that the cost-minimalization test is appropriate, but goes on to point out that in its application, costs other than those I have thus far considered must be taken into account. These are the *information costs* associated with varying standards—extra costs occasioned by the need to gather and assess specific information for *each* region, extra costs that are avoided by uniform standards. As one expert has noted, "It is possible, of course, for the Federal Government to establish standards according to regional or local situations. This, however, would be a large task and would require studies and hearings in local areas." Government proponents of uniform standards might have had in mind the higher information costs associated with varying standards during debates on the Clean Air Amendments. The Nixon administration offered as a "principal advantage" of uniform standards the fact that "the process of putting [such] air quality standards into effect would be accelerated, primarily because there would no longer be any time [and thus expense] consumed in reviewing and approving air quality standards for each air quality region."

The question then is whether the efficiency benefits of varying standards are greater than the information costs associated with setting them. This is an empirical issue for which we do not have hard data. We do, however, have some evidence bearing on the short-run answer; in addition, there is an observation one can make about the long run. The evidence is that the EPA already needs and acquires so much information about each basin in order to assess its implementation plan that it is doubtful that much more would be required to set appropriate individual standards for each. The observation is that even if this hypothesis is not true at present, in the longer run it might be. Individual standards might come to be justifiable in terms of total-cost minimization, not simply because the EPA would necessarily have more information in the longer run, but because in the future the air resource is likely to be more valuable than it is now, as a result of increasing demand brought on by population and economic growth. As a resource increases in value, the value of allocating the resource efficiently likewise increases. Since the information costs associated with varying standards remain more or less fixed over time, at some point they are likely to be surpassed by the efficiency gains such standards would yield. . . .

Three Objections to the Test and Its Application

A number of objections are typically raised against application of the cost-minimization test to a problem like air pollution, and it is time we confronted these. I will try to deal briefly with what seem to be the three most common objections, considering them in an order inverse to their persuasive force.

Objection 1: When it comes to health and well-being, costs should not be a concern. It should be clear that this objection amounts to patent nonsense. Behind the objection is the assumption that there is some absolute of "good health," and the further assumption that this absolute has infinite value, since it should be chosen no matter what cost that choice implies. Yet neither assumption seems reasonable. There is good health and there is better health. Moreover, we see that people regularly choose at some point not to opt for better health because they consider that the resulting benefits are not worth the costs of attaining them. People take dangerous jobs because they pay more; they purchase less than the best possible medical care because the best costs too much. Or, in the particular context of our problem, note that there are people who live in the smoggy San Fernando Valley area of Los Angeles because housing there is cheaper. To say individuals behave this way because they have limited resources and cannot afford any other course of action only underscores the point. Society too has limited resources, and it cannot afford to spend more on cleaner air than cleaner air is worth. At some point there arises a better bargain than obtaining still cleaner, more healthful air.

Objection 2: When it comes to the effects of air pollution, many of the costs cannot be calculated. The thrust of this objection is that the cost-minimization test is acceptable in principle only where all relevant costs can be calculated, and that in the case of air pollution such calculation is impossible. The weakness of this objection is that such calculation is unavoidable. In the abstract it probably is impracticable, perhaps impossible, to attach concrete dollar values to intangibles such as the aesthetic benefits of clean air. I would even agree that concrete cost figures for the health benefits of improved air quality cannot be calculated. However, these concessions have to do with valuations in the abstract. In the practical context of determining air quality standards, on the other hand, implicit valuations are necessarily made. The choice of a given standard, made through collective action, always reveals some conception and balancing of costs and benefits: A slack standard, with low control costs, is rejected as insufficient; a tight standard, with high control costs, is rejected as too expensive. In the process, an implicit value is attached to those effects not avoided by the more stringent standard. Otherwise there would be no reason to stop—to say that further avoidance would not be worth the additional expense of the more stringent standard.

This sort of implicit valuation is both common and necessary; indeed, it is unavoidable. The fact that we cannot make concrete valuations does not negate the further fact that we *must* choose either to act or not act. (In other words, not making a choice is itself a choice.) In choosing, we make a value judgment about the merits of the alternatives, and this judgment—an implicit calculation of the incalculable—is as unavoidable as the choice.

Objection 3: The cost-minimization test is aimed at efficiency; it ignores considerations of fairness, which are more important. To my mind, this objection has the most substance and it is the most difficult with which to deal. Although the cost-minimization test itself is concerned solely with efficiency, employing the test does not necessarily mean that considerations of fairness are ignored. When I suggested the test, I said it identified the desirable air quality goal *from the standpoint of efficiency.* Once so identified, the goal could still be rejected as unfair.

It appears to me that unfairness in this context could bear two rather different meanings. This first meaning must be that it is unjust for a society to live in an impure environment. This reasoning comes close to suffering the same faults as the first objection discussed. I suppose it is "unfair," in a very loose sense, if the air is not perfectly clean. It is a sad thing; it is a sorry state of affairs; it is just plain wrong. But it is also plain hard fact, even if we run our society solely in terms of fairness, that using limited resources to make the atmosphere cleaner and cleaner means diverting those resources from other worthwhile enterprises that should in fairness also be pursued. Choices among opportunities must always be made, whatever the rhetoric of resolution. I suspect that the most principled choices will be made when the full dimensions of both considerations—fairness and efficiency—are kept in mind, and when it is remembered that a step toward a fair result in one problem area, if accompanied by positive economic costs, *may* mean unfairness in another.

The second meaning of unfairness is more pointed. I take this meaning to be that injustice lies in the *distribution* of air quality among economic and social groups. The cost-minimization test might mean (in fact, as we saw, it almost surely would mean) clean air in some places and less clean air in others. The poor and disadvantaged might generally live in the less clean areas. The conclusion is that this inequality is unfair. The prescription is to make the air as clean for the poor and disadvantaged as for the more fortunate, or, if this is meteorologically impossible, as clean for them as the more fortunate demand for themselves, even if this means the air will be cleaner for the fortunate than the level for which they would be willing to pay. In other words, the prescription is to "protect the health of persons regardless of where such persons reside."

There are several problems with this position. First, it overlooks the likelihood that the costs of improved air quality will be borne disproportionately by the poor and disadvantaged in such forms as reduced income and higher prices; that is, they may be worse off with better air. A related problem is that the position assumes that the poor and disadvantaged prefer cleaner air to increased or maintained economic well-being, rather than leaving the choice to them. Since the cost-minimization test would aim at spending on air quality only to the efficient point, it would make available for distribution wealth that would have been spent by any more costly air quality policy—such as one calling for uniformly stringent standards. That wealth could be distributed on an equitable basis, thereafter leaving the poor and disadvantaged free to choose among a number of goods and services: cleaner air, better housing, more entertainment, warmer clothes, a newer car. Many, I suppose, would opt to stay where they were, preferring to enjoy things other than better air quality. To deny them that choice would appear to be paternalistic.

Perhaps an answer to this is that our society has shown little inclination toward equitable income redistribution, so that anyone concerned with the welfare of the poor and disadvantaged should take every opportunity to improve their lot by redistributing goods and services, even if such improvement is inefficient under the cost-minimization test. This is fair enough, if those who advocate this approach (a) are aware of its pitfalls (it may operate to make the intended beneficiaries worse off), and (b) spend at least equal time advocating more equitable income redistribution. The general point is that while considerations of fairness are important, so are those of efficiency. In a world of limited means the two are tied together. We should always consider not only whether an efficient result might be unfair, but also whether that unfairness might not be avoided by measures other than costly misallocations of resources. And, even if it cannot be, we must consider in each case whether it is fair to be inefficient, whether in fact an inefficient program might not bring with it effects which are themselves unacceptable in equity terms. The choice among competing environmental policies ultimately comes down to confronting just such issues as these, yet it seems they were ignored when Congress opted for the policy of uniform air quality standards.

The Confusion of Goals and Instruments:
The Explicit Consideration of Cost in Setting
National Ambient Air Quality Standards

GEORGE EADS

A number of important regulatory statutes—the Clean Air Act prominently among them—have been interpreted by the courts as precluding the explicit consideration of costs in setting health-based standards. Proposals that these statutes be amended

Reprinted by permission of Rowman & Littlefield Publishers, Inc., from Mary Gibson, ed., *To Breathe Freely: Risk, Consent, and Air* (1985).

to permit (or even require) the incorporation of explicit cost considerations or the balancing of benefits against costs have been intensely controversial. Several grounds for objection have been raised, two of which will concern me in this chapter.

The first is what some see as the ethical implications of such a change. On one view, "clean" air is a "right" that "ought not" be subject to cost-benefit trade-offs. Subjecting to such trade-offs important public decisions that determine the level of protection to be provided undercuts in a fundamental way the values that the programs designed to enforce these rights are intended to achieve. On this view, it is irrelevant whether introducing trade-offs might result in higher levels of health protection: it would be wrong and should be avoided regardless of the consequences.

A somewhat different but related source of opposition comes from those who hold strong pro-protection views and who believe that introducing cost considerations explicitly into basic standard-setting decisions would inevitably shift the outcome of the policy debate against these values. While this latter group sometimes uses the same vocabulary as those who hold "rights" views, their concerns are more tactical than ethical. Simply put, these concerns are with the consequences of the shift.

Supporters of both viewpoints draw comfort from a claimed ability to divide regulation into two distinct phases: "goal setting" and "goal implementing." In regulating airborne emissions from stationary sources the goal-setting phase encompasses the establishment of the National Ambient Air Quality Standards (NAAQS): the levels of pollutant concentration above which it is considered to be likely that some significant "sensitive population" will experience adverse health effects. The goal-implementing phase includes the establishment of the various source-specific emissions limits that determine when (and how) the NAAQS will actually be attained. In the goal-setting phase, issues of cost and technological feasibility are not to be considered. In the implementation phase, both considerations are permitted.

The attractiveness of this dichotomy to those who hold a "rights" view of airborne pollution control is easy to understand. Practicality may compel cost considerations to enter into decisions concerning how and how rapidly to implement standards, but the partitioning off of a goal-setting phase appears to permit an important element of the pollution control debate to be conducted in an atmosphere relatively free of their contaminating influence. The dichotomy's appeal to those who are concerned merely with the presumed consequences of permitting cost considerations to be explicitly considered is more complex and will be explored later.

In this chapter I advance two contentions: that the goal-setting/goal-implementing dichotomy is untenable as a matter of law, and that recent changes in our understanding of the physiological effects of air pollution effectively rule out setting the basic Clean Air Act standards without consideration of cost even if it were tenable to do so. If I am correct, then the controversy over whether the Clean Air Act ought to be amended to permit the explicit consideration of cost in the setting of the NAAQS ceases to be an argument over the ethical implications of including cost as one factor in setting these standards and instead becomes a strategic debate over the practical consequences of permitting explicit consideration of cost. . . .

The Process for Setting and Implementing the National Ambient Air Quality Standards

The Link Between Setting and Attaining the NAAQS

. . . [A] description of how the Clean Air Act is supposed to operate demonstrates that the distinction between goal setting and goal implementing is untenable. Once the NAAQS have been established (in theory, without regard to considerations of cost or technological feasibility), little or no leeway is open to the states under the provisions of the Clean Air Act. The District of Columbia Court of Appeals summarized the situation well in its recent decision in the case of *American Petroleum Institute v. Costle:*

> The goal of the Clean Air Act is to protect the public health and welfare by improving the quality of the nation's air. 42 U.S.C. Sec. 7401(b). Improved air quality is accomplished by the establishment of national ambient air quality standards (NAAQS) and by implementation thereof through state programs to control local sources of pollution. 42 U.S.C. Sec. 7410. The Act directs the Administrator to establish two types of NAAQS. Primary ambient air quality standards are "standards the attainment and maintenance of which in the judgment of the Administrator, based on such criteria and allowing an adequate margin of safety, are requisite to protect the public health." 42 U.S.C. Sec. 7409(b)(1). Secondary standards "specify a level of air quality the attainment and maintenance of which in the judgment of the Administrator, based on such criteria, is requisite to protect the public welfare from any known or anticipated adverse effects associated with the presence of such air pollutant in the ambient air." 42 U.S.C. Sec. 7409(b)(2). State control programs must provide for the attainment of primary standards "as expeditiously as practicable but . . . in no case later than three years from the date of approval of such plan. . . ." 42 U.S.C. Sec. 7410(a)(2)(A)(i). State programs that implement secondary standards must specify a "reasonable time at which such secondary standard will be attained." 42 U.S.C. Sec. 7410(a)(2)(A)(ii). *Thus, the ozone standards at issue in this case must be implemented through state plans within three years for the primary standard and within a reasonable time for the secondary standards.*

The "Goals" of the Clean Air Act

The Court's use of the word "goal" in its decision in *API v. Costle* is instructive. For the Clean Air Act does embody goals, though not in the way envisioned by those proposing the goal-setting/goal-implementing dichotomy. The goals of the act are contained in the section referred to by the Court, the statement of congressional findings and purpose. This statement reads, in part, that the purpose of the act is "to protect and enhance the quality of the nation's air resources so as to promote the public health and welfare and the productive capacity of its population."

This goal statement is similar in character to statements of other goals enunciated by Congress: the elimination of poverty, the ending of racial discrimination, the achievement of full employment and stable growth. Each of the pieces of legislation setting forth programs for purposes such as these contains broad statements of ideals toward which the nation should strive and a strategy for their attainment. Some of these strategies are precisely specified; others are left deliberately vague. However, when read as part of the statute as a whole, and when their legal interconnections are made clear, it becomes apparent that the NAAQS are not goals, with the emissions control programs means of implementing them; instead both the NAAQS and the emissions control programs are instruments for achieving the broader national goal of controlling air pollution.

Clean Air Regulation in Practice: The De Facto Enshrinement of the NAAQS as "Goals"

The above account describes the Clean Air Act as it is supposed to operate. It does not describe reality. The supposedly immutable deadlines have been pushed back by Congress once already and are certain to be pushed back once again. Protestations to the contrary notwithstanding, this has not been due primarily to the laziness of governmental officials or to the recalcitrance of industry, but to the growing realization that the original objectives as well as the dates for their attainment were hopelessly unrealistic. The Environmental Protection Agency, under both pro-environment and anti-environment presidents, has used whatever discretion it has—and sometimes then some—to introduce flexibility into compliance plans. Even where deadlines have not been extended, they have routinely been missed with impunity.

The menu of penalties available to EPA has not been invoked. There has, of course, been a long series of threat and counterthreat. But only one state, California, has ever had federal funds withheld. Occasions have occurred (and have been allowed to pass) where EPA could, in theory, have shut down entire major industries, or levied punitive fines having roughly the same effect. . . .

[E]nvironmental regulation in practice has been more like an extended and generally acrimonious negotiation involving EPA, the states, various industries, and Congress, with the courts often serving as the referee. In this complex and politically explosive negotiating process, the NAAQS have, in fact, taken on a life of their own and, in doing so, have become something very much like goals. . . .

Abandoning the Notion of a Health Effects Threshold: Enter Costs

. . . If the process of attaining the NAAQS bears little resemblance to the one described in the Clean Air Act, the same can be said of the process for establishing the NAAQS themselves. Again, the reason is pragmatic. In this case, the pragmatism arises from our growing understanding of the nature of the relationship between the exposure to pollutants and possible human health effects.

The original model that underlay the NAAQS—and that permitted at least the possibility of setting these standards without considering costs—was that distinct thresholds existed below which the exposure to pollutants would cause no discernible harm, even to sensitive populations. However, as we have become better able to detect physiological responses of organisms (including human beings) to low doses of pollution and as more and more large-scale epidemiological studies have been conducted, the notion of well-defined thresholds has had to be abandoned. This abandonment has been clearest in the case of ozone where, in revising the standard, the EPA administrator declared:

The criteria document confirms that no clear threshold can be identified for health effects due to ozone. Rather, there is a continuum consisting of ozone levels at which health effects are certain, through levels at which scientist [*sic*] can generally agree that health effects have been clearly demonstrated, and down to levels at which the indications of health effects are less certain and harder to identify. Selecting a standard from this continuum is a judgment of prudent public health practice, and does not imply some discrete or fixed margin of safety that is appended to a known "threshold."

And later:

[A review of the scientific studies] illustrate[s] several important points: (1) all alternative standard levels reflect some risk, (2) there is no sharp break in the probability estimates that would suggest selecting one alternative standard level over another, and (3) the choice of a standard between zero and a level at which health effects are virtually certain (0.15 ppm) is necessarily subjective.

How, then, was an ozone standard to be set? *No* level of ambient exposure above zero could be ruled out if consideration was given just to health effects. But an ambient standard of zero was clearly prohibitively expensive. The administrator's solution to this dilemma is worth quoting at length:

The Clean Air Act, as the Administrator interprets it, does not permit him to take factors such as cost or attainability into account in setting the standard; it is to be a standard that will adequately protect public health. He recognizes that controlling ozone to very low levels is a task that will have significant impact on economic and social activities. *This recognition* causes him to reject as an option the setting of a zero-level standard as an expedient way of protecting public health without having to decide among uncertainties. However, it is public health, and not economic impact, that must be the compelling [not sole?] factor in the decision. Thus, the decision as to what standard protects public health with an adequate margin of safety is based on the uncertainty that *any* given level is low enough to prevent health effects, and on the relative acceptability of various degrees of uncertainty, given the seriousness of the effects.

In short, costs weren't supposed to be considered, but they were—indirectly. Furthermore, this was not due to any willingness on the part of the EPA administrator to flout the law, but due to the impracticality of carrying out the law. But in order to develop a standard that would stand up in court, he was forced to pretend (though the pretense was relatively transparent in this case) that costs did not play an overt role in his decision. This he succeeded in doing, for, as we have seen, the courts upheld his decision. But in the process, the public lost the chance to examine the role that cost— as opposed to other factors—did play in influencing his judgment. All we know is that the importance of cost did not rise to the level where it offended the reviewing courts.

Is the Statutory Structure Really All That Important?

The preceding sections demonstrate that the structure of the Clean Air Act is not consistent with a goal-setting/goal-implementing dichotomy that would separate the process of setting the NAAQS from the processes designed to ensure their attainment. They also demonstrate the impossibility of excluding cost considerations, at least implicitly, from consideration in setting the NAAQS. But they also suggest that, where the realities of Clean Air Act administration clash in a serious way with the act's statutory structure, an accommodation is somehow reached. Thus, costs are considered, draconian penalties are not assessed, dates for attainment are slipped, both formally and informally—all without changing the act itself. The act thus becomes what I would term a "policy fiction," and arguments, intense though they may be, about changing the structure of the act to reflect these accommodations become arguments, at least in part, over the value of maintaining this policy fiction.

It might be contended that policy fictions are healthy—even necessary—to the smooth working of the policy-making process. By diverting conflict away from more substantive areas—like compliance methods and dates—they permit the unwieldy structure of the Clean Air Act to function. If some of the more symbolic (according to this view) features of the act were modified, attention would focus on the act's underlying structural flaws, and the whole consensus underlying the process of controlling air pollution through this structure might come unraveled. . . .

But the view that policy fictions are desirable and even necessary to the functioning of the act and therefore should be retained has its limits. For these fictions are not costless. To be sure, EPA administrators have found a way around the absolute prohibition on consideration of costs implied by the act (at least as the courts have interpreted it), and the courts, while continuing to cite precedents that would appear to compel overruling these administrators' decisions, have nevertheless upheld them as "supported by a rational basis on the record" and held that the process by which these decisions were reached, "[a]lthough not a model of regulatory action," was not so flawed as to warrant invalidation of the final standard. But to arrive at this outcome, decision-making processes have had to be artfully concealed and evidence that has figured prominently in decisions has had to be disowned. It has been impossible, for example, to adopt a decision-making procedure . . . in which the evidence upon which a NAAQS will be based, scientific and otherwise, is formally arrayed, evaluated, and presented to the administrator in a form amenable to the clear weighing of trade-offs. It also has been made impossible to demand of the administrator a rational, documented accounting of the weight he or she has given various factors in reaching the announced decision. We see only the shadow, not the substance. . . .

Prevention of Significant Deterioration: Control-Compelling Versus Site-Shifting

CRAIG N. OREN

The Clean Air Act is concerned largely with cleaning up areas with unsatisfactory air quality so that they meet minimum national standards. But what of areas that

Reprinted by permission of Craig N. Oren from 74 *Iowa Law Review* 1 (1988).

already meet these standards? Should it be enough that no new violations of the standards are created? Or is additional protection from air quality deterioration necessary, and if so, what kind? Should new factories and power plants in clean air areas be expected to install the best available pollution controls, even if not necessary to maintain compliance with the air quality standards? Should areas with clean air be required to maintain air quality better than the national standards—to be required, in effect, to meet a tighter standard than required in areas with dirty air? Should a nondegradation program be confined to national park and wilderness areas, or should its scope be nationwide? These questions have bedeviled the Act since 1970, when the statute's present structure took shape.

Congress sought to resolve the issue through the Prevention of Significant Deterioration (PSD) program included in the Clean Air Act Amendments of 1977. This program is intended to make sure that clean air stays clean—that areas with air quality better than the national ambient air quality standards (NAAQS) not be "degraded" to bare compliance with the standards. In a nutshell, PSD requires that each new or expanded "major emitting facility" in "clean air areas" use the "best available control technology" (BACT) for minimizing additional air pollution. The program also establishes "increments" that limit the cumulative increase in pollution levels over the "baseline concentrations" in clean air areas. . .

Control-Compelling and Site-Shifting

. . . This Article . . . suggests that a[n] . . . appropriate way to analyze the program is to see it as functioning in two different ways. First, the program aims at "control compelling": that is, it is intended to force pollution sources to improve their pollution controls or change their production processes to cut emissions. In this sense, the program is in effect a supplement to the nationally applicable standards imposed by new source performance standards (NSPS), which similarly compel new sources to cut emissions. In effect, BACT is a site-specific NSPS.

A control-compelling program may to some extent influence where sources decide to locate. For instance, the NSPS program not only reduces emissions from new sources, but also, by eliminating polluter havens—states with lax emission standards for new sources—sways decisions about where a plant should be sited. In doing so, the program overcomes potentially destructive competition among states that might make an effective pollution control program impossible. Similarly, the case-by-case BACT requirement that applies to each new or modified source seeking a PSD permit both lowers emissions and restrains competition among states for new sources.

The PSD program, though, influences siting in another way as well. The program is intended to prevent or discourage even well-controlled sources from locating in areas—park vicinities, or sites either with poor air pollution dispersion or where there has been substantial growth from other sources—in which they would cause an undesirable level of air pollution. I will refer to this second function as "site-shifting." The increments are the primary tools of site-shifting. Rather than attempting to overcome interstate competition, site-shifting tries to distribute emissions in the way that will cause the least harm to the environment.

The control-compelling and site-shifting aspects of PSD tend to blend together in practice, as we shall see, but they are analytically separate. A pollution control program need not both control-compel and site-shift. The NSPS program, for instance, does not site-shift, but rather applies the same level of control regardless of where the source is located. In contrast, an absolute ban on locating sources in certain areas would move sources away without necessarily having any effect on control technology; zoning laws, which traditionally confined industrial uses to certain zones, are an example. . . .

The Legitimacy of the Program's Goals

By its terms, the PSD program places constraints on growth in air pollution that are more stringent than the limits imposed by the ambient standards. The legislative sponsors of PSD were voluble in discussing the goals they believed these constraints would achieve. . . .

Protection of Public Health and Welfare from Effects that May Occur at Levels Below the Standards

The Clean Air Act commands that ambient standards be set at levels that will protect public health and welfare. Nevertheless, PSD's supporters urged in the Congressional debate that the program was necessary because the existing ambient standards were not sufficiently stringent to protect public health and welfare from all known or suspected effects. For instance, the House Report accompanying the provision argued that the existing standards did not take into account the possibility of additive or synergistic effects among pollutants and did not cover such possibly harmful air pollutants as sulfates. Thus PSD was claimed to be necessary not to prevent violations of the current ambient standards, but rather to provide needed protection in addition to the standards. . . .

Yet the PSD increment system seems a curiously inapt response to overly lax standards because it does not protect the dirty air areas where the need to safeguard public health is the greatest. Moreover, the language of the Act and its legislative history readily support the argument that EPA must set the standards on the basis of health and welfare alone. Thus one frequently suggested alternative to PSD that would protect dirty and clean air areas alike is that the ambient standards could be successfully challenged in the courts as insufficiently stringent under the statute. More directly, if Congress were truly convinced that the standards are inadequate, the Act could simply have been amended to establish tighter standards; this solution seems far from fanciful in a statute that sets precise emissions levels for new automobiles and exact quantitative levels of permissible deterioration.

PSD must be explained, therefore, as a reaction not to standards that are too lenient for all areas, but rather to standards that are considered too lenient for clean air areas only. In other words, the logic behind PSD is that lenient air quality standards are more defensible in dirty than in clean areas. Indeed, this is exactly the tack that PSD's supporters took, urging that clean areas should be given more protection than the ambient standards afforded. For instance, Senator Edmund Muskie, the floor manager of the PSD codification in the Senate, repeatedly expressed unwillingness to accept the ambient standards as the measure of acceptable air quality in clean areas. It is necessary, therefore, for the defender of PSD to explain why air quality standards should be tighter in clean than dirty areas—why clean air areas ought to be restricted by a tertiary standard, consisting of the baseline plus increment, stricter than the primary and secondary standards that govern dirty air areas. Only in this way would it be possible to justify the site-shifting aspect of the increments, which limits permissible pollutant concentrations below the levels of the standards.

There are a number of possible rationales for such a differential. One argument, vociferously expressed by PSD's proponents, is that tighter standards in urban areas would impose unacceptable administrative and compliance costs. This has considerable basis. For instance, EPA estimates that a one-hour sulfur dioxide standard of 0.25 parts per million would necessitate an eight-million-ton reduction in sulfur dioxide emissions in the East and Midwest at an annual cost of $3.5 to $5 billion. A similar reduction, when proposed as a means to curb acid rain, has been projected to cause the loss or relocation of thousands of jobs of high-sulfur coal miners in the Midwest and to cause utility rate increases of close to ten percent in the same region. Ozone presents an even more drastic case. EPA currently estimates that one out of three Americans lives in an area where the present ambient standard for ozone is sometimes violated, and that about twenty-five urban areas will not attain the standard in the foreseeable future. Tightening the standard from 0.12 to 0.08 parts per million would more than triple the population of nonattainment areas. Under these circumstances, reluctance to tighten the standard becomes understandable.

It might be possible to disregard high compliance costs if it were clear that health and welfare effects are occurring at the levels of the present standards. But considerable uncertainties exist. One difficulty is that the scientific evidence that a particular effect occurs is often open to challenge. . . .

More fundamentally, there are no clear thresholds at which health effects begin to occur. This is true in two different senses. First, a given air pollutant may cause a broad spectrum of effects. A very low amount of lead, for instance, will increase blood lead levels. A higher amount may increase the concentration in the blood of an enzyme that is used by the body to make hemoglobin. At a still higher amount, synthesis of heme, a prime ingredient of hemoglobin, is decreased. All of these are effects; but there is no objective way to decide which are "health" effects which ought to be prevented through a nationwide ambient standard.

Second, there is tremendous variability in individual susceptibility to pollution; some persons may experience effects at very low concentrations that would not affect the vast majority of individuals. For example, the short-term sulfur dioxide effects recounted earlier affect only a minority of asthmatics; only about 5,000 to 50,000 sensitive persons live in areas where these effects are likely to occur despite compliance with the present ambient standards. The Act's legislative history calls for protection of sensitive populations, but stops short of defining the term. Thus there is no indication whether the most sensitive individual who breathes the ambient air must be protected, or if EPA can choose, as it did in setting both the lead and ozone standards, to leave some exceptionally susceptible persons unprotected. . . .

The fact that the standards cannot prevent all effects in dirty air areas is not necessarily an argument for protection beyond the standards in clean air areas. Rather, it implies a need for the same kind of "pragmatic judgment" about the appropriate level of protection in clean air areas as is necessary in dirty air areas. Perhaps the balance should be struck differently in clean than in dirty areas. But the mere need to strike a balance in dirty areas is not necessarily a reason to strike a more protective balance in clean areas. Just as the argument that the standards are too weak does not necessarily justify more protection for clean air than for dirty air areas, the lack of thresholds does not necessarily call for a differential between clean and dirty areas. . .

We have already seen a rationale that might be used to justify the distinction between clean and dirty areas: that the costs of attaining tight ambient standards in dirty air areas are too great to justify additional stringency. Conceivably, though, this balance might be struck differently in clean air areas, where there would not be the difficulty of having to control existing sources. For instance, suppose that reducing ozone from 0.12 to 0.08 parts per million would cost urban areas a billion dollars. This might be far greater than the cost of keeping a given rural area at its present concentration of 0.08 or below. Thus a pragmatic judgment, to use Senator Muskie's phrase, might be that the cost of reducing ozone to 0.08 does not justify the standard to dirty air areas, but does justify keeping clean areas at 0.08.

There is, however, a difficulty in applying this reasoning to the health standards, or to welfare standards that are intended to curb effects on humans, rather than on attributes of the natural environment. While the cost of enforcing a stringent ozone standard in a clean area may be less than the cost in a dirty area, the benefits of implementation in a dirty air area are bound to be greater given the higher population of dirty areas. Moreover, the marginal health damage from pollution generally increases as concentrations rise and cause ever more incapacitating health effects. Thus it might be more sensible to pay the cost of implementing a more rigorous ozone standard in an area barely meeting the present 0.12 standard, than to expend resources controlling new hydrocarbon sources cleaner areas.

It is, therefore, by no means obvious that the implementation cost differential between clean and dirty areas justifies a program like PSD that applies more stringent ambient standards to clean than to dirty areas. Rather, the cost-benefit balance may be the same in both clean and dirty areas, or may actually justify greater stringency in dirty air than clean air areas. Similar reasoning applies to the argument from insurance. While the cost of buying insurance for clean air areas to protect against undiscovered effects may be less than the cost of insurance for dirty air areas, the payoff in health protection is bound to be far less in clean air than dirty air areas. Hence, there may be no basis for using PSD in effect to buy insurance for clean areas, but not for dirty areas.

If cost differentials are an uncertain basis for a distinction between standards in clean and dirty areas, perhaps a more supportable explanation is "hysteresis": the tendency to put a higher value on preserving an existing possession than would be attached to acquiring that possession. Such a tendency can be seen in economists' surveys on the willingness of individuals to pay for improving air quality or to accept compensation for worsened air. Often persons who are willing to pay only modest amounts for better air quality refuse to name a price at which they would be willing to see visibility obscured. Thus individuals treat the benefit of keeping an area at a certain air quality level as being greater than the benefit of restoring an area to that level. This may justify the differential between standards in clean and dirty areas, since it implies that the perceived benefits of keeping an area at say, 0.06 ppm of ozone are greater than that which would be realized from cleaning up a 0.08 area to that level. . . .

Consequently, the weaknesses of the ambient standards do not appear to provide a clear-cut case for a program like PSD that focuses on protecting areas where the standards are being met. Moreover, PSD seems poorly designed as a health protection program. Here, as elsewhere, it is necessary to distinguish between the control-compelling and site-shifting elements of PSD.

The control-compelling elements protect health by curbing emissions from new sources, and therefore minimizing growth in ambient concentrations of pollution. The site-shifting elements, by contrast, are intended to prevent or discourage sources from locating in certain areas. Ideally, a site-shifting program would protect health by focusing control on sources that wish to locate in areas where ambient concentrations are high and where, therefore, the risk to health posed by a new source would be at its greatest. Such a standard also would tend to deter sources from locating in areas in which the standard is close to being violated, since control technology requirements in those areas would be relatively onerous.

This could be done, for instance, by a uniform standard that covers old and new pollution alike. The site-shifting portions of PSD, though, do not work in this way, but rather rely on an increment system that limits the amount of new pollution allowed in an area after a certain date. The amount of air pollution that existed on that date is largely irrelevant to determining whether a source may locate. Such a system seems based primarily on hysteresis in that it focuses on preventing the loss of existing air quality, not on the total level of pollution.

The increment system has considerable potential to misallocate public health and welfare protection beyond the standards. Hypothesize, for instance, an increment of twenty micrograms and area "Pure" with baseline of ten and area "Sullied" with baseline of fifty. The increment system treats the two areas equally, even though, with Pure's lower baseline concentration, less marginal damage can be expected to occur there from increasing air pollution concentrations. If Pure grows more quickly than Sullied and uses up its increment, the increment system allocates growth to Sullied even though Pure's air quality is better than Sullied. This result is difficult to justify as protecting health, since, after all, baseline and increment-consuming sources are indistinguishable in their effects. Moreover, it is likely that Pure, with the lower baseline, has fewer people than Sullied. Yet because the increment system disregards the baseline concentration, it affords the greater overall protection to Pure. The site-shifting aspect of the increments tends to push sources from Pure to Sullied; similarly, the control-compelling aspect of the increments puts emphasis on controlling the source in Pure, even though the source in Sullied poses the greater health threat.

Notes and Questions

1. Evaluate the strength of the following the arguments for uniform NAAQS:
   (a) to prevent unequal conditions of competition among the states

(b) to ensure every individual in the country an equal level of air quality

(c) to ensure every individual in the country a minimum level of air quality

What other arguments are plausible? What are the strongest arguments for uniform NAAQS?

2. What conceptions of fairness would justify uniform NAAQS? Is there a coherent conception that would call for uniform NAAQS but accept that other components of individual welfare should be allocated through markets rather than guaranteed by the government?

NAAQS are supposed to be set at a level above which the most sensitive population would experience the first adverse health effects (a "critical population, critical effects" methodology). What is the argument for requiring, on fairness grounds, such a stringent standard, given that in other redistribution programs the government generally guarantees only a minimum level of welfare? How would you assess the desirability of criterion for setting NAAQS if it turned out that the health of the most disadvantaged members of society would be improved much more dramatically and at much lower social cost through upgrading the health care system? For discussion of the distributional effects of air quality regulation, see the Peskin article excerpted in Chapter V.

3. Section 116 of the Clean Air Act gives the states the authority to set state ambient standards that are more stringent than the federal standards. Is this authority desirable given the various justifications for uniform NAAQS?

4. How are these arguments affected by the existence of the PSD program, which imposes more stringent ambient standards in certain areas, and by the nonattainment provisions, which effectively allow, at least in the short and medium terms, less stringent ambient standards in other areas?

5. Would Krier favor more stringent ambient standards in densely populated areas or in relatively unpopulated areas? Krier's article was written before the 1977 amendments. How are his arguments affected by the multiplicity of ambient standards introduced by the PSD program and the nonattainment provisions?

Consider the following perspective:

> Note . . . the peculiar result of the approach that Congress has taken: under the Clean Air Act, the highest air pollution concentrations are permitted in those areas where the most people live, while remote areas are afforded extra protection!
>
> This is not a result one would arrive at if standard-setting were based on a calculation of where scarce pollution resources would do the most good. The latter approach would seem to dictate that the tightest standards should exist where pollution would be the most damaging—that is, in heavily populated areas. (Portney, "Air Pollution Policy," in Paul R. Portney, ed., *Public Policies for Environmental Protection* [Washington, D.C.: Resources for the Future, 1990], at 27, 78)

How compelling is this view? In light of the possible mobility of capital and population, what might be the consequences of the disuniform ambient standards?

6. Recall the discussion of federalism in Chapter VII. Why should the NAAQS be set at the federal level? Are they a good tool for dealing with interstate spillovers? What independent justifications might exist for giving the federal government the authority to impose these standards?

7. Eads states that "[t]he original model that underlay the Clean Air Act . . . was that distinct thresholds existed below which the exposure to pollutants would cause no discernible harm." According to him, post-1970 scientific advances have rejected the existence of such

thresholds. Even in 1970, however, Congress was aware that thresholds did not exist. For example, in his testimony before the Senate Subcommitte on Air and Water Pollution, which was considering the bill that led to the Clean Air Act of 1970, the highest ranking federal official with responsibility for air quality, stated that it was "virtually impossible to state . . . that there is a no-effects level" (Legislative History of the Clean Air Act of 1970, at 1489). Why would Congress have enshrined into the NAAQS the fiction concerning the existence of thresholds?

In setting the NAAQS, should EPA continue to pretend that thresholds exist? If it does not, on what basis should it justify its choice of standards? For a discussion of EPA's difficulties in setting NAAQS, see R. Shep Melnick, *Regulation and the Courts: The Case of the Clean Air Act* (Washington, D.C.: Brookings Institution, 1983), at 247-48.

EPA's experience in setting the NAAQS for lead is instructive. EPA determined that a standard of 30 micrograms of lead per deciliter of blood would protect the most sensitive population (children aged 1–5). It then determined that a target mean concentration for the whole population of 15 micrograms would protect 99.5 percent of the sensitive population. Nonetheless, the approach would leave about 25,000 children, mostly in inner cities, inadequately protected. On what basis could EPA have decided not to protect 99.9 percent of the sensitive population? For further discussion, see id. at 271–76; Finamore and Simpson, "Ambient Air Standards for Lead and Ozone: Scientific Problems and Economic Pressures," 3 *Harvard Environmental Law Review* 261 (1979); Melnick, supra, at 281–84.

8. In 1981, Eads presented the following testimony before the Senate Committee on Environment and Public Works:

> When decisions having such enormous potential economic consequences for the country are being made, it is foolish to pretend that economic concerns will not enter into the decision-making process. Indeed, it is positively deceitful to require that the economic considerations which do influence the Administrator's decision be hidden from public view. (Clean Air Act Oversight: Hearings Before the Committee on Environment and Public Works, U.S. Senate, 97th Cong., 1st Sess. [1981], at 199)

How compelling is this position?

Consider a different perspective. Guido Calabresi defines the "costs of costing" as "the external costs—moralisms and affront to values, for example—of market determinations that say or imply that the value of a life or of some precious activity integral to life in reducible to a money figure" (*Tragic Choices* [New York: W. W. Norton, 1978], at 32). Which perspective do you find more attractive?

In the late 1970s, the United States Court of Appeals for the District of Columbia Circuit developed the "hard look" doctrine, primarily in the context of the review of regulatory decisions by EPA. The court required that the agency explain in particularity all the steps that led to its decision. This requirement is designed to improve the quality of agency decision making and to make the agency more accountable to the public. What should courts do in light of the conflict between the decision-making process described by Eads and the "hard look" doctrine?

9. How compelling is Oren's distinction between the "control-compelling" and "site-shifting" components of PSD? For example, Oren states that "the NSPS program . . . does not site-shift, but rather applies the same level of control regardless of where the source is located." Does a uniform standard, however, not shift sources to where raw materials or labor might be cheapest? In contrast, ambient standards (and the increments in the case of the PSD program) appear to be described as "site shifting," but do they not compel more stringent controls for sources that would like to locate in areas in which the ambient air quality levels are

only slightly lower than the applicable ambient standards? Oren endorses the control-compelling rationale of the PSD program but not its site-shifting rationale. What type of non-degradation program would be consistent with his views?

10. Is the PSD program more than an overbroad and poorly targeted way to protect parks and other areas of exceptional environmental value? For discussion of the PSD program's effects on parks, see Oren, "The Protection of Parklands from Air Pollution: A Look at Current Policy," 13 *Harvard Environmental Law Review* 313 (1989).

11. In the case of sulfur dioxide, the NAAQS is 80 micrograms per cubic meter (annual mean) and the increment for a Class II PSD area is 20 micrograms per cubic meter. Under the PSD program, the baseline is determined at the time that the first major emitting facility applies for a permit. Consider the impact of the program on three areas, each of which at the time of the 1977 amendments has ambient air quality levels for sulfur dioxide of 40 micrograms per cubic meter:

> (a) The first major emitting facility applies for a permit soon after the enactment of the 1977 amendments and consumes the full increment; as a result, nonmajor facilities are subsequently prevented from locating in the area.
>
> (b) Between 1977 and 1985, several nonmajor facilities locate in the area and degrade the ambient air quality levels by 20 micrograms per cubic meter. The first major emitting facility applies for a permit in 1985, when the baseline concentration is 60 micrograms per cubic meter, and consumes the full increment.
>
> (c) The area cannot attract any industry because it is not in attainment of the particulate standards, even though it is classified as PSD for sulfur dioxide. Between 1977 and 1995, emissions from a neighboring state degrade its ambient air quality levels by 40 micrograms per cubic meter (consistent with the prevailing case law, the impacted state is unsuccessful at securing a legal remedy against the upwind pollution).

In light of these examples, evaluate the strength of Oren's hysteresis argument.

12. Consider an ambient standard that applies uniformly in two areas, one of which has a population density that is one hundred times higher than the other. Assume that the costs of control are the same in both areas. What discrepancy between wilingness-to-pay (WTP) and willingness-to-accept (WTA) measures is necessary to support Oren's hysteresis argument? Is such a discrepancy plausible? Might the PSD program have been chosen over more stringent uniform NAAQS simply to avoid economic dislocations, even if such dislocations could be justified on cost-benefit grounds?

13. Richard Stewart explained as follows the judicial decisions from the early 1970s finding that the Clean Air Act embodies a nondegradation requirement:

> [T]heir justification is found in principles of diversity that underlie the first amend-ment and the federal structure of our political system. This justification is associated with the impact of experience on individual tastes and values. Environmental degra-dation in general, and air pollution in particular, are associated with industrialization and urbanization, processes that together shape outlooks and expectations in charac-teristic ways. At least since the advent of the romantic movement, relatively pristine environments and the individual and social life associated with them have been rec-ognized as fostering conditions that not only furnish mental and spiritual refreshment for temporary refuge from urbanized life, but also provide fertile soil for the devel-opment of aesthetic, social, and political views distinctly different from those fos-tered by urban industrialism. (Stewart, "The Development of Administrative and

Quasi-Constitutional Law in Judicial Review of Environmental Decisionmaking: Lessons from the Clean Air Act," 62 *Iowa Law Review* 713, 750 [1977])

How persuasive is Stewart's argument? If diversity (as opposed to better quality) is desirable, should some low-quality areas purposely be kept that way?

14. No areas have been redesignated under the PSD program from Class II to Class III. See Robert V. Percival, Alan S. Miller, Christopher H. Schroeder, and William P. Leape, *Environmental Regulation: Law, Science, and Policy* (Boston, Mass.: Little, Brown, 1992), at 853–54. How do you explain this phenomenon? For some speculation, see *To Breathe Clean Air* (Washington, D.C.: National Commission on Air Quality, 1981) Chapter 5. Consider the following factors:

   (a) the role of the NAAQS as constraints on ambient air quality deterioration
   (b) the process by which baselines are set
   (c) the interaction between the nonattainment and the PSD program
   (d) long-distance transport of pollutants
   (e) regional distribution of the demand for industrial location
   (f) procedural difficulties

How does the lack of redesignations affect your views about interstate competition for industry?

15. Evaluate the strength of the following arguments for uniform NSPS standards:

   (a) to create incentives for the introduction of new technology
   (b) to prevent unequal conditions of competition among the states

What are the strongest justifications for uniform NSPS standards? For a critique of BAT standards generally, and NSPS standards in particular, see the Ackerman and Stewart article excerpted in Chapter VI.

16. Consider the following definitions in the Clean Air Act, from sections 111(a), 169(3), and 171(3), respectively:

   (a) NSPS: "a standard for emissions of air pollutants which reflects the degree of emission limitation achievable through the application of the best system of emission reduction which (taking into account the cost of achieving such reduction and any nonair quality health and environmental impact and energy requirements) the Administrator determines has been adequately demonstrated";
   (b) BACT: "an emission limitation based on the maximum degree . . . which the permitting authority, . . . taking into account energy, environmental, and economic impacts and other costs, determines is achievable for such facility through application of production processes and available methods, systems, and techniques, including fuel cleaning, clean fuels, or treatment or innovative fuel combustion techniques . . .";
   (c) LAER: "rate of emissions which reflects . . . the most stringent emission limitation which is contained in the implementation plan of any State . . ., unless the owner . . . demonstrates that such limitations are not achievable, or . . . the most stringent emission limitation which is achieved in practice by such class or category of source, whichever is more stringent."

Can you explain the differences among these standards? Given that the states have some discretion in determining BACT, is it likely that the PSD program has much (in Oren's terminology) control-compelling effect? Is it likely that state implementation plans would contain limitations more stringent than NSPS?

17. Why should the NSPS standards be set at the federal level? Given that the increments under the PSD program are set by EPA, what are the arguments for imposing a federal BACT

requirement in clean air areas? For discussion of this issue, see a portion of the Oren article (pp. 97–104) that is not excerpted. Does it make sense to place responsibility for regulating new sources in the hands of the federal government and for regulating existing sources in the hands of the states?

18. It is instructive to compare the structure of the Clean Air Act with that of the Clean Water Act. As indicated in the introduction to this chapter, under the Clean Air Act, the federal government sets ambient standards and emission standards for new sources, whereas the states set emission standards for existing sources.

In contrast, under the Clean Water Act, the federal government sets effluent limitations (the water equivalent of emission standards). Water quality standards (the water equivalent of ambient standards) are set by the states, though subject to some federal supervision. The principal federal requirement is that the states engage in "use attainability analysis" if the uses that they designate for their waters do not meet the federal goal of fishable/swimmable water. Use attainability analysis is defined as "a structured scientific assessment of the factors affecting the attainment of the use which may include physical, chemical, biological, and economic factors." See 40 CFR §131.1 et seq. How do you explain the differences in the structures of the two statutes? How do you explain the permissibility of using economic factors for use attainability analysis under the Clean Water Act but not for the NAAQS under the Clean Air Act? For a plea for a greater role for water quality standards, see Pedersen, "Turning the Tide on Water Quality," 15 *Ecology Law Quarterly* 69 (1988).

The scientific consensus is that thresholds are more likely to exist for water than for air. For example, aquatic life can be sustained if the level of dissolved oxygen is above a floor and, consequently, if the level of biochemical oxygen demand (BOD)—a commonly used measure of pollution—is below a particular amount. How does this fact affect your answers to the preceding questions?

X

Liability for the Cleanup of Hazardous Waste Sites

The readings in this chapter examine the Comprehensive Environmental Response, Compensation, and Liability Act (CERCLA), popularly known as Superfund, which imposes liability for environmental problems associated with the disposal of hazardous substances. Superfund is among the most controversial of the federal environmental statutes. One set of complaints centers around the breadth of its liability scheme: strict liability, joint and several liability, weak causation requirements, and limited defenses. Superfund is also criticized for the high cost of cleanups, which average about $30 million for sites on the National Priorities List (NPL)—the list of the most hazardous sites, now numbering about 1200.

The article by Richard L. Revesz and Richard B. Stewart provides an overview of the structure of the statute. The central component of Superfund is a set of provisions imposing liability on a large numbers of potentially responsible parties (PRPs) at each site, under expansive liability rules with weak causation requirements and limited defenses. The statute also contains taxing provisions, which raise money used to finance cleanups at sites at which all the PRPs are insolvent or unknown. The authors also review the process for cleaning up contaminated sites, which is cumbersome and time-consuming. On average, sites are cleaned up more than a decade after the time that the Environmental Protection Agency (EPA) becomes aware of their existence. Moreover, the statute gives imprecise and mutually contradictory commands on the choice of cleanup standards. Revesz and Stewart also show how the Superfund statute has had significant effects on the real estate, banking, and insurance industries, as well as on municipalities and the legal profession.

The article by James T. Hamilton and W. Kip Viscusi analyzes the risks addressed by the remedies chosen at Superfund sites. Based on an analysis of EPA's choice of remedy at seventy-eight sites, the authors report that a large majority of risk pathways (mechanisms by which the hazardous substances at the site can endanger populations) is attributable to future land uses: 70 percent of the cancer pathways and 79 percent of the noncancer pathways fall in this category. The authors also show that three-quarters of all the pathways affect residential popula- tions; here too, the bulk of the risk is to future populations (individuals who will move to sites that are currently undeveloped or nonresidential). Simply counting the number of pathways, however, does not give a fair measure of the total risks to which the various populations are exposed in the present and future because some pathways may give rise to more risk than others. Thus, the authors weight each path- way by the total magnitude of the risk estimated for that pathway. They show that 82 percent of the risks addressed by Superfund cleanups are to future residential populations at the site. The study thus raises the important question whether expen- sive, permanent cleanups should be replaced by far less costly containment mea- sures coupled with restrictions on the future uses of the property.

The article by Lewis A. Kornhauser and Richard L. Revesz examines the rela- tive desirability of holding PRPs jointly and severally liable, rather than nonjointly liable. This choice, which has been the focus of considerable attention in discussions about Superfund reform, is particularly salient in the context of Superfund because, as a result of the long time that elapses between the disposal of hazardous sub- stances and the discovery of an environmental problem, liable parties might be dif- ficult to locate or be insolvent. Under joint and several liability, the remaining liable parties are responsible for these orphan shares. In contrast, under nonjoint liability, the plaintiff (usually EPA or a state) bears the shortfall.

Opponents of joint and several liability claim that it overdeters, discourages set- tlements, and produces unfair allocations of responsibility. The article by Korn- hauser and Revesz shows that, on each of these three criteria, neither rule dominates the other (the portion excerpted in this chapter deals with the first of these issues). For example, in the case of two generators sending waste to a single site, when one of them is insolvent, the relative desirability of the rules depends on the other gen- erator's solvency. Joint and several liability performs better when this solvency is high, nonjoint liability performs better when it is in an intermediate range, and both rules produce identical results when it is low. While the analysis proceeds from a simplified economic model and does not explicitly account for the many real-world complications that arise in every Superfund case, its major insights are applicable to more complex situations.

The Superfund Debate

RICHARD L. REVESZ AND RICHARD B. STEWART

During the last decade, the Superfund approach to environmental liability and remediation has become highly controversial. The costs of remedying the environmental problems caused by hazardous substances are great, although Superfund is far from the most costly U.S. environmental program. Its annual costs are in the range of $3 to $5 billion—a fraction of the costs of the federal air or water pollution regulation programs.

Much of the controversy generated by Superfund stems from its far-reaching statutory system of liabilities, which goes far beyond that of the common law. . . . The broad net of Superfund liability includes current and past owners and operators of waste sites and waste generators and transporters. Defendants at Superfund sites include not only large industrial firms, but also a broad array of other entities—municipalities, local dry cleaners, hospitals, and a myriad of small businesses. As a result of this expansive liability regime, Superfund also has had significant effects on the real estate, banking, and insurance industries, as well as on the legal profession.

Defendants have criticized the cleanup levels demanded by the U.S. Environmental Protection Agency (EPA) as excessively stringent and costly. Superfund is also widely regarded as a wasteful and inefficient program, plagued by high transaction costs, serious administrative deficiencies, and long delays in cleaning up sites.

The Liability and Taxing Regimes

Under Superfund, the cleanup of hazardous waste sites is funded by two separate sources: a liability regime and a taxing regime.

The Liability Regime

Superfund contains an extensive and far-reaching liability scheme. Liability is triggered when the government or a private party incurs response costs in dealing with a release or threatened release of hazardous substances into the groundwater, surface water, soil, or air. In the event of such a release, the following categories of parties, often referred to as *potentially responsible parties* (PRPs), are liable: the current owner or operator of the site at which the release occurs; prior owners and operators during whose period of ownership there was disposal of hazardous substances at the site; generators of the hazardous substances; and transporters of the hazardous substances who had responsibility for selecting the site. Liability is also imposed on an owner of the site who, though not otherwise liable, obtains knowledge of the release or threatened release and subsequently transfers the property without disclosing such knowledge.

Reprinted by permission from Richard L. Revesz and Richard B. Stewart, eds., *Analyzing Superfund: Economics, Science, and Law* (Washington, D.C.: Resources for the Future, 1995).

The liability standard under the statute is strict liability, rather than negligence. Thus, a PRP cannot avoid liability by showing that it met the regulatory or common law standards of care applicable at the time that it engaged in the activity, or even that it was also complying with hazardous waste regulatory standards currently in force.

Liability under Superfund is both retroactive and prospective. In addition to imposing liability for cleanup costs attributable to generation, transportation, treatment, storage, or disposal of hazardous substances undertaken before the passage of the statute, it also attaches cleanup costs for wastes disposed after the passage of the statute.

The defenses to liability are extremely limited. A PRP can escape liability only if it can show that the release or threatened release was caused solely by an act of God, an act of war, an act or omission of a third party, or a combination of these causes. Not surprisingly, only the third-party defense has been of practical significance, and the Superfund statute imposes significant limitations on it. Although this defense is the source of considerable litigation, very few PRPs have successfully established it.

A PRP seeking to defend on this ground must show that the third party was the sole cause of the harm, and that the third party was not the PRPs employee or agent. Moreover, these acts or omissions cannot occur in connection with a direct or indirect contractual relationship between the third party and the PRP asserting the defense. Thus, for example, a generator cannot raise the defense if the need for a cleanup arose as a result of the actions of either its transporter or the operator of the site where the wastes were eventually deposited. The party raising the defense must also show that it exercised due care with respect to the hazardous substances and that it took precautions against foreseeable acts or omissions of the third party.

Moreover, the courts have held that PRPs are jointly-and-severally liable if the harm at the site is indivisible—that is, if the wastes are sufficiently commingled that it is not possible to determine which wastes were responsible for the release. PRPs have the burden of showing that the waste and corresponding cleanup costs for which they are responsible are divisible from those attributable to other parties.

Joint-and-several liability is coupled with a *right of contribution*, so that if one PRP had to pay the full cleanup costs at a site, it could require other PRPs to pay their equitable shares of the liability. The right of contribution, however, is unavailing if other PRPs are insolvent or cannot be located. Accordingly, the solvent PRPs, as a group, must absorb the "orphan" shares of insolvent or absent PRPs. The existence of joint-and-several liability is especially significant in the Superfund context. Because significant periods of time—often several decades—can elapse between the disposal of hazardous substances and the cleanup, it is particularly likely that some PRPs will not be found or will be insolvent once they are found.

The Superfund statute also has causation requirements that are highly attenuated. Thus, a PRP can be liable even if it cannot be shown that its hazardous substances were the ones implicated in the release or threatened release that gave rise to cleanup costs. Liability can be imposed upon a PRP if its hazardous substances at some point were present at a site at which there was later a release or threatened release of the same or even another hazardous substance.

In addition to its cleanup provisions, CERCLA authorizes federal, state, and Indian tribe authorities who manage or control natural resources to sue for damages to such resources resulting from the release of a hazardous substance. The categories of persons liable are the same, and the principles of liability are generally the same, as in the cleanup program. Thus, for example, contamination at a site might lead to the impairment or destruction of wetlands. Following a cleanup that removes the hazardous substances from the site, the PRPs could remain liable for *natural resource damages* (NRD) in connection with any residual damage to the wetlands.

The Taxing Regime

The Superfund taxing provisions are an adjunct to the liability scheme. Currently, three separate taxes are levied on chemicals, petroleum products, and general corporate profits to finance the Hazardous Substances Superfund (the trust fund), which gives the statute its popular name. This fund is used for two primary purposes: to pay for cleanups at sites at which all the PRPs are either insolvent or unknown, and to advance money for EPA cleanups at other sites pending EPA's recovery of cleanup costs from the PRPs. Thus, the fund is a revolving as well as residual form of financing, which covers cleanup costs that cannot be recovered through the liability scheme. At the time of the passage of CERCLA in 1980, Congress authorized a $1.6 billion fund, with the money to be raised over five years. The 1986 SARA amendments provided for an additional $8.5 billion, also to be raised over five years. When Congress reauthorized the statute in 1990, it provided for a funding level of $5.1 billion between October 1, 1991, and September 30, 1994. The total costs of cleaning up sites potentially subject to Superfund have been estimated at $100 billion or more.

Impact of the Liability Regime

The Superfund liability scheme has transformed vast sectors of the U.S. economy and has had effects far beyond the PRPs at Superfund sites. Our discussion in this regard focuses on the real estate, banking, and insurance industries, as well as on municipalities and the legal profession, where the impact of Superfund has been particularly significant.

The Real Estate Industry

Purchasers of real estate face the threat that they will buy contaminated land and that, at some point in the future, they will face liability as the current owners of the land. CERCLA recognizes an "innocent landowner" defense to owner liability. In order to assert this defense, an owner must establish that at the time it acquired the facility it did not know and had no reason to know about the hazardous substances responsible for the release or threatened release. Having "no reason to know" is further defined as undertaking "all appropriate inquiry . . . consistent with good commercial or customary practice." In order for purchasers to take advantage of this defense and avoid potentially far-reaching liabilities, it is now customary in the context of transfers of commercial real estate for purchasers to undertake environmental assessments, which, depending on the circumstances, can include extensive testing of soil and groundwater.

The Banking Industry

With respect to the banking industry, the statute, somewhat confusingly, exempts from liability "a person, who, without participating in the management of a [site], holds indicia of ownership primarily to protect his security interest." Two categories of cases involving mortgage lenders are relevant. First, if the borrower defaults and the lender forecloses, taking title to the property, logic would suggest that because the lender then acquires full indicia of ownership it can no longer qualify for the exemption and would face liability as a current owner. Second, preforeclosure liability arises when the bank becomes sufficiently involved in the activities of the debtor—for example, by monitoring the debtor's operations—that it is deemed to participate in the debtor's management. Unfortunately, the courts have been quite divided about what constitutes too much involvement, and a regulation by EPA attempting to clarify the issue was recently struck down in the courts as beyond EPA's authority.

As a result, banks routinely perform, or require the performance of, environmental assessments before they approve mortgages for commercial real estate. Moreover, critics of Superfund claim that the potential liability of banks, and, perhaps more importantly, the considerable uncertainty surrounding the scope of such liability, has undesirably increased the cost and reduced the availability of credit, especially for small businesses.

The Insurance Industry

The insurance industry also has been centrally affected by the Superfund statute. Between 1973 and 1986, the standard *comprehensive general liability* (CGL) policy held by individuals and corporations included a pollution exclusion clause, which provided that insurance would not cover bodily injury or property damage arising out of pollution except if the release was "sudden and accidental." In large part as a result of Superfund, insurers amended this clause in 1986, explicitly excluding any pollution-related liability. The impact of Superfund on the insurance industry is manifested in two distinct ways.

First, firms interested in protecting against liability for pollution must now purchase specialized insurance, which, to the extent it is available at all, carries high premiums, high deductibles, high coinsurance rates, and low caps. Moreover, the availability of such insurance is quite limited. Thus, many firms have had little option but to self-insure, sometimes risking bankruptcy in the event of an environmental accident.

Second, Superfund has raised an enormous amount of contentious litigation concerning the liability of insurers under policies written before the 1986 change in the pollution exclusion clause. PRPs in Superfund actions routinely seek indemnification from their CGL insurers. The interpretation of insurance contracts is a matter of state law, and the state supreme courts that have addressed the issue have split almost evenly on whether the release or threatened release of hazardous substances at Superfund sites is "sudden and accidental" and meets the other terms of policy coverage. The litigation on this matter between insureds and insurers has consumed exceedingly high transaction costs and has led to proposals for the establishment of a fund (to supplement the existing trust fund), financed by assessments on insurance companies, to pay for a portion of Superfund cleanup costs in place of case-by-case litigation between insureds and insurers.

Municipalities

Municipalities also have been caught in the Superfund web. Typically, municipal solid waste contains a small percentage of hazardous substances. Some municipalities disposed of this waste at sites also used by industrial generators. If liability is apportioned proportionally to the aggregate amount of waste contributed by each PRP, on the premise that cleanup costs are roughly proportional to the volume of waste to be cleaned up, the municipalities will generally bear a high percentage of the costs. If, instead, the relevant criterion is the amount of hazardous substances in the waste contributed by each PRP, the bulk of the burden will be placed on the industrial generators. Judicial decisions adopting the former approach have threatened to imperil the financial stability of small towns.

The Legal Profession

The legal profession has also been powerfully affected by the Superfund liability scheme. In the 1970s, the bulk of environmental law practice consisted in large part of challenging EPA and state command-and-control regulations and the implementation of those regulations. Typical lawsuits pitted industrial firms or environmental groups on one side against the federal government on the other. The specialized environmental bar was disproportionately located in Washington, D.C. Largely as a result of Superfund, environmental disputes now routinely involve controversies among industrial and commercial firms that are PRPs at the same site, and between such firms and insurers and banks. The federal government is sometimes both the enforcer of the law and a polluter responsible for cleanup costs at a site. As a result of the broad scope of Superfund liability, environmental law has become a standard component of legal practice nationwide.

The Cleanup Process

The process leading to the cleanup of Superfund sites is cumbersome and slow, and consists of several stages. First, EPA must become aware of a site's existence. Generally, a site is brought to the agency's attention by a state or municipality, or by citizen complaints; there is no federal discovery program. EPA then places the site in the

CERCLA Information System (CERCLIS)—the inventory of locations that potentially require cleanup. To date, over 36,000 sites have entered the CERCLIS database.

Second, EPA conducts a Preliminary Assessment (PA) to ascertain the risks posed by the site. If warranted, a Site Inspection (SI) then follows. At each of these stages, many sites are classified as sufficiently harmless to warrant no further attention.

Third, EPA ranks sites under the Hazard Ranking System (HRS). The HRS is a composite score that measures the risk of the site by reference to three possible routes of human exposure: groundwater, surface water, and air.

Fourth, sites that receive a score above a given cut-off are placed on the National Priorities List (NPL); currently there are over 1,200 sites on the NPL. Only sites listed on the NPL are eligible for the expenditure of money by EPA for long-term remedial action from the trust fund. This limitation, however, does not apply to EPA removal actions (quicker and less extensive measures often undertaken in the face of emergencies).

For sites on the NPL, the fifth stage of the process involves the preparation of a Remedial Investigation/Feasibility Study (RI/FS). This stage consists of a more detailed examination of the site and a preliminary study of possible remedies.

Sixth, EPA issues a Record of Decision (ROD). This document contains an analysis of alternative remedies, with their expected costs, and selects the remedy that will be implemented at the site.

Seventh, comes the Remedial Design/Remedial Action (RD/RA). The former is a more detailed design of the remediation technique chosen in the ROD; the latter is the actual cleanup of the site.

The process, however, does not always occur in this linear fashion. Cleanup activities at a site are often divided into separate parcels known as operable units; one unit, for example, might involve soil removal and another, groundwater treatment. The different operable units may progress at different rates: one might be at the RD/RA stage whereas the other might be at the RI/FS stage.

A RAND study completed in 1989 . . . showed that, for a site that ultimately gets listed on the NPL, it takes on average forty-three months between the time EPA becomes aware of a site's existence and its listing. Twenty months then elapse until the beginning of the RI/FS, thirty-eight additional months until the issuance of the ROD; the RD/RA takes an additional forty-three months. Thus, on average, the time elapsed between listing on CERCLIS and the completion of the RD/RA is twelve years; eight-and-a-half years elapse between the listing on the NPL and the completion of the RD/RA.

Typically, EPA or a state in which a site is located is responsible for the stages leading to listing on the NPL. Of the later stages, the RI/FS and the RD/RA can be conducted by EPA or the state, or by a group of PRPs. In contrast, the issuance of the ROD is the sole responsibility of EPA.

In the early years of the Superfund program, EPA followed a "fund lead" strategy for cleanup. It hired contractors to carry out cleanup activities, paid them out of the fund, and then sought reimbursement from PRPs. The limited size of the fund and the delays and difficulties in obtaining reimbursement led EPA to make increasing use of an "enforcement lead" approach. Under this approach, EPA uses its CERCLA authority to issue an administrative order to PRPs or seek a court order requiring the PRPs to undertake the cleanup. Currently, PRPs are undertaking the bulk of RI/FSs and RD/RAs, typically as a result of settlements with EPA. Evidence suggests that the cost of a given cleanup is about 20 percent lower when it is undertaken by the PRPs rather than by EPA, presumably because private PRPs have stronger incentives to minimize costs and can supervise contractors more effectively.

The Determination of Cleanup Standards

The most important decision at any NPL site is the determination of the extent of the cleanup and the choice of cleanup technology. If the site's soil is contaminated, should the site simply be capped to reduce the probability of releases into the groundwater, or should the soil be removed and incinerated off-site? The first option will typically be a great deal cheaper, but might pose some long-term risks. Similarly, in the face of groundwater contamination, is it sufficient to prevent migration of the contaminated groundwater through containment measures and secure an alternative source of drinking water (or do nothing at all if the contaminated groundwater is not used for drinking), or instead should one undertake a "pump and treat" program? The latter course of action will be far more expensive, and there is substantial question about its long-term effectiveness. Unfortunately, the statute says little that is helpful in answering these questions, and a wide range of remedies has been used in actual cleanups at NPL sites.

CERCLA contains two sets of provisions dealing with cleanup standards. First, it directs EPA to select remedies protective of "human health and the environment." In making this determination, EPA is to consider a wide range of factors. For example, remedial actions must be "cost effective," but must also "to the maximum extent practicable" utilize "permanent solutions" and technologies that will result "in a permanent and significant decrease" in the volume, toxicity, and mobility of contaminants. These provisions leave EPA with considerable discretion. EPA has tended to emphasize more permanent and costly remedies, such as treating contaminated groundwater rather than simply taking steps to prevent its migration.

Second, CERCLA requires that sites be cleaned in accordance with any "legally applicable" or "relevant and appropriate" standards (ARARs or applicable or relevant and appropriate requirements), where such standards exist. Any standard promulgated under a federal environmental law is "legally applicable" and therefore automatically an ARAR; more stringent state standards are ARARs, if certain procedural conditions are met. The statute, however, does not define when standards are "relevant and appropriate," and therefore also ARARs.

The ARAR prescription is particularly problematic in the case of groundwater contamination. The Safe Drinking Water Act (SDWA) defines permissible levels of various pollutants in publicly supplied drinking water. If groundwater at a Superfund site is contaminated by such a pollutant, the SDWA standard probably will not be deemed "legally applicable" if this groundwater is not used as the source of publicly supplied drinking water, or if it is treated before its distribution to households. The standard might, however, be deemed "relevant and appropriate" and therefore qualify as an ARAR. The Superfund statute provides that where SDWA standards are "relevant and appropriate," the cleanups must at least achieve the Maximum Contaminant Level Goals (MCLGs) under the SDWA. (The SDWA provides that

MCLGs "be set at the level at which no known or anticipated adverse effects on the health of persons occur and which allows an adequate margin of safety." MCLGs are aspirational goals; the enforceable standards or Maximum Contaminant Levels (MCLs) must be set as close to MCLGs "as is feasible.") In the case of known or probable carcinogens, MCLGs require a zero concentration of pollution—probably an unattainable objective in groundwater remediation.

The requirement that Superfund cleanups satisfy ARARs, however, is subject to significant exceptions. In the case of groundwater remediation, ARARs need not be used in a variety of circumstances, including if "the remedial action includes enforceable measures that will preclude human exposure to the contaminated groundwater." More generally, for any remediation financed solely by the trust fund, exceptions from ARARs are appropriate on the basis of a balance between "the need for protection of public health and welfare and the environment at the facility," and the availability of amounts from the trust fund to respond to other sites. Moreover, ARARs generally do not exist for soil remediation, which is a major element in many cleanups.

The Magnitude and Policy Implications of Health Risks from Hazardous Waste Sites

JAMES T. HAMILTON AND W. KIP VISCUSI

Our analysis represents the first systematic effort to document the character of the risks addressed by Superfund. Which population groups are most affected? How do these risks arise? What is the magnitude of the risks that are present? Are the risks in fact trivial, or are there serious threats to public health?

Most important from a policy standpoint, should we expand the range of policy options being considered? For instance, should we continue to favor the more "permanent" options of hazard treatment and removal over on-site containment and land-use restrictions? Although our analysis does not assess all attributes of hazardous waste policies, it does highlight a key dimension—the pathway by which the risks arise. Limited policy options, such as capping and fencing a site or restricting land use, can eliminate some mechanisms by which risks arise. Our analysis documents the nature of these risk pathways and provides startling evidence on the way in which these risks arise.

The character of the risks is in many respects quite surprising. Whereas risks to current residents have played a pivotal role in generating political support for the Superfund program, *the overwhelming preponderance of the risks is to future populations* for land uses that represent departures from current behavior. In our database of seventy-eight sites, there are thirty-five sites where future residential pathway risks occur despite the absence of any current residential risks exceeding

Reprinted by permission of the authors and publisher from Richard L. Revesz and Richard B. Stewart eds., *Analyzing Superfund: Economics, Science, and Law* (Washington, D.C.: Resources for the Future, 1995).

the 10^{-6} cutoff level used in this analysis. The detailed analysis of the character of the risks, which leads to a wide variety of similar insights, is particularly instructive in highlighting how different policy mechanisms can influence the pathways responsible for generating the risk. . . .

To address these issues, we analyze the human health risk assessments conducted at seventy-eight Superfund sites that had Records of Decision (RODs) signed in 1991 or 1992. We focus on the distribution of these risks across different categories of analysis in risk assessments. These categories of analysis include:

time frame of exposure (current use or potential future uses)
location of exposure (on-site or off-site)
population type (residential, worker, recreational, trespasser)
exposure medium (such as soil or groundwater)
exposure route (such as ingestion, dermal contact, inhalation)

. . . Part of the decisions involved in conducting the baseline risk assessments lies in determining what pathways to evaluate to derive quantitative estimates of cancer and noncancer risks. Our database defines the pathways in risk assessments by a number of different category variables: time scenario of exposure, exposed population, age group, location of population, location of medium, exposure medium, and exposure route.

The *time scenario* variable refers generally to whether land use envisioned in the risk assessment corresponds to the current use or is related to a projected use in the future. Current land use is determined by the risk assessor according to site inspection data, zoning information, census data, and aerial photographs. Our designation of a pathway as a current or future scenario is determined by whether the risk assessment defined the pathway as current or future. Note that not all current risks are risk pathways that actually represent a risk today. Some assessments are based on current potential scenarios where the land use in an area does not change but other things may change, such as the size of a groundwater contamination plume so that wells not currently contaminated are assumed to become contaminated. These "current potential" risks are defined as current risks in our analysis if the risk assessors described them as such risks.

Future risks are generally those associated with changes in land use or activities. The guidance provided by the RAGS [Risk Assessment Guidance for Superfund] encourages risk assessors to consider a scenario where land that is currently not residential is brought into residential use in the future. The guidance document states:

> Because residential land use is most often associated with the greatest exposures, it is generally the most conservative choice to make when deciding what type of alternative land use may occur in the future. Assume future residential land use if it seems possible based on the evaluation of the available information.

Thus, future residential risks may be estimated at sites that are currently undeveloped or industrial and that have a low probability of future residential use. In our database of seventy-eight sites, there are thirty-five where future residential pathway risks occur despite the absence of any current residential risks exceeding the 10^{-6} cutoff level used in this analysis.

Exposed populations for which pathways are estimated include residents; workers; recreational users, such as swimmers or hunters; and trespassers. Though risk assessments are often conducted with very specific age group designations for the particular pathway described, we have (for this analysis) generally collapsed the different age groupings into adult (ages eighteen and higher) and child (ages less than eighteen).

The risk assessment category for the *location of population* generally refers to where the particular population is exposed to the contaminant (for residents, location of population refers to where they live).

Location of medium refers to whether the contaminant for which the pathway is estimated is on-site or off-site.

Exposure medium describes in what medium the individual is exposed to the contaminant (such as air, groundwater, soil, or biota, that is, plants or animals containing chemicals that are later consumed by humans).

Exposure route details how a person comes into contact with the chemical. For example, soil contaminants may enter the body through ingestion or through dermal contact or through inhalation. . . .

In our analysis, we break down the description of cancer and noncancer pathways by risk assessment categories. Determining the relative magnitudes of current versus future risks is important in distinguishing how estimates of human health risks at Superfund sites are affected by assumptions about future land use. Designating whether risks involve residents, workers, recreational users, or trespassers is a necessary step in analyzing the efficacy of different policy options for reducing human health risks. Similarly, analyzing whether the populations exposed are on-site or off-site and whether the contaminants are on-site or off-site is a necessary part of evaluating the impact of remedies at Superfund sites.

We also analyze the contribution of specific chemicals to the risks posed. Since uncertainty may exist over the toxicity of particular chemicals, consideration of the relative frequency of these chemicals at sites and their estimated contribution to pathway risks may help determine where additional resources could be devoted to defining the risks of these chemicals or developing remedies to deal with particular types of contaminants. . . .

An examination of the distribution of pathways is instructive to get a sense of the frequency with which alternative risk exposure mechanisms are operative. However, one should be cautious in proceeding from a pathway count to making inferences about the total level of the risk associated with a particular grouping of pathways. The risk associated with a set of pathways is governed not only by the number of such pathways but also by the magnitude of the risk associated with them. Pathways for which there is a high probability of an adverse outcome consequently pose greater risk than those with a lower probability. . . .

Table X.1 provides a comprehensive overview of the distribution of the risk pathways by various categories of analysis. The columns of statistics in the table provide the pertinent breakdowns within the risk assessment categories for all 1,430 pathways, for the 1,015 cancer pathways, and for the 415 noncancer pathways in the sample.

The first distinction in the table, which is perhaps the most salient result of the study, pertains to the breakdown between risks arising from current uses of the land

and risks arising from future uses. This distinction pertains not to the time period of the risk but rather to the nature of the context in which the risks will arise. For example, future risks to current residents are generally captured under the "current" timeframe designation, but new uses, such as the decision to build a residential area on land that is now a Superfund site, would be a "future" risk. The striking result of Table X.1 is that the great majority of the risk pathways pertain to such future risk exposures as opposed to risks associated with current uses. Overall, 70 percent of the cancer pathways, 79 percent of the noncancer pathways, and 72 percent of the total pathways pertain to future as opposed to current uses.

Table X.1. Distribution of Pathways by Risk Assessment Categories (%)

Risk Assessment Category	Total Pathways ($N = 1,430$)	Cancer Pathways ($N = 1,015$)	Noncancer Pathways ($N = 415$)
Scenario			
current	27.8	30.5	21.0
future	72.2	69.5	79.0
Exposed population type			
residential	73.2	71.2	78.1
worker	17.4	17.8	16.4
recreational	3.6	3.8	3.1
trespasser	5.8	7.2	2.4
Age group			
adult	62.7	65.3	56.4
child	37.3	34.7	43.6
Location of population			
on-site	69.2	70.2	66.7
off-site	23.2	23.3	22.9
not indicated	7.6	6.5	10.4
Location of medium			
on-site	79.6	80.3	77.8
off-site	13.7	14.3	12.3
not indicated	6.7	5.4	9.9
Exposure medium			
air (from soil)	9.0	9.0	8.9
air (from water)	9.0	10.4	5.3
soil	33.6	38.2	22.2
groundwater	37.2	30.8	52.8
surface water	1.0	1.1	0.7
sediment	5.2	5.7	3.9
biota	3.6	2.9	5.3
structures	0.1	0.2	—
sludge	0.8	0.9	0.5
combination	0.3	0.4	—
leachate	0.1	0.2	—
mothers' milk	0.3	0.2	0.5
Exposure route			
ingestion	58.4	53.7	69.9
dermal contact	22.6	25.7	14.9
inhalation (vapor phase chemicals)	13.0	14.6	9.4
inhalation (dust)	5.7	5.8	5.3
dermal contact and inhalation	0.3	0.2	0.5

Of the exposed population types, the most important in terms of the risk pathways is that of residential populations. Approximately three-fourths of all pathways pertain to residential populations, with the next most important group being workers, for whom only 17 percent of the pathways are pertinent. Recreational users, such as those who fish in streams on Superfund sites, account for a very small fraction of all the risk pathways.

In terms of the age distribution of those affected by the risk pathways, most of the risk pathways (over 60 percent) are to adult populations, and just over one-third of the risk pathways pertain to children (that is, those under eighteen years of age). The main difference between these figures and the overall age distribution of the U.S. population is that whereas 37 percent of the risks are to the child age population, this group comprises only 26 percent of the U.S. population overall. Thus, the pathways affecting children occur almost 1.5 times as often as the representation of children in the population.

The location where the risks arise is also of substantial interest, particularly as it relates to the potential efficacy of policy options that limit future uses of land at or near Superfund sites. Both the location of the populations and the location of the medium (that is, the location of the medium from which the risk arises) are heavily concentrated toward on-site risks. Of the total pathways, 69 percent pertain to risks to on-site populations; 80 percent of the media associated with the pathways pertain to on-site media. The particular media that appear to be most prominent are soil and groundwater, as each of these accounts for over one-third of all pathways. The other relatively important exposure media are air (from soil), air (from water), and sediment, each of which accounts for 5 percent to 10 percent of all pathways. If the two air pathway mechanisms are aggregated, they account for almost one-fifth of all pathways.

The final component of Table X.1 lists the exposure route by which the risk arises. The dominant exposure route is that of ingestion, such as drinking contaminated groundwater or ingesting dirt, where this category gives rise to 58 percent of all pathways. Dermal contact accounts for 23 percent of all of the different exposure routes, and inhalation of vapor phase chemicals and dust are next in importance.

For the exposure routes as well as for most of the other components of the table, the distribution of pathways is fairly similar for both cancer pathways and noncancer pathways. The major distinctions are that the noncancer pathways play a more prominent role in the future risk scenarios, are more likely to affect residential populations, are less likely to affect adults, are more likely to involve groundwater exposure rather than soil, and are more likely to arise from ingestion rather than dermal contact. . . .

Table X.2 analyzes the distribution of the exposed population and the location of the exposed population for each of the two land use scenarios. Overall, the great majority of the pathways are accounted for by residents based on future risk scenarios, which account for 59 percent of the pathways. In contrast, current risks to current residents, as well as future populations in current residential areas, account for only 14 percent of the risk. The next most prevalent category, that of workers, also has a greater number of pathways for future scenarios as opposed to current time frames, but the difference is not as stark as in the case of the residential risks.

Table X.2. Distribution of Pathways

Distribution by Exposed Population Type				
Scenario	Residential	Worker	Recreational	Trespasser
Current	13.85	6.43	1.96	5.52
Future	59.37	10.91	1.68	0.28

Distribution by Population Location			
Scenario	On-site	Off-site	Not Indicated
Current	15.38	10.91	1.47
Future	53.78	12.31	6.15

Note: All figures are a percentage of total pathways, where $N = 1,430$.

There is also a substantial difference in the character of the risks with respect to their population location and the time frame for the analysis. On-site risks under current scenarios account for 15 percent of the pathways, which is only somewhat greater than the current off-site risks of 11 percent. For the future based scenarios, however, on-site risks escalate to account for 54 percent of the pathways, which is more than four times as great as the 12 percent of the pathways due to future off-site risks. Overall, 90 percent of current pathways are on-site residential pathways, 59 percent of future pathways are on-site residential, and future on-site residential pathways account for 43 percent of all pathways in the sample. *The dominant exposure to risks consequently arises from the expected future residential exposures on Superfund sites.*

In the case of residents, the chief risks arise from ingestion of either groundwater or soil. Resident ingestion of groundwater accounts for a quarter of all total pathways. Although Superfund anecdotes frequently highlight the importance of children who eat dirt, it is noteworthy that ingestion of soil plays a much greater proportional role in the risk pathways for workers than it does for residents. Dermal contact with soil and ingestion of groundwater also account for a substantial share of the risks to workers and a significant share of all total pathways.

The risk pathways for recreational users and trespassers account for a very small percentage of all pathways in the sample, but the distribution within these groups is nevertheless of interest. The primary risk to recreational users is that from dermal contact with soil, with ingestion of soil and inhalation of the vapor phase chemicals from soil being next in importance. For trespassers, the major risks are from ingestion of soil and ingestion of sediment, which together account for almost half of all the risks to trespassers. . . .

Analysis of the frequency of risk pathways gives a sense of how often the pathways are pertinent; consideration of the risk levels associated with the pathways indicates the magnitudes of the risk per pathway. However, the overall level of the risk that will be generated at a Superfund site will reflect the combined influence of the frequency with which particular types of risk pathways occur as well as the levels of the risk associated with different types of pathways. . . .

To convey information concerning both the frequency and magnitude of pathway cancer risks, Table X.3 provides statistics on the risk-weighted shares of the different cancer risk pathways. Rather than simply determining the fraction of the pathways represented by particular types of exposures, such as risk to future generations, each of these pathways is weighted by the total magnitude of the risk estimated for that pathway and the risk-weighted pathways are then summed for the entire sample. The statistics in Table X.3 provide information on the percentage of the total risk-weighted pathways accounted for by each pathway type.

The principal purpose of combining the influence of the frequency of pathway occurrence with the magnitude of the risk is to generate a hybrid of the two influences discussed above. For example, in the case of future risk scenarios, we found

Table X.3. Risk-Weighted Shares of Cancer Pathway Risks (%)

Risk Assessment Category	Total Cancer Pathway Risk	Future Cancer Pathway Risk	Current Cancer Pathway Risk
Scenario			
current	8.8	—	100.0
future	91.2	100.0	—
Age group			
adult	74.9	74.5	78.5
child	25.1	25.5	21.5
Exposed population type			
residential	87.3	89.4	65.6
worker	11.0	9.4	27.7
recreational	1.4	1.2	3.6
trespasser	0.2	0.0	3.1
Location of population			
on-site	77.5	81.0	40.9
off-site	19.9	17.4	45.4
not indicated	2.7	1.6	13.7
Exposure medium			
air (from soil)	4.5	4.2	7.3
air (from water)	2.1	2.0	3.1
soil	32.9	34.2	19.4
ground water	47.7	49.1	32.8
surface water	0.0	0.0	0.0
sediment	0.5	0.5	1.0
biota	10.0	7.5	36.5
structures	0.0	0.0	0.0
sludge	0.3	0.3	0.0
combination	0.0	0.0	0.0
leachate	0.0	0.0	0.0
mothers' milk	1.8	2.0	0.0
Exposure route			
ingestion	65.4	64.6	74.1
dermal exposure	28.0	29.2	15.6
inhalation (vapor phase chemicals)	6.2	5.9	9.4
inhalation (dust)	0.3	0.3	0.9
ingestion and dermal	—	—	—
inhalation and dermal	0.0	—	—

that future risk pathways were not only more prevalent than pathways based on current risk scenarios, but that these pathways posed a greater risk level per pathway as well. The compounding of these influences is borne out in the statistics in Table X.3, as 91 percent of all total cancer pathway risks are attributable to future risk scenarios. This emphasis on future risks is much greater than the unweighted share of future pathways, which we found in Table X.1 to be only 72 percent.

The other statistics in the table are presented for total cancer pathway risks, future cancer pathway risks, and current cancer pathway risks. In terms of the distribution of risks, by far the largest risk share is for adults for current pathways and for future pathways.

A very strong contrast arises with respect to risks to the various exposed populations. Residential populations account for 66 percent of the current cancer pathway risks, and this figure escalates to 89 percent for future risk pathways. Similarly, exposures to workers account for 28 percent of current cancer pathway risks, and this figure drops to only 9 percent for future risk pathways.

The location of the populations affected by these risks also changes dramatically depending on the time frame for the risk scenario. The percentage role of on-site population risks rises from 41 percent to 81 percent when one moves to the future risk scenarios, and the role of off-site drops from 45 percent to 17 percent. The implications of the exposed population and location of population results is that future risk scenarios put a much greater weight on risks posed to on-site residential areas. Of the exposure media listed in Table X.3, the most noteworthy pattern is that groundwater risks account for almost half of the future cancer risk pathways, as contrasted with about one-third of these pathways for current cancer risk assessments. . . .

Most of the political pressures that generated the impetus for the Superfund program arose because of the concern of existing populations for the risks that they believe these sites currently pose. Consideration of the risk assessments for Superfund sites indicates, however, that it is not the existing risks that are most salient. Rather, the dominant risks arise from future risk scenarios that generally involve alternative uses of the land. Indeed, these future risks account for 90 percent of all the risk-weighted pathways for the Superfund sites in our sample. Chief among these future risks is that of future residents living on-site. The underlying assumption driving the EPA risk analyses is that there will be new residential areas on existing future Superfund sites where there are currently no such residential areas.

Analysis of the structure of risks is of fundamental importance with respect to the choice of different possible modes of government intervention. If some mechanism were available that could eliminate these future risks, such as the use of various use restrictions and containment options, then the great preponderance of the risks analyzed in human health assessments at Superfund sites would be eliminated. Indeed, examination of the risk pathways suggested that many of the risks likely to remain with such containment and land-use restriction options, such as that to trespassers, are very low even without adopting policies, such as fencing, to reduce these risks.

Evaluating the Effects of Alternative Superfund Liability Rules

LEWIS A. KORNHAUSER AND RICHARD L. REVESZ

One of the central issues in the Superfund . . . debate concerns the fate of joint-and-several liability. . . . [We seek] to inform th[is] debate by comparing the effects of joint-and-several liability to those of nonjoint liability. . . . Our major conclusion is that . . . neither rule dominates the other.

Our discussion . . . focuses on a simple situation: a single plaintiff, whether the U.S. Environmental Protection Agency (EPA) or a state, seeks to recover cleanup costs from two defendants; we therefore do not focus on many of the real-world complications posed by Superfund. We proceed in this fashion because it is important to understand the basic questions raised by joint-and-several liability before dealing with additional complications. The major insights of this chapter are applicable to more complex situations as well. . . .

We consider . . . the relative effects of joint-and-several liability and nonjoint liability in transmitting to generators of hazardous wastes appropriate incentives for waste reduction and care. The first section of our discussion on this topic covers the role of liability rules in transmitting incentives for desirable conduct concerning wastes generated both before and after the passage of Superfund. The second section presents the model that guides our inquiry. The third section analyzes the situation in which the generators have sufficient solvency to satisfy the judgments entered against them; it shows that joint-and-several liability and nonjoint liability have identical effects but that both, when coupled with strict liability, produce under-deterrence. The fourth and final section studies how the situation is different when one of the defendants has limited solvency; it shows that, under these circumstances, it is not possible to draw any general conclusion about whether joint-and-several liability is preferable to nonjoint liability. . . .

The Role of Liability Rules

The relative desirability of competing regimes' imposing liability on generators of hazardous wastes should be evaluated by reference to the incentives that the regimes transmit for desirable conduct. Generators make several relevant decisions that can be influenced by liability rules; the following list is by no means exhaustive.

First, a generator must determine the volume of hazardous wastes that it will produce. Because the wastes are generated as a byproduct of profitable economic activity, this decision is affected by the costs of different production processes as well as the expected liability associated with the generation of hazardous wastes. So, for example, a generator must trade off the costs of more expensive production

Reprinted by permission from Richard L. Revesz and Richard B. Stewart, eds., *Analyzing Superfund: Economics, Science, and Law* (Washington, D.C.: Resources for the Future, 1995)

processes that would yield a smaller amount of wastes per unit of useful output against the higher costs, including the expected liability, of disposing a larger amount of wastes.

Second, a generator must determine whether to recycle, to treat (that is, render nonhazardous), or to dispose of the wastes. Each of these processes has certain immediate costs and gives rise to different levels of expected liability.

Third, a generator must choose a level of care for the handling of the wastes in the predisposal phase. More care raises the cost of disposal but decreases the generator's expected liability.

Fourth, for wastes that it chooses to dispose, a generator must choose a disposal site. Different sites might charge different fees, present different risks that at some point hazardous wastes will be released into the environment, and give rise to different cleanup costs in the event of a release.

Fifth, both in the case of on-site and off-site disposal, a generator must decide on the effort that it will expend in monitoring the disposal site to detect releases of the wastes into the environment. In fact, such monitoring has led to the cleanup of a number of sites that are not on the National Priorities List (NPL). Once a release occurs, cleanup costs might rise quickly as a function of the time that the problem is left unattended. Thus, by monitoring a site and perhaps undertaking a cleanup before the site has come to the attention of EPA, a generator can reduce its liability. Here, a trade-off exists between the monitoring costs and the expected liability.

Sixth, once a release is detected—whether by the generator, by another PRP, or by EPA or a state—a cleanup has to be undertaken. A liability scheme can provide incentives for the generator to act to ensure that the cleanup is performed in a cost-effective manner.

In analyzing the effects of liability rules, the Superfund program should be separated into two distinct components: a prospective component, for wastes generated and disposed after the passage of the statute in 1980; and a retrospective component, for wastes generated and disposed prior to the passage of the statute.

Obviously, with respect to the retrospective component, a liability scheme cannot create incentives for actions that have already been completed. They can, however, provide incentives for monitoring the site and for ensuring that a cleanup is performed in a cost-effective manner. Opponents of retroactive Superfund liability typically overlook these incentives.

With respect to monitoring, it would be desirable if PRPs undertook cleanups at non-NPL sites before the site came to the attention of EPA or the state. Not only would EPA then be able to better deploy its scarce enforcement resources, but environmental problems could be addressed earlier, before the cleanup costs escalated. This problem of cost escalation is particularly acute under the Superfund program, where groundwater remediation comprises a large part of cleanup costs. Such remediation is less likely to be necessary, or its extent reduced, if the cleanup is undertaken soon after the problem is discovered.

With respect to cleanup decisions, there is strong evidence suggesting that, for a given cleanup level, the cleanup costs are lower when the actions are undertaken by PRPs rather than by EPA; current estimates of this differential run at around 20 percent. Moreover, EPA is constrained in the number of sites at which it can supervise cleanups. It has no alternative but to do so at sites that have no PRPs with the resources or expertise to perform cleanups. For other sites, it relies heavily on PRPs, which now perform about 70 percent of the cleanups. The vast majority of the PRPs currently undertaking cleanups generated hazardous wastes before the passage of Superfund. If these parties did not face liability, the cleanups would have to be performed by EPA. Thus, in the absence of retroactive liability, not only would cleanup costs be higher, but our ability to address the problem of contaminated sites would be impaired.

The Model

In our analysis we model a situation in which two manufacturers, Row and Column, dispose of hazardous wastes at a single landfill. These actors benefit from the dumping because the wastes are the byproduct of profitable economic activity. At some time in the future, these wastes may leak into the environment and cause serious damage, including, perhaps, the contamination of groundwater supplies. For ease of exposition, we think of this damage as the cost of cleaning up the landfill and the surrounding area affected by the release. We assume initially that, as the volume of wastes at the landfill grows, this cost increases more rapidly than linearly. We then show that this assumption of convex cleanup costs is not necessary for our central result, which holds even if the cleanup costs are linear or concave.

The expected damage from a release does not fall directly on the generators unless a legal provision shifts the liability to them. Instead, it falls on those who would either undertake a cleanup or, alternatively, suffer the consequences if the problem were left unattended. In our discussion, we assume that EPA initially bears the cleanup costs and then seeks reimbursement to the extent allowed by the liability regime.

The efficient amount of wastes is that which maximizes the social objective function: the sum of the benefits derived by the manufacturers minus the expected damages. An economically rational generator, however, does not make its decision based on the social objective function. Instead, it seeks to maximize its private objective function: the benefit that it derives from the activity that leads to the production of the wastes minus whatever share of the damage is allocated to it. This share depends on the liability regime that applies in the case of a release of hazardous wastes. For expositional convenience, we assume that the only way in which an actor can affect its expected liability is through the quantity of wastes that it chooses to generate. The results developed below, however, extend to the more complex situation discussed in the previous section on the role of liability rules. . . .

The argument can best be developed by reference to a simple example. Let x and y be the amount of wastes generated by Row and Column, respectively. Let $(100 + 20x)$ and $(100 + 20y)$ be the benefits that Row and Column, respectively, obtain from engaging in the economic activity that produces wastes as a byproduct. (If either Row or Column exits the market, it does not receive the fixed component of $100, but the remaining actor receives that additional amount; one can thus think of the market as guaranteeing a $200 profit, which is either split by two firms or captured by a single firm.) Let $(x + y)^2$ be the damage from the disposal of these hazardous wastes. Net social benefits are maximized where $(x + y) = 10$; thus, one efficient outcome is for Row and Column to generate 5 units of wastes each. . . .

Under strict liability, as long as the defendants are sufficiently solvent, joint-and-several liability transmits the same incentives as nonjoint liability. Under either rule, each defendant would pay its share of the total damage. So, for example, if Row generated 4 units of wastes and Column generated 6 units, Row would pay 40 percent of the total damage of $100, or $40, and Column would pay 60 percent of this amount, or $60. Joint-and-several liability and nonjoint liability produce the same consequences because there is no scenario under which either defendant would be called upon to pay costs attributable to the other.

In contrast, the choice between joint-and-several liability and nonjoint liability matters under strict liability when at least one of the defendants has limited solvency. If Row were wholly insolvent, Column's liability would be unaffected under nonjoint liability; it would still pay $60. Under joint-and-several liability, Column would instead pay the full damage of $100. . . .

Full Solvency

Having established that, for strict liability, joint-and-several liability has the same effect as nonjoint liability—regardless of whether it is coupled with joint-and-several liability or nonjoint liability—fails to transmit desirable incentives: it leads to the overproduction of wastes. Recall that social welfare is maximized when Row and Column each generates 5 units of wastes. Assume that Row has tentatively decided to generate 5 units and that Column, without consulting Row, is trying to figure out how much to generate. If it also generated 5 units, it would accrue benefits of $200 and face a liability of one-half the total damage of $100, or $50; its net benefits would therefore be $150. What would happen, however, if Column generated 6 units rather than 5? Its benefits would rise from $200 to $220, its share of the damages would rise from one-half to six-elevenths (6/11), and the total damages would rise from $100 to $121. In sum, Column's net benefits would be $154 rather than $150.

In turn, Column will have imposed a cost on Row. Row's share of the damages will fall from one-half to five-elevenths (5/11), but, as a result in the increase in damages, it will have to pay $55, rather than $50. Its net benefits will fall from $150 to $145. Thus, by deciding to generate 6 units of wastes rather than 5, Column captures an additional $4 in net benefits, but imposes costs of $5 on Row. As a result, Column's action decreases the aggregate level of net social benefits.

Of course, Row can play the same game too. The symmetric Nash equilibrium of this game (the point at which neither Row nor Column has an incentive unilaterally to change its strategy) occurs when each of these actors generates 6.67 units of wastes. We will denote this level as $x(\infty)$, which is larger than x^* (the socially optimal level of wastes generated). The net benefits of each of these parties is then $144.42, rather than the $150 that each would have accrued if they both had acted in the socially optimal fashion.

Limited Solvency

We assume here that each actor has a fixed solvency, which is available to pay the actor's share of the social loss. We assume that the benefits that the actor derives from engaging in the activity leading to the production of hazardous wastes are not included in this solvency. We offer two interpretations for this assumption. First, one might think of an actor's solvency as a bond that it must post in order to engage in the activity. Recourse against the actor is then restricted to the size of the bond. Alternatively, one might interpret the model as containing an implicit time structure. Benefits accrue to the actor in the present while the social loss occurs (and responsibility for it is apportioned) in the future. If current profits are distributed in the present (when they accrue), then the actor's solvency has the traditional interpretation of the difference between its assets and liabilities.

To analyze the problem of limited solvency, we modify in one respect the model presented previously: we assume that the actors derive benefits from the level of hazardous wastes generated only up to a technological limit, which we call x^H. Beyond x^H, no further benefits accrue from additional waste generation. Let x^H be equal to 9 units. We look only at the particular case in which Row's solvency is zero and examine the properties of strict liability for different values of Column's solvency. We restrict our inquiry in this fashion because it is sufficient to generate our central conclusion: that it is not possible to make general comparisons between joint-and-several liability and nonjoint liability.

Joint-and-Several Liability Equilibria

Recalling that Row's solvency is zero, consider first a case in which Column is fully solvent. Under joint-and-several liability, Column will be responsible for the whole liability. If Row were generating x^*, Column would generate x^* as well, as it would incur the full social cost of departing from x^*. (As in the prior section on full solvency, we defined x^* as the socially optimal amount of waste generation; in our example, x^* is equal to 5 units.) But because Row is generating at the technological limit of x^H, which is more than x^*, Column's best response is to generate less than x^*—an amount that we shall call a. (The first symbol is Row's level of waste generation and the second is Column's.) Therefore, the resulting equilibrium is (x^H, a). In our example, a is the value of y that maximizes $100 + 20y - (y + 9)^2$ and is equal to 1 unit. Column's net benefits are therefore $20.

If Column is not fully solvent, there are two possible equilibria: (x^H, a) or (x^H, x^H). Column will generate a (the equilibrium will be at (x^H, a)) if Column has sufficient solvency to pay for the liability attributable to it if it generated x^H, which is the full damage caused by both parties' generating a total of $2x^H$. In contrast, the equilibrium is at (x^H, x^H) if Column does not have sufficient solvency to pay for the liability attributable to it if it generated a, which is the full damage caused when both parties generate a total of $(x^H + a)$.

But the equilibrium can be at (x^H, x^H) even if Column has sufficient solvency to pay the liability that results from the equilibrium (x^H, a). Column's additional liability caused by generating more than a is greater than its corresponding additional benefit, up to the point at which it becomes insolvent. Beyond its point of insolvency, however, it continues to accrue benefits, but its liability does not increase.

We define Column's *critical solvency* under a strict liability coupled with joint-and-several liability (given that Row is insolvent) as s_{Csj}; this is the lowest solvency for which Column would choose a (1 unit) rather than x^H (9 units) in light of Row's choice of x^H. In our example, if Column generates 9 units of wastes, its benefits are $280. Provided that its solvency is less than $260, its net benefits from generating 9 units are greater than its net benefits from generating 1 unit; s_{Csj} is therefore equal to $260 ($280 minus $20). . . .

Nonjoint Liability Equilibria

Under nonjoint liability, Column would only be responsible for its apportioned share of the liability, and the plaintiff would not be compensated for the share attributable to Row if Row is insolvent. Even under nonjoint liability, however, Row's insolvency affects Column's incentives. Where Row is insolvent and generates at the technological limit of x^H, Column faces higher costs than where Row is solvent and generates $x(\infty)$. This result follows from the assumption that cleanup costs, as a function of the volume of wastes, increase more rapidly than linearly. Consequently, if Column were fully solvent, it would, in response to Row's insolvency, restrict the amount of wastes that it generates to b, which is less than $x(\infty)$. In our example, Column is responsible for a share of $y/(y + 9)$ of the total damages $(y + 9)^2$. Then b, the value of y that maximizes $[100 + 20y - [y/(y + 9)](y + 9)^2]$, is equal to 5.5 units. At this level of waste generation, Column accrues benefits of $210 and must pay $79.75; its net benefits are therefore $130.25.

By analogy to the analysis performed for joint-and-several liability, it follows that under nonjoint liability, there is either an equilibrium at (x^H, b), where Column remains solvent, or (x^H, x^H), where Column is insolvent. Our example illustrates a general proposition: because for a given level of waste generation, Column bears less liability under nonjoint liability than under joint-and-several liability, it follows that b is always greater than a.

We define Column's critical solvency under strict liability, this time coupled with nonjoint liability, given that Row is insolvent, as s_{Csn}. In our example, if Column generates 9 units, its benefits are $280. Provided that Column's solvency is less than $149.75 ($280 minus $130.25), its net benefits from generating 9 units are greater than its net benefits from generating 5.5 units; s_{Csn} is therefore equal to $149.75. . . .

Table X.4. Equilibria Generated by Joint-and-Several and Nonjoint Liabilities

Region	Column's Solvency	Equilibria	
		Joint-and-Several Liability	Nonjoint Liability
A	$0-s_{Csn}$	(x^H, x^H)	(x^H, x^H)
B	$s_{Csn}-s_{Csj}$	(x^H, x^H)	(x^H, b)
C	$s_{Csj}-\infty$	(x^H, a)	(x^H, b)

The critical solvency s_{Csn} is smaller than s_{Csj}, Column's critical solvency (given that Row is insolvent) under strict liability coupled with joint-and-several liability. The reason is that if Row is insolvent and Column is solvent, Column faces a smaller liability under nonjoint liability, and the prospect of remaining solvent is correspondingly more attractive. Thus, there is a range of solvencies—the range between s_{Csn} and s_{Csj}, between \$149.75 and \$260 in our example—in which Column would choose to remain solvent under nonjoint liability but would become insolvent under joint-and-several liability.

Comparison of Equilibria Under Joint-and-Several Liability and Nonjoint Liability

Table X.4 summarizes the equilibria generated by joint-and-several liability and nonjoint liability. Recall that Row's solvency is zero. This table defines three regions for different ranges of Column's solvency. It reveals that in region A, the two rules perform identically, yielding an equilibrium at (x^H, x^H). In region B, joint-and-several liability produces an equilibrium at (x^H, x^H) whereas nonjoint liability does so at (x^H, b). Finally, in region C, the equilibrium under joint-and-several liability is at (x^H, a), whereas under nonjoint liability it is at (x^H, b).

We use three measures to compare the performance of the two rules. We determine, first, which rule leads to higher social welfare; second, which rule results in less unfunded liability; and, third, which rule leads to the generation of less waste. Obviously, in region A, both rules produce identical results.

In region, C, we ascertain whether an equilibrium at (x^H, a) under joint-and-several liability is preferable to an equilibrium at (x^H, b) under nonjoint liability. From a social welfare perspective, (x^H, a) is preferable. Where one actor is generating x^H, joint-and-several liability makes the other actor see the full social cost of its actions, whereas nonjoint liability does not. Thus, a is the optimal response by Column to Row's choice of x^H.

Joint-and-several liability also results in less unfunded liability. When the equilibrium is at (x^H, a), Column is solvent and pays the full liability, leaving no unfunded liability. At (x^H, b), Column is also solvent, but under nonjoint liability, it pays only its share of the liability, leaving Row's share unfunded.

Finally, the (x^H, a) equilibrium results in an amount $(x^H + a)$ of wastes, whereas the (x^H, x^H), equilibrium results in $2x^H$ of wastes. Because a is smaller than x^H, joint-and-several liability is preferable to nonjoint liability. Thus, in region C, joint-and-several liability is preferable under all three criteria.

The comparison between the equilibria in region B follows by analogy. From a social welfare perspective, the equilibrium under nonjoint liability, (x^H, b), is preferable to the equilibrium under joint-and-several liability, (x^H, x^H), because, as indicated, given that Row has chosen x^H, the socially optimal response by Column is a; of course, b is closer to a than is x^H. In terms of amount of waste generated, the (x^H, b) equilibrium is also preferable, as $(x^H + b)$ is smaller than $2x^H$.

With respect to unfunded liability, if Column's solvency is only s_{Csn}, which is the lower bound of the range defined in region B, the equilibrium under nonjoint liability, (x^H, b), will result in less unfunded liability. Indeed, as Column expands its production of wastes from b to x^H, the total liability increases, but Column's contribution to that liability does not, as its full solvency is consumed at the (x^H, b) equilibrium. Where Column has a solvency higher than s_{Csn}, however, it is not possible to make any general comparison between the rules: the equilibrium under nonjoint liability, (x^H, b), will result in less unfunded liability for certain benefit-and-loss functions, but in more unfunded liability for other functions.

In summary, neither rule dominates the other. In region C, joint-and-several liability is always preferable under all three criteria. In region B, nonjoint liability is always preferable under two criteria—social welfare and waste generated—but is only sometimes preferable under the criterion of unfunded liability.

Notes and Questions

1. Is it desirable for Superfund to impose liability on individuals who unknowingly purchase contaminated real estate? Consider the following issues:
    (a) incentives for detection of the environmental problem
    (b) likelihood that detection will lead to a cleanup
    (c) potential for unjust enrichment
    (d) avoidance of loopholes that would frustrate the statutory scheme
For a discussion of the relationship between causation and responsibility under Superfund, see Nagle, "CERCLA, Causation and Responsibility," 78 *Minnesota Law Review* 1493 (1994).

2. Is it desirable for Superfund to impose liability on banks? Consider the two ways in which banks can be liable: following foreclosure and before foreclosure if they become sufficiently involved in the activities of the debtor. Consider whether the banks or the purchasers of the property might have a comparative advantage in conducting environmental audits of the property.

3. Consider different ways in which Superfund might treat municipalities that disposed of their wastes at sites also used by industrial generators:
    (a) exempt from liability
    (b) subject to liability, with damages apportioned proportionally to the total amount of waste dumped
    (c) subject to liability, with damages apportioned proportionally to the amount of hazardous waste dumped
Which rule is most desirable from a fairness perspective? Which is more desirable from an efficiency perspective? Consider separately the questions of prospective and retroactive liability.

4. What risk management criterion can you discern from the Superfund cleanup standards? To what extent are they health-based standards set without regard to cost? Recall the discussion of risk management criteria in Chapter IV.

5. The most expensive part of cleanups is generally groundwater remediation. Should such remediation be undertaken if the surrounding communities do not currently use the water for drinking and if there exist abundant alternative supplies to meet future demand? Is it desirable to undertake such cleanups if they can be justified only by reference to existence values? Is it likely that such valuations would be high in the case of groundwater?

6. How should the Hamilton and Viscusi study influence the selection of Superfund remedies? Is it likely that deed restrictions will be respected in the future? What is the likely impact of containment and fencing (rather than a permanent cleanup) on land use development around

the site? Would the racial and demographic characteristics of the surrounding communities affect your views on this matter? How would you view the choice between extensive cleanup that would make the site suitable for future residential use and less extensive cleanup that would permit only industrial use? Does your answer depend on how the surrounding land is currently used?

7. Hamilton and Viscusi did not consider the size of the populations affected at a Superfund site or the costs of achieving risk reductions. Another empirical study found that EPA did not accept higher risks to surrounding populations at sites with higher cleanup costs. Gupta, Van Houtven, and Cropper, "Do Benefits and Costs Matter in Environmental Regulation?: An Analysis of EPA Decisions Under Superfund," in Richard L. Revesz and Richard B. Stewart, eds., *Analyzing Superfund: Economics, Science, and Law* (Washington, D.C.: Resources for the Future, 1995), at 83. This study also found, somewhat paradoxically, that higher risks are selected in urban areas, which are more densely populated, than in rural areas. The authors speculate that EPA may be reluctant to excavate soils in urban areas and that sites in urban areas might be intended for industrial use, for which extensive cleanups might be less necessary. Finally, as is explained more fully in Chapter V, the study also found that the racial and economic composition of the surrounding population is not a factor in the choice of target risk.

Should factors such the size of the affected population or the magnitude of cleanup costs be considered in deciding how much to spend on cleanups? Are the arguments raised by this question comparable to those relevant to the assessment of the desirability of uniform ambient standards under the Clean Air Act, which are raised in Chapter IX? Consider in this connection the more localized nature of the effects of hazardous waste contamination.

8. Should communities surrounding a Superfund site be able to agree to less extensive cleanups than would otherwise be required in return for the financing of other projects that reduce health risks? For example, consider two possible uses for $30 million in an isolated community:

    (a) a $30 million cleanup, which would reduce to one in a million the probability that a resident would contract cancer from exposure to the site

    (b) a cleanup costing $5 million, which would reduce this probability to one in a hundred thousand, coupled with $25 million for a clinic

If it is undisputed that the latter option produces a larger improvement in the health of the community, should EPA be able to defer to the community's preference for the latter option? For further discussion, see Graham and Sadowitz, Superfund Reform: Reducing Risks Through Community Choice, *Issues in Science and Technology*, Summer 1994, at 35. Should EPA be able to choose the latter option if the community desires the former? How should the analysis be affected if, instead of using the $25 million for a clinic, the community wishes to use the money for a school or to reduce taxes?

9. Which of the incentives transmitted by a liability rule could also be transmitted through a tax system? How would taxes have to be structured? From the perspective of transmitting desirable incentives, are taxes a feasible alternative? Recall the discussion of regulatory tools in Chapter VI.

10. Superfund is an *ex post* liability scheme. In addition, activities connected with the disposal of hazardous wastes are covered by the *ex ante* regulatory scheme of the Resource Conservation and Recovery Act (RCRA). RCRA creates a tracking system for hazardous wastes and imposes regulatory standards on operators of hazardous waste sites. Could appropriate incentives be transmitted solely through a liability system? Could appropriate incentives be transmitted solely through a regulatory system? For general discussions of different policy mechanisms in this area, see Hahn, "An Evaluation of Options for Reducing Hazardous

Waste," 12 *Harvard Environmental Law Review* 201 (1988); Lyons, "Deep Pockets and CER-CLA: Should Superfund Liability Be Abolished," 6 *Stanford Environmental Law Journal* 271 (1987); Russell, "Economic Incentives in the Management of Hazardous Wastes," 13 *Columbia Journal of Environmental Law* 257 (1988).

11. Given the conclusions of the study by Kornhauser and Revesz, on what basis should one choose between joint and several liability and nonjoint liability? Which institution is best able to make that choice?

12. There is widespread agreement that joint and several liability has negative effects on the availability of insurance. For example, Kenneth Abraham notes: "The threat of joint and several liability creates special uncertainty, because the probability of liability—and of consequent loss for the insurer—is affected by the behavior of nonpolicyholders whom the insurer cannot necessarily identify in advance" ("Environmental Liability and the Limits of Insurance," 88 *Columbia Law Review* 942, 959 [1988]).

Evaluate this statement in light of the argument in the Kornhauser and Revesz article. Consider the following factors: First, under both joint and several liability and nonjoint liability, as long as the damages of waste disposal are not a linear function of the amount of waste dumped, the amount of waste generated by one actor affects the liability of other actors sending wastes to the same site. Second, under either rule, if the damage function is convex (it increases more than linearly), the insolvency of Row can be the but-for cause of Column's insolvency.

13. In a portion of the article that is not excerpted, Kornhauser and Revesz study the settlement-inducing properties of joint and several liability. They analyze this issue for the pro tanto set-off rule, which applies in Superfund actions in which EPA or a state is the plaintiff. Under this rule, in the event that the plaintiff settles with one defendant, its claim against the nonsettling defendants is reduced by the amount of the settlement. Kornhauser and Revesz show that for this rule joint and several liability deters settlements if the plaintiff's probabilities of success against the defendants are independent (in the case of two defendants, if the probability that the plaintiff will prevail against one defendant does not depend on whether the plaintiff prevails against the other). In contrast, joint and several liability encourages settlements if these probabilities are perfectly correlated (if the plaintiff either prevails against both defendants or loses to both defendants). Which situation is most likely in a Superfund case? In this connection, consider whether the legal dispute concerns the plaintiff's choice of cleanup or the factual responsibility of the defendants.

14. Consider the fairness implications of the choice between joint and several liability, and nonjoint liability. Under joint and several liability, solvent PRPs are responsible for the shares attributable to their nonsolvent counterparts. Under nonjoint liability, the plaintiff, generally EPA or a state, would fail to recover its full cleanup costs, and the shortfall would be financed by the Hazardous Substances Superfund.

An empirical study compared the financial impact on various sectors of the economy of three different Superfund policies:

    (a) the status quo

    (b) a waiver of liability for wastes disposed of before 1980 at multiparty sites

    (c) a limitation on the scope of joint and several liability such that solvent PRPs
        would not have to pay for the shares attributable to insolvent PRPs (though they
        would remain responsible for the shares attributable to unidentifiable PRPs)

It found that the percentage of total cleanup costs borne by each industry remains almost constant under each liability option (Probst, "Evaluating the Impact of Alternative Superfund Financing Schemes," in Richard L. Revesz and Richard B. Stewart, eds., *Analyzing Superfund: Economics, Science, and Law* [Washington, D.C.: Resources for the Future, 1995], at 145).

Consider how the two liability rules would allocate costs among different firms in an industry. Which is fairer? For a more extensive study of the impact of Superfund on different industries, see Katherine N. Probst, Don Fullerton, Robert E. Litan, and Paul R. Portney, *Footing the Bill for Superfund Cleanups: Who Pays and How* (Washington, D.C.: Brookings Institution and Resources for the Future, 1995).

15. Superfund is often criticized on the ground that it gives rise to excessive transaction costs—costs incurred in the process of determining cleanup remedies and imposing financial liability. An empirical study performed by RAND found that, for the private sector, Superfund transaction costs are between 23 and 31 percent of total Superfund outlays. (RAND did not seek to determine the government's transaction costs); Dixon, "The Transaction Costs Generated by Superfund's Liability Approach," in Richard L. Revesz and Richard B. Stewart, eds., *Analyzing Superfund: Economics, Science, and Law* (Washington, D.C.: Resources for the Future, 1995), at 171. Are such costs excessive?

Consider the following comparisons: defendant transaction costs as a percentage of total outlays are about 35 percent for tort litigation, 50 percent for asbestos claim litigation, and 14 percent for airline crash litigation. Which, if any, of these benchmarks is appropriate? Consider, alternatively, the transaction cost as a percentage of total outlays in administrative compensation schemes. For example, for workers' compensation and social security disability, these shares can be as low as 20 percent and 10 percent, respectively. Are these comparisons more appropriate? Evaluate the significance of the following factors:

  (a) recency of the statutory scheme
  (b) breadth of the liability system
  (c) nature of the cleanup standards
  (d) administrative, rather than judicial, remedy selection

16. The Dixon study showed that transaction cost are disproportionately high for PRPs that are responsible for only a small share of the liability at a site. Congress sought to remedy this problem in 1986 by requiring EPA, "[w]henever practicable and in the public interest," to enter into settlements with *de minimis* PRPs "as promptly as possible" (42 U.S.C. §9622[g]). An empirical study of de minimis settlements found that EPA has vastly underutilized this settlement tool and that, even when it has entered into such settlements, it has done so late in the cleanup process, after the expenditure of considerable transaction costs. See Kornhauser and Revesz, "De Minimis Settlements Under Superfund: An Empirical Study," in Richard L. Revesz and Richard B. Stewart, eds., *Analyzing Superfund: Economics, Science, and Law* (Washington, D.C.: Resources for the Future, 1995), at 187. The problem might be that EPA, as a litigant, concerns itself with maximizing its expected recovery in litigation and minimizing its transaction costs. Should it view its mission as maximizing social welfare and therefore strive to minimize the transaction costs that it imposes on third parties? How can appropriate institutional incentives be transmitted?

17. Compare the Superfund liability rules with those of the European Union's proposed directive on civil liability for damage caused by waste. The proposed directive imposes liability on the "producer" of the waste, which it defines, in Article 2.2, to include "the person who had actual control of the waste when the incident giving rise to the damage . . . occurred, if he is not able within a reasonable period to identify the producer" and "the person responsible for the installation, establishment or undertaking, where the waste was lawfully transferred to such installation, establishment or undertaking." Under Article 3, the producer shall be liable "irrespective of fault on his part"; under Article 5, "[w]here . . . two or more persons are liable for the same damage or the same impairment of the environment, they shall be liable jointly and severally;" and under Article 6, "[n]o liability shall attach to any person if

he proves that, in the absence of fault on his part . . . the damage or impairment of the environment was caused by an act or omission of a third party with the intent to cause such damage or impairment." Article 13 provides that the proposed directive "shall not apply to damage or impairment of the environment arising from an incident which occurred before the date on which its provisions are implemented." What are the similarities between this proposed directive and Superfund? What are the salient differences?

18. PRPs at Superfund sites are liable not only for cleanup costs but also for the damage to natural resources under the management or control of the federal or state governments or of Indian tribes. Under what circumstances could the cleanup of a site nonetheless leave natural resource damages?

One measure of damages is the cost of restoration of the injured resource. If restoration of the particular injured resources is inordinately costly, should the purchase of a comparable resource be acceptable? For example, consider a scenario under which, following the completion of a cleanup, the cost of restoring the injured wetland would be $1 million. A similar wetland, currently in private hands and only five miles away, could be purchased for $10,000. Should the purchase of the latter be considered an adequate substitute? Should it make a difference if the substitute is instead fifty miles away? Should it make a difference if the substitute is not a wetland but, instead, a forest that is equally valuable from an environmental standpoint? There is controversy under Superfund about the resolution of these questions.

19. Article 4.2 of the European Union's proposed directive on civil liability for damage caused by waste provides that costs incurred in restoration of the environment can be reimbursed except if they "substantially exceed" the corresponding benefit or "other alternative measures to the restoration of the environment may be undertaken at a substantially lower cost." In the latter case, the plaintiff can be reimbursed for the implementation of these alternative measures. Is this approach desirable?

20. A different measure of damages is the diminished value of the injured resource. In many cases, existence values are an important component of the total value of environmental resources. For a criticism of the use of the contingent valuation methodology (CVM) to value such resources, see Stewart, "Liability for Natural Resource Injury: Beyond Tort," in Richard L. Revesz and Richard B. Stewart, eds., *Analyzing Superfund: Economics, Science, and Law* (Washington, D.C.: Resources for the Future, 1995), at 219. Stewart argues that the valuation of natural resources should not be undertaken by the tort system, but instead should be based on an administrative system of scheduled assessments. If CVM is not used, on what basis should this inquiry be performed?

Environmental Law
in an International Community

XI

Environmental Regulation and International Trade

The interaction between environmental regulation and international trade played an important role over the last decade in discussions concerning the future of the General Agreement on Tariffs and Trade (GATT), which was established to police the system of multilateral free trade, and it is now a salient issue on the agenda of the GATT's successor, the World Trade Organization (WTO). Environmental issues were also at the forefront of the debate about whether the United States Senate should ratify the North American Free Trade Agreement (NAFTA).

The article by Richard Stewart explores the lessons that the international community can draw from the experience of two federal political systems: the United States and the European Union. Stewart distinguishes between two analytically distinct categories: (1) product-based regulation, under which a nation imposes restrictions on the importation of a product on the ground that the product poses an undue environmental risk; and (2) process-based regulation, under which a nation imposes restrictions on the importation of an otherwise unobjectionable product on the grounds that it was manufactured by an environmentally undesirable process. The federal systems studied by Stewart outlaw product-based regulations that are facially discriminatory, under which restrictions are imposed only on out-of-state products. As to facially neutral measures, the legality of the restriction generally turns on a comparison of the environmental benefit to the state imposing the restriction and the detriment to the state that is the target of the restriction. Stewart notes that the GATT standards are probably more deferential to the jurisdiction imposing the restriction than the standards in the United States and the European Union.

Process regulation has not played an important role in either of the federal systems that Stewart studied. It is, however, a central concern in the international arena.

Stewart notes that one nation's environmental policies can have four different types of impacts on another: (1) pollution spillovers, under which pollution generated in A is deposited in B; (2) resource externalities (use values), under which A overexploits a resource, thereby depriving consumers or producers in B of the opportunity to exploit the resource in the future; (3) preservation externalities (nonuse values), under which A's exploitation of the resource deprives individuals in B of an existence value (the value that comes from knowing of the existence of the natural resource to individuals not intending to use it); and (4) competitiveness externalities, under which B's industries are competitively disadvantaged as a result of A's failure to adopt more stringent environmental regulation, or under which a race to the bottom ensues. Stewart is skeptical of process-based regulation except where there are interjurisdictional pollution spillovers or effects on the global commons.

The article by Howard Chang criticizes the resolution of the what is undoubtedly the best-known trade-and-environment case: the 1991 decision of a GATT panel finding that a process-based ban by the United States on the importation of tuna fish caught by Mexico ran afoul of the GATT's prohibition on quantitative import restrictions. Pursuant to the Marine Mammal Protection Act, the United States had banned the importation of tuna caught with "purse seine" nets in the eastern tropical Pacific Ocean on the ground that such nets kill dolphins. Chang notes that the GATT panel's decision was consistent with the GATT secretariat's general preference for a "carrots only" approach, under which the only option open to a nation interested in improving another nation's treatment of the environment is to pay the latter to alter its environmental practices. Chang maintains that, from the perspective of economic efficiency, the use of trade sanctions, or "sticks," is likely to be desirable because sticks deter overuse, whereas carrots provide incentives for countries to pollute more in order to obtain larger payments.

The article by Jagdish Bhagwati and T. N. Srinivasan argues against the widely held view favoring the international harmonization of environmental standards. Different nations may value environmental protection differently; moreover, the effects of pollution are generally dependent on physical characteristics, such as climate or landscape, that might vary across nations. As a result, uniform standards would be undesirable from the perspective of maximizing social welfare. The authors take issue with two central objections to the diversity of environmental standards: (1) that low-standard countries are engaging in "unfair trade," and (2) that high-standard countries will abandon those high standards unless they can force the upward harmonization of low standards in other countries. As to the first objection, the authors argue that differences in environmental standards are like differences in wages, capital costs, skills, infrastructure, or weather, all of which affect the cost of doing business and do not give rise to unfair trade practice claims. With respect to the second, the authors doubt that a race to the bottom is likely to occur.

International Trade and Environment: Lessons from the Federal Experience

RICHARD B. STEWART

This Article has two objectives. First, the Article will develop a conceptual framework for analyzing the interrelationship between trade restraints and environmental protection policy. Second, it will draw from the experiences of two federal-type political systems—the United States and the European Community—potential lessons for the international institutional treatment of trade and environment issues. . . .

The Free Trade Regime

The cornerstone of the case for free trade is the mutual economic benefit resulting from trade among nations with differences in comparative advantage in producing goods and services. . . . In the classic Ricardian conception, comparative advantage was based on relative differences in factor endowments—such as the character of agricultural land, climate, timber, and mineral resources—among nations. But an enlarged conception of comparative advantage has come to include differences in human capital and industrial and technological infrastructure. No reason exists in principle why comparative advantage should not also encompass differences in national economic, social, and regulatory policies and legal and administrative systems. Economists also regard national differences in the ability of ecosystems and populations to assimilate pollution as an element of comparative advantage.

There are additional reasons, beyond comparative advantage, why free trade should enhance the welfare of all nations engaging in trade. A wider market enhances the opportunity to realize economies of scale. It also promotes specialization, with attendant gains in productivity. A greater array of suppliers stiffens the efficiency-promoting discipline of competition. The wider network of international contacts accelerates the diffusion of knowledge and technological innovation.

Experience confirms the economic benefits of a free trade regime (FTR). Empirical studies show a strong correlation between the degree of trade liberalization and economic growth rates among different nations and a similar correlation between changes in trade policy and growth in individual nations. . . .

Given the benefits of free trade, why should nations ever seek to impose barriers to it? In most cases, restrictions are imposed to protect the interests of producers and workers who would be injured by greater competition. In many of these cases, however, a nation's consumers would suffer greater losses in welfare from trade barriers than the benefits that these barriers afford to producer interests. . . .

Trade barriers may also reflect a demand for national autonomy in social and economic policy. Sentiment exists that a political community should be able to set its own priorities and principles, rather than have them dictated by remote economic

Reprinted by permission of the author and publisher from 49 *Washington and Lee Law Review* 1329 (1992).

forces. Some states with centrally-planned economies have adopted trade restrictions for fear that a FTR would disrupt their ability to carry out central planning. A related concern is that if political integration follows economic integration, a nation's political autonomy will be further compromised. . . . This concern for policy independence has become especially prominent in the environmental context, as nations have sought increasingly to impose restrictions on trade in the name of health, safety and ecological protection.

Environmental Protection in the Context of Economic Integration: The Federal Experience

Products

. . . Assume that producers in state or nation A make a product which they wish to sell in B. B, however, prohibits or (more often) imposes restrictions or in some cases a tax on the import and sale of such products in B on the ground that the product poses an undue risk to health, safety, or the environment, or that such risks are not adequately disclosed in the product's labelling. . . .

Assuming that the product is sold in A without such restrictions, why should B prohibit or restrict a product that A allows? B's citizens may place a greater value on health, safety, and environmental protection than those of A. This decision may reflect a stronger preference for environmental quality relative to other goods and services. B's citizens also may be wealthier than A's; the demand for environmental protection typically rises with income. Institutional structure also influences environmental policy. Even if preferences and wealth of consumers in the two states are similar, B's political and administrative institutions may, for various reasons, give greater weight to preferences for environmental protection than those of A. Alternatively, A may allow marketing of the product because of a political judgment that the benefits to its producers outweigh the environmental risks. B, however, has no interest in the welfare of A's producers and judges that the risks to its consumers outweighs the benefits. This conclusion may be especially likely if the product is not produced at all in B or is produced only in small quantities. B may also impose restrictions on imports in order to protect its producers. This restriction may take the form of a discriminatory ban or other regulation of imported but not domestically manufactured products or, more often, regulatory requirements and procedures that impose a greater effective burden on imports than on similar domestic products or domestic substitutes for the imported product. The regulatory measure may be neutral on its face, but have the practical effect of favoring domestic producers. Examples include the ban by Minnesota of nonrecyclable plastic milk cartons but not paperboard cartons (Minnesota has a substantial timber products industry) and Canada's heavy tax on nonrecyclable beer containers, which has a disproportionate impact on U.S. producers. The resulting disproportionate impact, however, does not rule out the possibility that the measure will achieve significant environmental protection benefits.

Whatever the reason for such restrictions—and a given restriction may have several explanations—they undermine the FTR by harming foreign producers as well as both domestic and foreign consumers. Foreign producers are disadvantaged relative to B's producers, who can reap economies of scale in complying with B's regulatory restrictions because their sales in B are likely to be large compared to those of foreign producers. If different states adopt different restrictions, the ability of all producers to realize scale economies will be diminished, the transaction costs of achieving compliance with different requirements will be increased, and other benefits of free trade will be reduced. This will result in economic harm to consumers in all states, including those in B. Nonetheless, these detriments may well be outweighed by the protection afforded to B's consumers and its environment by the restriction. . . .

The Supreme Court found in the Commerce Clause the implied power for the Court to invalidate state measures that unduly burdened the functioning of the common market among the states. The [European] Court of Justice enjoys a more explicit constitutional power to the same end. These courts' task has been to determine whether the local benefits of a product restriction and the interest in local autonomy outweigh the detriment to the FTR.

Both courts have outlawed facially discriminatory state measures that impose restrictions on imported products but do not impose those same restrictions on domestic products. In the absence of facial discrimination, the analysis is more contextual. Several questions are relevant. How great is the environmental justification for the measure? How great is the detriment to trade? Are there ways of achieving a similar benefit through means that are less disruptive of trade? Is the measure framed in such a way as to give competitive advantage to local producers, and what is the justification for tailoring the restriction in this way as compared to other means of protecting the environment? What is the justification for special tests or inspection of imported goods, which may duplicate those already imposed by the exporting state? In these contextual evaluations, the ultimate judgment is one of proportionality: taking into account the importing state's sovereignty interest, is the restraint on trade (including protectionist effects) manifestly disproportionate to the environmental benefits achieved, considering the availability of other means of securing those benefits. In making such judgments it is a crucial question, which court decisions do not resolve, whether the court is to weigh the detriment to the welfare of consumers in the state imposing the restriction, or only the welfare of out-of-state producers and consumers. The interests of in-state consumers are surely among those protected by the FTR; it would seem that detriments to their interests should accordingly be included in assessing the detriment to human welfare caused by trade restrictions. On the other hand, given the premise of political decentralization on which federal-type systems are founded, it would seem inappropriate for a suprastate tribunal to invalidate state legislation on the ground that the legislature had erred in determining which regulations would best advance the net interests of its citizens. In addition, it is difficult for a court to measure the relevant benefits and detriments. As a result, courts have invalidated such regulations only when they lack significant environmental justification, or a less trade restrictive alternative with comparable regulatory benefits is available, or the measure is clearly protectionist, or the detriments substantially outweigh the benefits. The ultimate test is a protectionist standard of net proportionality, deferentially applied: the extent of otherwise unobtainable environmental benefits must be weighed against the welfare detriments of trade restrictions, giving due respect to state political authority. . . .

Processes

Regulation of in-state manufacturing and resource harvesting or extracting processes by different states is not so clearly offensive to the FTR as the measures surveyed above because it does not directly obstruct the free flow of goods and services among states. But pollution and other forms of environmental degradation caused by processes and process-based environmental regulation can create several different types of troublesome externalities. These externalities include the following:

Pollution spillovers. If some of the pollution generated in A is deposited in B, A's regulation of pollution by its producers is likely to be inadequate because A will give little or no weight to the interests of B's residents. If pollution is exactly reciprocal— if B pollutes A just as much as A pollutes B—there may be incentives for cooperative approaches to regulation, but such reciprocity is rare. The externality created by pollution generally consists of negative use value; pollution increases the economic costs or diminishes the benefits associated with resource use. Examples would include the costs of purifying polluted water for drinking or reduced crop yields or diminished recreational fishing opportunities due to air and water pollution.

Resource externalities (use values). A may be exploiting its resources in a wasteful fashion, running down the resource stock at an excessive rate and depriving consumers and producers in B of the benefit of future use of those resources. For example, A may be overcutting its forests, driving up the future cost of timber, or recklessly destroying pristine areas that residents of B would spend money to visit. On standard economic assumptions, it is difficult to understand why A would engage in such behavior, assuming that the lost future benefits are greater than the benefits of present use; shortsightedness, corruption, or some other form of institutional failure must be invoked to explain such behavior.

Preservation externalities (nonuse values). A may be exploiting its resources in ways that deprive citizens in other states of the satisfaction of knowing that those resources are preserved, independent of any use that might be made of them. Examples include the destruction of pristine environments and the eradication of endangered species. Again, the diminution in welfare associated with the loss of these nonuse values may exceed the benefits from current resource exploitation. Unlike the case with use values, however, it is quite easy to understand why this form of welfare impairment might arise. Preservation of natural resources for nonuse values is a collective good. A and its citizens can not selectively provide the benefits of preservation to those outside the state who would be willing to pay to have those resources preserved. Because of free-rider effects, those outside the state are unlikely to bond together and pay for such preservation. Accordingly, the economic incentives of A and its citizens to preserve such resources will not be adequate.

Competitiveness externalities. A may fail to adopt strong environmentally protective process regulations for fear that its industries will be competitively disadvantaged in relation to industries in B, who may fail to adopt similarly strong policies. B and others may reason likewise. The result may be a "race to the bottom" that leads everywhere to lower levels of environmental protection than all states would prefer. The externality here consists of uncertainty regarding the reaction of each state to the environmental policy of every other state and the resultant tendency to take a risk averse approach to the threat of job loss and industrial dislocation by adopting less stringent environmental measures.[1] . . .

Conceivably B could seek to deal with such externalities by excluding imports of products manufactured by laxly controlled plants in A. B could also impose a tariff on such imports equal to the difference in pollution control compliance costs incurred by industries in A and those in B. Such measures might, however, be challenged as violative of the Commerce Clause or the Treaty of Rome. The question, however, is moot, for no state in the United States or the European Community has apparently attempted to impose such measures, and no law exists on the issue. . . .

Principles of Governance for Trade-Related Environmental Regulation in the International Setting

What considerations, then, should guide determination by GATT or its equivalent of the validity of national measures with asserted environmental justifications that restrict trade? . . .

Products. Consider first, unilateral restrictions on product imports based on regulations designed to prevent environmental, health, and safety risks in the importing nation. The issues presented in assessing the justifications for such regulations, their competitive impact, and the appropriate balance between free trade and regulatory restrictions on products are very similar to these presented in federal-type systems. In principle, there appears to be no reason why a GATT-type tribunal should not be able to evolve a satisfactory body of international law to govern the subject. In practice, however, trade rivalries are likely to be sharper, and the difficulty of resolving controversy over the scientific analysis of risk and appropriate risk management policies greater, in the international context than in a federal-type system. United States-European Community disputes over EC exclusion of meat from U.S. cattle

---

[1]The "race to the bottom" argument is carefully examined and criticized in Richard Revesz, *Rehabilitating Interstate Competition: Rethinking the "Race to the Bottom" Rationale for Federal Environmental Regulation.* His basic critique of the argument is that just as consumer welfare is promoted by competition among different sellers of goods and services, so is it also enhanced by competition among states in providing different environmental regulatory regimes for industry and residents. Because states will differ in their geography, ecology, state of development, and citizen preferences, there is no reason to suppose that such competition will result in uniform and unduly lax regulation. The other forms of externalities discussed in this article are nonpecuniary externalities—environmental and health harms or risks that are not reflected in the market prices of factor inputs or goods and services. However, competitiveness externalities are pecuniary externalities. According to standard economic theory, pecuniary externalities, which are reflected in market prices, should not result in market failures. One would probably have to invoke game theory and problems of uncertainty and strategic interdependence in order to explain how such competition might lead states to adopt laxer environmental regulation than states would otherwise prefer.

which have received bovine growth hormone (BGH) and U.S. exclusion of EC wine with trace pesticide residues illustrate how contentious issues of science and risk management can be. Science often tells us that the level of risk is highly uncertain. Also, differences in risk management approaches cannot be dissolved by appeals to "sound science."

The dispute between the United States and Canada over Canada's heavy tax on nonusable beer containers also illustrates the difficulty of evaluating the extent of environmental justification for measures that have the effect of protecting domestic producers. U.S. environmentalists are fearful that the GATT and other free trade agreements will result in wholesale invalidation of U.S. product regulations, such as those aimed at pesticide residues. While some U.S. regulation, such as the zero-tolerance standard for carcinogens imposed by the Delaney Amendment, might be suspect, knowledgeable observers believe that the GATT standards are substantially more deferential to national regulation than those applied by the United States Supreme Court and the Court of Justice to state regulation. . . .

Processes . . . Restrictions on product imports aimed at adverse environmental effects associated with the processes by which those products are produced present difficult issues, which have become a storm center of controversy following a GATT panel's decision invalidating U.S. prohibitions on the import of tuna caught through fishing practices that cause the incidental taking of porpoise in greater numbers than allowed by U.S. law. Other examples include existing or proposed bans on imports of tropical hardwood timber, Canadian sealskin and Canadian lobster. Many of these prohibitions have been imposed by the United States, which has also asserted uni-lateral authority to ban imports of product X from a country because of opposition to the process by which that country produces an unrelated product Y. Examples include threatened prohibition of imports of fish products and pearls from Japan because of opposition to Japanese whaling practices and imports of tortoise shell products respectively. There is also growing talk in the United States of imposing countervailing duties on imports of products manufactured through industrial processes that do not meet U.S. environmental standards, either on the grounds that the less stringent exporting state's environmental standards represent an export sub-sidy that is countervailable, or violate antidumping laws because the total social costs borne by consumers in the producing state, including pollution as well as ordi-nary product costs, are greater than the costs charged consumers in the importing state (ecodumping), or are a general form of unfair competition. These proposals would seek to tax the "embedded pollution" in product imports.

As previously noted, there is no precedent in federal-type systems for such mea-sures. This circumstance may reflect a number of factors including the relatively recent rise of environmental concerns, the fact that process spillovers have been more readily addressed by legislation in the United States and the European Com-munity than internationally, and the relatively limited leverage that a single state would exercise. In analyzing the validity and wisdom of such measures in the absence of precedent, it is important to distinguish among the different types of environmental externalities generated by different process and regulatory practices in different nations.

Competitiveness spillovers have been a particular focus of attention in the United States. Some assert that unless other nations adopt relatively stringent U.S. environmental standards, U.S. industry will suffer a serious competitive disadvantage, leading to "industrial flight," and job losses. These assertions raise difficult and controversial issues. Most empirical studies by economists conclude that environmental regulatory costs are generally not a significant factor in industrial location decisions or trade performance. But, there are clear examples of "industrial flight" in a few limited contexts—such as the relocation of furniture finishing operations from Los Angeles to Tiajuana, Mexico. In addition, there are indications of some displacement of pollution-intensive industry in the chemical sector from North America to Southeast Asia. Even in the absence of significant relocation, U.S. industry may suffer some comparative competitive disadvantage, particularly if the calculus of costs includes not only capital outlays and operating costs incurred in order to comply with regulatory requirements, but also includes the invisible costs imposed by liability risk and regulatory constraints, delays and uncertainty, which are especially high in the U.S. legal and administrative system. These costs are far greater for new products and processes than existing ones, and may cause long-term impairment of comparative advantage even without any relocation of existing capacity. On the other hand, environmental regulations may create some competitive advantages by encouraging the development of environmentally superior processes that can be sold in the export market or that improve efficiency in resource use.

Because of the difficulties in establishing the extent of competitive disadvantage that may result from a nation's adoption of more stringent environmental regulation, disputes over the justification for import duties and other trade restrictions assertedly designed to offset such disadvantage will depend on who has the burden of proof. Given the clear potential for serious protectionist abuse of such measures, the burden of proving serious competitive disadvantage should rest with those who would restrict trade.

There is, however, a more fundamental objection to such measures. As the opinion of the GATT panel in the tuna/dolphin case correctly observes, no reason exists why differences in environmental conditions and preferences among different nations and consequent differences in process regulations should not be regarded as an appropriate aspect of comparative advantage. The "level playing field" principle underlying proposals to equalize environmental compliance costs admits of no stopping point. It would equally apply to labor and wage policies, education policy, tax policy, and the entire array of government social and regulatory policies that affect production costs. In the contemporary world, any effort to distinguish "natural" advantages from those created by government policy is bootless. Equalizing all costs of production would produce a world without any trade at all. Moreover, problems of establishing cost differentials and allocations raise such intractable problems that unilateral tariffs or other measures assertedly designed to correct competitiveness differentials would in practice be extraordinarily arbitrary. . . .

Exploitation of resources in the global commons. The matter stands somewhat differently, however, in the case of resources that are part of the commons, wholly outside any nation's territory. Examples include the oceans, Antarctica, and the stratosphere. For resources within national territory, the interests of the citizenry, reinforced by the prospect of trade (including tourism), provide at least some incentives for the establishment of systems of property rights and regulation to protect the resource. In addition, important considerations of national sovereignty cut strongly against the unilateral use by other nations of trade restrictions to induce the host nation to adopt more protective environmental measures domestically. Both of these considerations are absent in the case of the resources of the ocean and Antarctica. Without an international agreement, there is no system of property rights or regulation protecting resources such as whales or stratospheric ozone. All too often, the result is a textbook example of the tragedy of the global commons. An unregulated regime of free trade will simply hasten the tragedy rather than advance the common welfare. In these circumstances, it is far easier to justify restrictions on products from other nations that have been produced through means that destroy the commons resource base. Use by nations of self-help to preserve that resource base is, in principle, a fair response to those who would use self-help to destroy it. But, this principle does not provide carte blanche for trade restrictions. Such restrictions must, as in the context of restrictions on product imports based on the risks posed by the products themselves, advance environmental rather than protectionist goals and do so through means whose adverse effects on trade are not unduly disproportionate to the environmental benefits obtained. . . .

Process-based pollution spillovers. Still different considerations are raised by import product bans aimed at the pollution generated by the processes of their production, where such pollution causes or threatens significant environmental harm in the importing state. Examples might include regional ozone transport or acid deposition, depletion of stratospheric ozone through CFC and halon emissions, and global climate change resulting from GHG [greenhouse gas] emissions. In this instance a product ban could be understood as a form of self-help against deliberate injury. Alternatively, the injured importing state might impose a tariff based on the additional production costs that the exporting state would have to incur in order to prevent the injury, although as a practical matter this would be very difficult to determine. Given the lack of other remedies available under international law, such forms of self-help do not seem objectionable in principle, but there are very significant problems in ensuring that such measures are justified by the end asserted—prevention of serious environmental injury—and not exercised in a protectionist or otherwise arbitrary fashion. These problems raise questions such as how serious is the pollution spillover? Would the restriction be effective in preventing it? Are the preventive requirements that it would impose on the polluting country reasonable, judged in light of the importing nation's practices and those of other nations? What competitive advantage would a ban or tariff give to domestic manufacturers? Is there evidence that the measure is aimed at economic rather than pollution spillovers? Is the environmental benefit to the importing nation reasonably proportionate to the trade detriments imposed? It would not appear beyond the competence of a specialized tribunal to work out a satisfactory resolution of these matters. In addition, there is the problem of distinguishing transboundary physical spillovers from those that are merely local. Many environmentalists believe that everything is connected to everything else, but in a regime of sovereign states, distinctions of degree, if not of kind, must be recognized. The standards for upholding product import restrictions assertedly aimed at pollution spillovers should be reasonably demanding in order to rule out the dangers of arbitrary or protectionist actions, and might only be met in the case of regional pollution spillovers causing serious current injury.

An Economic Analysis of Trade Measures to Protect the Global Environment

HOWARD F. CHANG

This article addresses the role of trade restrictions in supporting policies to protect the global environment. The issue of environmental trade measures was raised most prominently in 1991 and in 1994, each time by a controversial decision by a dispute-settlement panel of the General Agreement on Tariffs and Trade (GATT). Both GATT panels held that a ban by the United States on imports of tuna from specified countries violated GATT Article XI, which prohibits quantitative import restrictions. The United States had banned these tuna imports pursuant to national legislation, the Marine Mammal Protection Act of 1972 (MMPA), which includes provisions limiting the number of dolphins that may be killed through tuna fishing. The reasoning advanced by each GATT panel has potentially sweeping implications for a wide variety of U.S. environmental laws and international environmental treaties that rely on trade restrictions for enforcement. The looming conflicts between trade liberalization and environmental protection has placed these problems high on the GATT agenda. . . .

In both the 1991 case and the 1994 case, the United States invoked GATT Article XX, which provides a list of general exceptions to all GATT obligations. Article XX lists a variety of national measures that are recognized as directed toward legitimate goals. In particular, the United States claimed that the ban on tuna imports was a measure "necessary to protect human, animal or plant life or health" within the meaning of Article XX(b) and "relating to the conservation of exhaustible natural resources" within the meaning of Article XX(g). Although the language in these provisions is broad, the 1991 panel held that these exceptions were intended to apply only to measures to protect animal life and natural resources *within* the jurisdiction of the party applying them. The 1994 panel rejected the geographic restriction adopted by the 1991 panel, but nevertheless ruled against the United States on the ground that the U.S. ban on tuna imports was a trade measure that would succeed in protecting dolphins only by changing the policies of other countries.

This article draws on economic theory to support a more liberal reading of Article XX that includes neither geographic restrictions nor an absolute prohibition on

Reprinted by permission of Howard F. Chang from 83 *Georgetown Law Journal* 2131 (1995).

trade sanctions. A more appropriate rule would also permit, for example, unilateral trade restrictions that protect either the global commons or endangered species found abroad. This critique of the panels' interpretations of the GATT focuses on the criterion of economic efficiency: the analysis evaluates alternative interpretations of the GATT largely on the basis of aggregate costs and benefits, measured in terms of what individuals are willing to pay to avoid particular costs or to gain particular benefits. . . .

The GATT Secretariat . . . has explained that it favors the use of subsidies as "carrots" over the use of trade measures as "sticks." . . . I offer a critique of the "carrots only" approach endorsed by the GATT Secretariat. Using concepts from game theory, I defend unilateral sticks (and multilateral sticks against countries not parties to the environmental protection agreement) as necessary to restrain more effectively exploitation of the environment pending a multilateral agreement. Finally, I address some possible problems posed by the use of sticks in general, and I advance some reasons to think that the use of sticks is nevertheless more likely to promote global economic welfare than the "carrots only" approach. . . .

An Economic Critique of the "Carrots Only" Solution

. . . The "carrots only" contractual approach to the problem of negative externalities corresponds with the type of solution indicated by a naive reading of the Coase theorem, which suggests that as long as parties can bargain with one another, they will reach an efficient solution regardless of the initial allocation of legal rights. Changing the legal rights alters the welfare each party expects to enjoy in the absence of an agreement—that is, it moves the "threat point" in the negotiations. This shift in the default payoffs, however, merely reallocates wealth; it does not render the outcome inefficient. In the international environmental context, the "carrots only" solution gives countries the right to harm the global environment and puts the burden on others to offer concessions sufficiently valuable to the offending nations to induce them to stop. This solution amounts to an endorsement of the "victim pays" principle rather than the "polluter pays" principle.

The Coase theorem, however, assumes no transaction costs, so that there are no barriers to parties reaching these efficient agreements. In reality, transaction costs will make these agreements difficult to reach: agreements will take real time and effort to negotiate, and the parties may sometimes fail to reach agreement altogether. . . .

Market failures are particularly acute in the context of the global environment. The large number of countries with a stake in the global environment will lead to free-rider problems: each country has the incentive to wait for others to offer the carrots necessary to forge an agreement. The free-rider problem exists even if the true preferences of each party are common knowledge among all the other parties. Under conditions of "symmetric information," the problem amounts to a multilateral version of the classic bilateral prisoner's dilemma game, under which the dominant strategy of each party is to refuse to cooperate, even if each player would be better off under the cooperative solution than under the noncooperative equilibrium.

Imperfections in information compound the free-rider problem among countries: in reality, each government participating in the negotiations will be uncertain about the preferences of the other governments. Given these asymmetries in information, each country will have an incentive to understate its interest in protecting the global environment (and to overstate its interest in exploiting it) in order to win a better deal for itself in the negotiations. This "preference revelation" problem makes it difficult to induce each country to bear its fair share of the costs of environmental protection, as would be the case with any global public good.

If the parties reach an agreement at all, it will take time. . . . Over time, haggling provides the parties with further information about one another's true preferences, and it is the very costliness of the passage of time without an agreement that provides this information. Countries most impatient for an agreement make greater concessions in less time; those with the least to lose from delay hold out longer.

For all these reasons, it is naive to cite the mere possibility of international agreements as a panacea for global environmental problems. Because strategic behavior can cause bargaining to fail, we cannot rely on multilateral agreements alone to protect the global environment. Even when bargaining eventually succeeds, harm to the environment will occur prior to the conclusion of an agreement. Given the reality of strategic behavior, the allocation of legal rights (the threat point) is no longer a matter of indifference from the perspective of efficiency. In a regime in which unilateral sticks are prohibited, countries would be more inclined to engage in environmentally harmful actions. We would expect the level of environmentally harmful activities to rise for two distinct reasons: first, sticks deter overuse of the environment, and second, the use of carrots alone creates perverse incentives.

Sticks Deter Overuse of the Environment

Under the reasoning of the GATT panels, countries would be shielded from sanctions, and (at least under the reasoning of the 1991 GATT panel) foreign producers would be shielded from extrajurisdictional trade measures, which otherwise would discourage each from harming the environment. These prohibitions on the use of sticks would bring forth more environmental harm even in the absence of any prospect of negotiations toward a multilateral agreement. Insofar as the environment is a public good among multiple countries, we would expect to observe inefficiently high levels of environmentally harmful behavior, because each party fails to internalize the negative externalities associated with its behavior. A prohibition on sticks would remove an effective deterrent to excessive exploitation of the environment. If we cannot use sticks to protect the global commons, for example, then unless and until we obtain a multilateral agreement, we are left with the usual free-rider problems that cause each party to overuse the natural resources held in common. Even if some countries were to reach an agreement to protect the environment, other countries will have an incentive not to sign the agreement—they would prefer to "free ride" on the restraint exercised by signatories to the agreement.

Carrots Create Perverse Incentives

The prospect of being the beneficiary of carrots in a multilateral agreement would create additional *positive* incentives to harm the environment. In the absence of sticks to induce cooperation, a multilateral agreement must offer the polluting countries carrots: concessions by those countries that value the environment and must secure the cooperation of other countries without resort to unilateral sanctions. By moving the threat point in the bargaining game away from environmentally friendly countries and toward those that harm the environment, we reduce still further the incentives to exercise restraint in exploiting the global environment. The use of subsidies or other rewards to encourage pollution abatement has a number of perverse incentives.

More countries will pollute. In regulating a polluting industry in the domestic context, a government agency can use a system of subsidies rather than Pigouvian taxes. Either instrument would encourage individual firms to choose the efficient level of pollution abatement, but the subsidy would make the industry in question more profitable than it would be otherwise. Lured by the prospect of these profits, more firms would enter the industry than would otherwise, and the net result could be more pollution rather than less. The subsidy can create incentives that are identical to those of the Pigouvian tax only if the government pays the subsidy not only to actual polluters, but also to potential entrants into the polluting industry and to those polluters who exit the industry. These potential entrants may be infinite in number, however, and even if they are finite in number, they be may difficult to identify.

Similarly, a multilateral agreement that relies on subsidies rewards countries for harming the environment; only those who pollute receive the subsidy. This prospect makes environmentally harmful activity rational for countries that would otherwise be indifferent or even disinclined to harm the environment. Countries would be encouraged to "enter" the polluting industry by offering producers lax environmental regulations. . . . Countries will have reduced incentives to regulate their own producers: rather than regulating spontaneously without getting a carrot, some countries will be induced to delay in order to receive a carrot in exchange for restricting pollution later. . . .

Countries will pollute more. Furthermore, the fact that the size of these carrots will be determined by a bargaining process will encourage each polluting country to pollute still more. Those that would already be inclined to harm the environment would be encouraged by the bargaining process to harm it to an even greater extent to qualify for larger carrots. These strategies would yield positive payoffs for the offending countries because they could use the threat of continuing their environmentally harmful activities to extort carrots from other countries.

Why would other countries pay this "ransom" if they know that these activities are not what the "blackmailing" countries would consider optimal in the absence of the prospect of a ransom? If we assume conditions of symmetric information, the true preferences of each party would be common knowledge among all parties. In such a model, it must be in a blackmailing country's interest to carry out its threat ex post if the threat is to be credible ex ante. If a blackmailing country can make such a threat credible, then game theory suggests that it may indeed be able to extract concessions from another country by taking actions that harm the interests of the other country, even if that action is also costly to the blackmailing country. . . .

Bargaining will delay agreement. The same dynamic continues throughout the bargaining process, leading low-cost countries to turn down offers of carrots they in fact consider more valuable than their environmentally harmful behavior. They do so to mimic high-cost countries, in the hope of obtaining a better offer in the future. In a "bluffing equilibrium," low-cost countries hold out for larger carrots and thereby delay resolution of the environmental problem. Thus, a "carrots only" regime may not only increase the level of pollution, but also extend the period during which inefficiently high levels of pollution persist.

Advantages of sticks. If countries that value the environment are permitted to use sticks rather than carrots, they can avoid these perverse incentives. If a particular level of environmentally harmful behavior triggers trade sanctions, and environmental regulation leads to the removal of those sanctions, for example, then a country will have nothing to gain by pretending to find such levels of environmentally harmful behavior in its interests. To the extent that the severity of these sanctions turns on observed levels of environmental harm, the use of sanctions will discourage lax environmental regulations. First, greater levels of harm will increase the stake of the country employing sanctions in preventing the harmful activity. This effect will increase the costs that this country is willing to bear, including the costs of more draconian sanctions. Second, if the sanctioning country infers, from either the intransigence of the polluting country or its levels of pollution, that this "target" country enjoys large benefits from its own environmentally harmful activity, then it will believe harsher sanctions are needed to achieve its objective. With sticks, a country that signals an inclination to harm the environment can bring greater penalties upon itself; with carrots, the same signal can yield greater rewards. . . .

Sticks: Possible Disadvantages

The use of sticks rather than carrots is not without its own risks. There is, in theory, a corresponding risk that, if allowed to use sticks, countries will use them opportunistically, simply to extract more favorable terms from target countries in multilateral agreements. That is, countries may take advantage of trade instruments by employing them as strategic bargaining chips, just as countries may use environmentally harmful policies to extort carrots from those that value the environment. In this sense, a rule allowing the use of trade measures does not eliminate strategic behavior in the bargaining process. Given that strategic behavior is inherent in the bargaining process under conditions of asymmetric information, why might we be worried about the use of sticks? Sticks may raise issues of distributive justice and of economic efficiency.

Distributive Justice

Advocates of the "carrots only" approach point out that a ban on unilateral sanctions would redistribute wealth from the nations using sanctions toward the targets of those sanctions. They tout this effect as a reason to favor such a ban, not only on grounds of distributive justice, but also on environmental grounds: they point to evidence that "environmental quality and income levels are highly correlated." Wealth transfers to developing countries raise their income levels and thereby increase their interest in environmental protection. If we assume that the countries wielding sanctions tend to be affluent countries with large economies and that the targets of sanctions are often smaller and poorer countries, then the use of sanctions will have unfortunate redistributive effects compared to the "carrots only" approach.

Concerns about distributive justice per se, however, do not lend support for directing carrots at those countries that harm the environment. Transfers from those countries that most value the environment to those that harm the environment are extremely clumsy instruments for redistributing wealth. These transfers redistribute wealth on grounds imperfectly correlated with the affluence or poverty of the countries in question. They will include wealth transfers to affluent countries that harm the environment (such as Norway, which has resumed its hunting of minke whales) and will exclude transfers to poor countries that refrain from harming the environment. This erratic policy not only offends notions of horizontal equity and fairness, but also creates perverse incentives to harm the environment.

A superior policy response would be to transfer resources from affluent countries to poor countries generally, because such transfers can target those countries with the greatest need without creating . . . the perverse incentives described here. . . .

Economic Efficiency

Should the strategic use of sticks to shift the threat point in favor of the countries that wield them be a source of concern from an efficiency perspective? The fact that sticks reduce the payoffs of polluting countries does not in and of itself pose a problem: it is precisely this mechanism that discourages excessive harm to the environment. Nevertheless, the heavy-handed use of sticks could raise problems for global economic efficiency by going too far in protecting the environment. Sticks that impose large costs on the targeted countries might induce them to forgo environmentally harmful activities even when the economic benefits that they derive from these activities outweigh the costs. Over time, bargaining may allow these countries to resume these activities without provoking sanctions: if their activities are indeed efficient, then they should be able to offer other countries concessions sufficiently valuable to pay for the right to pollute. As already discussed, however, bargaining may fail to bring about efficient outcomes. Is the stick cure therefore likely to be worse than the carrot disease? There are two reasons to think that the use of sticks is still likely to effect an improvement over the "carrots only" regime.

First, the transparently opportunistic use of sticks to extort concessions is unlikely, given the adverse impact these tactics would have on a sanctioning country's foreign relations. The doctrine of proportionality in international law requires that any sanctions a country employs should be proportionate to the interests to be protected. This well-established principle implies that economic sanctions causing effects disproportionate to the environmental interests at stake would violate international law. The use of sanctions is constrained not only by international law generally, but also by the realities of the international political landscape. No country will resort to sanctions without some hesitation, because they can erode precious political capital in the realm of international relations. Even if all governments agreed that trade sticks are consistent with the GATT, they would be unlikely to use sanctions often. The threat of a hostile country, makes such an undertaking a risky and serious matter. Because trade sanctions and other threats are available to all parties, each party is likely to exercise restraint in employing them.

For these reasons, we rarely observe the blatant use of either sanctions or environmentally harmful activities simply to extort concessions from others. A rule allowing the use of trade sanctions offers the prospect that the potential for opportunistic behavior on each side will serve to inhibit abuse on the other side. Allowing countries to respond to the environmental threats posed by polluting countries with threats of their own preserves some symmetry that would be lacking under the "carrots only" approach. Given the potential for mutual threats, if any abuse of sanctions occurs at all, it is unlikely to take the form of naked blackmail. Any abuse of sanctions is more likely to be rather subtle, so as to appear proportionate to the legitimate interests of the country wielding the sanction.

Second, the use of sticks assures that the bargaining process is not biased against environmental interests. The "carrots only" approach leads inevitably to inefficiently high levels of environmental harm because it rewards rather than penalizes harmful behavior. Whereas the "carrots only" approach guarantees perverse incentives for harmful behavior, the use of trade sticks creates a mere theoretical possibility of excessive deterrence. The realities of international politics that inhibit the abuse of trade sanctions also greatly reduce the risk that countries will deter too much environmental harm.

Trade and the Environment: Does Environmental Diversity Detract from the Case for Free Trade?

JAGDISH BHAGWATI AND T. N. SRINIVASAN

The potency of the contention that fair trade or level playing fields constitute a precondition for free trade and that, therefore, harmonization of domestic policies across trading countries is necessary before free trade can be embraced to one's advantage, should not be underestimated today. It is nowhere more manifest or more compelling in its policy appeal than in the area of environmental standards.

Reprinted by permission of the authors and MIT Press from Jagdish N. Bhagwati and Robert E. Hudec, *Fair Trade and Harmonization: Prerequisites for Free Trade?* Volume 1, chapter 4, copyright 1996 MIT Press.

Both the *general* view that cross-country intra-industry (CCII) harmonization of environmental standards is required if free trade is to be implemented and the *specific* proposals currently in vogue to implement this view are therefore in need of analytical scrutiny.[1]. . .

In reviewing and assessing the demands for CCII harmonization of environmental standards, it is customary now to make a distinction of analytical importance between (1) environmental problems that are intrinsically *domestic* in nature (though they may be "internationalized" for reasons we will discuss); and (2) those that are intrinsically *international* in nature because they inherently involve "physical" spillovers across national borders. . . .

Objections to Diversity of Standards

It would seem, at first glance, that at least the intrinsically domestic environmental problems should be matters best left by governments to domestic solutions and within domestic jurisdiction. . . . Why should anyone object to the conduct of free trade with any country on the ground that its preferred environmental choices and solutions (by way of setting pollution standards and taxes) to intrinsically domestic questions are unacceptable because they are incompatible with the case for (gains from) free trade? Yet, the fact is that they do. . . .

Unfair Trade

If you do something different, and especially if you do what appears to be less, concerning environment than I do in the same industry or sector, this difference is considered to be tantamount to lack of "level playing fields" and therefore amounts to "unfair trade" by you. Free trade, according to this doctrine, is then unacceptable because it requires, as a precondition, "fair trade."

Losing Higher Standards

Then again, the flip side of the "fair trade" argument is the environmentalists' fear that if free trade occurs with countries having "lower" environmental standards, no matter what the justification for this situation, the effect will be to lower their own standards. This will follow from the political pressure brought to bear on governments to lower standards to ensure the survival of their industry.

An associated argument is that capital will move to countries with lower standards, so that countries will engage in a "race to the bottom," each winding up with lower standards than desired because standards are lowered to attract capital from each other. . . .

---

[1] By CCII, we mean harmonization of standards within the *same* industry across different trading countries.

Does Diversity of Environmental Standards Imply That Low-Standard Countries Are Indulging in "Unfair Trade"?

The theoretical analysis clearly shows that the basic presumption is that different countries will have *legitimate diversity of CCII environmental taxes and standards.* This diversity will arise even if they share the same "utility function" with associated trade-offs between income and different types of pollution: the diverse tax rates can come from differences in technology and in endowments in the broadest sense (so as to include weather, demography, geography, inherited abatement policies, etc.).

As it happens, there is also no compelling reason to think that every society must share the same utility function. It is perfectly appropriate, and not an indulgence of willful "sovereignty," for Mexico to value clean water higher than clean air, compared to the United States, because a dollar expended on the former instead of the latter will produce greater health gains for Mexicans, whereas the situation would be the reverse for the United States.

The overall trade-off between income and (some generalized index of) pollution will also be different between societies: income may be more valuable at the margin when societies are poor and poverty takes people close to malnutrition than when societies are rich and malnutrition results from overindulgence rather than deprivation. A clear example again is the emphasis on saving dolphins rather than increasing productivity in tuna fishing in the United States and the contrasting emphasis on ameliorating poverty instead in Mexico by using purse seines that kill dolphins while fishing for tuna.

The notion, therefore, that the diversity of CCII pollution standards and taxes is illegitimate and constitutes "unfair trade" or "unfair competition" is itself illegitimate. So is the consequent demand, following from this notion, that CCII harmonization is necessary for "free and fair trade," in the absence of which CCII differences must be treated as eco-dumping and be countervailed.

In fact, since the effect of such policies would be to force (at least some) countries to harmonize up their preferred lower CCII standards, the consequence would equally be to inflict a welfare loss on them. We might even argue that, while we advocate free trade traditionally, with diversity of domestic standards, on the presumption that voluntary trade is beneficial (relative to autarky) for every trading nation, and hence it is a mutual-gain policy prescription, the opposite is true for CCII harmonization to be superimposed on free trade: it will amount to immiserization of the trading nation whose standards are being "distorted" up.

This basic case against CCII harmonization can be challenged on grounds that we now examine and mainly find unpersuasive.

Objection 1: Competing with Foreign Firms That Do Not Bear Equal Burdens Is Unfair

This competitiveness argument is common, especially on the part of some business groups and also some unions. As notions of unfairness are expressed by them, and as implied by proposed legislation to equalize burdens, this is certainly a strongly felt belief. Underlying it is the sense of outrage that one's ability to hold on to an industry is compromised by the fact that one's rivals abroad do not carry the same burdens. The contrary arguments, which reject this competitiveness argument, are as follows:

(a) The fact that others abroad do not carry the same burdens is symmetric with the fact that these countries have different wages, capital costs, skills, infrastructure, weather, and what have you: all of which lead to differential advantages of production and trade competitiveness. Diversity of environmental tax burdens is thus no ground for complaints of unfairness.

(b) Losing competitive advantage because we put a larger negative value on a certain kind of pollution than others do is simply the flip side of the differential valuations. To object to that implication of the differential valuation is to object to the differential valuation itself, and hence to our own larger negative valuation. . . .

(c) Besides, attributing competitive disadvantage to differential pollution tax burdens in the fashion of CCII comparisons for individual industries confuses absolute with comparative advantage. Thus, for example, in a two-industry world, if both industries abroad have lower pollution tax rates than at home, both will not contract at home. Rather, the industry with the *comparatively* higher tax rate will.

Objection 2: Others' Lower Standards Do Not Reflect Their Citizens' Preferences Correctly

In turn, some environmentalist critics argue that the foreign governments do not reflect their citizens' "true preferences," and therefore in relation to these true preferences that would lead to higher valuation of pollution, the governments have unduly low standards, implying "unfair" competition.

There are counterarguments, in turn:

(a) Similar arguments, about failure of "political markets," apply to most countries, including high-standard countries, and to many areas of governmental regulation. It is commonly argued that the earliest legislations mandated "too-high" environmental standards that went beyond the "optimal" levels because costs were ignored and virtually limitless gains were assumed from the regulations. Now, in the United States for sure, cost-benefit considerations are steadily being introduced into the legislative process; and even the judiciary seems to have turned increasingly to this type of analysis, which then tends to weaken the bite of the standards legislatively laid down.

Since arguments can be made persuasively that all legislation strays from the optimal because of political market failures endemic to any political system, however democratic, objecting only to lower environmental standards as reflecting such political market failure is to be arbitrary. It is also to open a Pandora's box, in favor of the more powerful countries that can then throw stones at others' glass houses while building a fortress around their own.

(b) Again, even if one argues that the decisions made undemocratically by a dictatorship or an oligarchy are vitiated, there is no reason to believe that the higher standards being pursued by a foreign country representing the competitive interests of a foreign industry or labor union in an industry are what a more democratic process would yield. The correct approach should rather be to encourage a shift to more democratic procedures in arriving at social and economic legislation, including environmental policy. Process, not outcomes (especially outcomes sought by self-serving groups elsewhere), is what we should aim at in countries that lack democratic ways.

Should High-Standards Countries Force Low-Standards Countries into Upward Harmonization to Preserve Their High Standards?

In political-economy theory there are two forms of argument for CCII harmonization that take the high standards themselves to be at risk under free trade. Consider each, in turn.

1. The less common argument is simply that, under pressure of competition from the low-standard countries, the political equilibrium will shift in favor of those who oppose high standards.

But this argument suffers from the fallacy of misplaced concreteness. Intensified international competition, *no matter why it arises*, will put such pressure on governments to reduce business costs. Why pick on lower standards elsewhere, even assuming that they are contributing to the problem?

2. Far more worrisome to environmentalists than the simple effects of trade competition are the fears that "capital and jobs" will move to countries with lower standards, triggering a "race to the bottom". . . . where countries lower their standards in an interjurisdictional contest, below a level that some or all would like, in order to attract capital and jobs. So, a cooperative solution that would *coordinate* the setting of standards would, generally speaking, be a better solution. This coordinated solution, however, need not be characterized by harmonization at the level of the standards in the high-standard country or, in fact, by harmonization at all.

What we have here is a valid theoretical argument. It is stated with analytical rigor as follows: independent governments (or jurisdictions), setting public policy for environmental protection (via taxes and abatement) and competing for investment by reducing environmental standards in a world of mobile and scarce capital, will set these standards at levels that are "too low," that is, that are inefficient for the world economy (composed of the nations whose governments compete in this way). The inefficiency is to be construed as usual: alternative policies exist that make at least one jurisdiction better off and no other jurisdiction worse off.

The question that now arises is whether this theoretical possibility of the "race to the bottom" is an empirical possibility of any significance. . . . [W]e may ask whether there is any empirical support anyway for the propositions that (1) capital is in fact responsive to the differences in environmental standards and (2) different countries and jurisdictions actually play the game then of competitive lowering of standards to attract capital. Without both these phenomena holding in a significant fashion in reality, the "race to the bottom" could be a theoretical curiosity.

As it happens, systematic evidence is available for the former proposition alone, but the finding is that the proposition is not supported by the studies to date: at best, there is very weak evidence in favor of inter-jurisdictional mobility in response to CCII differences in environmental standards. . . .

Of course, there are many ways to interpret this finding of an extremely weak effect of CCII differences in environmental standards on industry location. There are three classes of explanation for the finding: (1) that the differences in standards are not significant and are outweighed by other factors that affect locational decisions; (2) that exploiting differences in standards is not a good strategy relative to not exploiting them; and (3) that lower standards may paradoxically even repel, instead of attracting, direct foreign investment.

Explanation 1

(a) The obvious, and most cited, explanation is that the standards differences are a small factor in the location decision because they are dominated by other more important factors such as tax breaks, infrastructure facilities, and proximity to markets.

(b) Industry location may be seen to be more sensitive to CCII differences in standards if executive enforcement and voters-cum-NGO [nongovernmental organization] activism are taken into account as well. The de facto differences in standards may then be more acute than assumed in many studies.

Explanation 2

(c) Another (static) explanation is that when multiplant firms, such as most multinationals, invest in different locations, they tend to work uniformly with the most stringent standards they face among these locations, to reduce the transaction costs involved in making diverse choices.

(d) Another (dynamic) explanation is that, faced with divergent standards, firms extrapolate that all countries are on an escalator to similar higher standards and therefore decide that it is best to be "ahead of the curve" in the countries that currently have lower standards and to conform to higher standards even though not required. In this case, again, convergence of standards adhered to will emerge, as in the preceding (static) argument, and differences in (required) standards across different jurisdictions will become moot, showing little relationship in practice between such differences and industry-location choices.

(e) Another (dynamic) explanation is that firms may argue that the higher-standard countries are the ones that innovate, that many innovations lead to embodied technical change, that such innovations are likely to be embodied (only) in recent vintages of capital goods that already meet the higher standards, and therefore the important benefit of significant technical change will accrue to a firm only insofar as processes and capital goods using higher-standard technology at present are being used by it.

Explanation 3

(f) An ingenious explanation of a different analytical variety is that multinationals are discouraged from investing in low-standard countries because local firms have comparative advantage in using pollution-intensive technology that conforms to lower standards. Hence, direct foreign investment (DFI) is likely to be less, not more, when CCII differences in standards are greater between countries!

A possible underlying explanation is that firms in the higher-standard countries are likely to scrap their earlier-vintage lower-standard equipment and sell it to the lower-standard countries for the local firms to use, instead of undertaking DFI themselves with such discontinued technology. In short, arm's-length sale of lower-standard equipment to local manufacturers may be preferred to DFI with such equipment, because the local firms are more likely to be able to work with this technology than the multinationals that have moved on to higher-standard, newer-vintage technology—engineering and maintenance know-how tend to be specific to the technology one is working with.

Notes and Questions

1. Suppose that state or nation B bans the importation of product X, from state or nation A, unless the product complies with stringent environmental standards. Is this restriction appropriate

(a) if product X is not manufactured in B?

(b) if product X is manufactured in B and domestic producers are held to the same standards as the foreign producers?

To what extent should it matter whether B generally has stringent environmental standards? To what extent should one inquire about legislative intent? In the second scenario, is the market share of the domestic producers in B relevant?

Should a Justice of the Supreme Court of the United States interpreting the dormant commerce clause apply different standards than a member of a GATT panel? For a detailed comparative analysis of product restrictions under the dormant commerce clause and the GATT, see Farber and Hudec, "Free Trade and the Regulatory State: A GATT's-Eye View of the Dormant Commerce Clause," 47 *Vanderbilt Law Review* 1401 (1994). For an extensive comparison of the treatment of environmental issues in the European Union and the GATT, see Petersmann, "Settlement of International Environmental Disputes in GATT and the EC," in *Towards More Effective Supervision by International Organizations* (Niels Blokker and Sam Muller, eds., Dordrecht: Martinus Nijhoff, 1994), at 165.

2. Stewart explains that the validity of product restrictions turns on a comparison of the importing state's environmental benefits against the costs imposed by the restraint on trade. He notes that "[i]n making such judgments it is a crucial question, which court decisions do not resolve, whether the court is to weigh the detriment to the welfare of consumers in the state imposing the restriction, or only the welfare of out-of-state producers and consumers." Assume that the restriction produces environmental benefits in B of 10 units, detriment of consumers in B of 5 units, and detriment to producers and consumers in A of 8 units. Can a sensible inquiry about the impact of the restriction on social welfare be performed without considering the detriment to the consumers in B? Are the federal courts or GATT panels well equipped to make this sort of inquiry? What alternative test could be used to determine the validity of the restriction? How would the analysis be different if producers in B obtained a benefit of 5 units from the trade restriction?

3. Article XX of the GATT provides:

> Subject to the requirement that such measures are not applied in a manner which would constitute a means of arbitrary or unjustifiable discrimination between countries where the same conditions prevail, or a disguised restriction on international trade, nothing in this Agreement shall be construed to prevent the adoption or enforcement by any contracting party of measures . . . necessary to protect human, animal or plant life or health; . . . relating to the conservation of exhaustible natural resources if such measures are made effective in conjunction with restrictions on domestic production or consumption . . .

The term "necessary" in Article XX(b) has been defined to mean "least GATT-inconsistent." Are the standards that you chose in your responses to the preceding questions consistent with Article XX?

4. The standards of Article XX were modified by the recent Uruguay Round of trade negotiations, conducted under the auspices of the GATT and completed in 1993. The Uruguay Round led to the Agreement on Sanitary and Phytosanitary Measures (SPS), which deals with trade restrictions concerning food safety and products that spread diseases, and the Agreement on Technical Barriers to Trade (TBT), which deals with issues not covered by the SPS Agreement.

Article 2.4 of the TBT Agreement provides that restrictions based on "relevant international standards" are presumptively appropriate. Under Article 2.2, to be acceptable, other restrictions "shall not be more trade-restrictive than necessary to fulfill a legitimate objective, taking account of the risks non-fulfillment would create." What types of cases might be decided differently as a result of this new standard?

Article 904 of NAFTA provides that trade restrictions are appropriate if "the demonstrable purpose of the measure is to achieve a legitimate objective." What are the salient differences between the GATT and NAFTA regimes? For a discussion of the environmental aspects of NAFTA, see Stewart, "The NAFTA: Trade, Competition, Environmental Protection," 27 *International Lawyer* 751 (1993).

The SPS Agreement provides, in Article 3, that, to form the basis for trade restrictions, standards must be "necessary," "based on scientific principles," "not maintained without sufficient scientific evidence," and based on a risk assessment "appropriate to the circumstances." Moreover, under Article 5, restrictions must be avoided if "there is another measure reasonably available, taking into account technical and economic feasibility, that achieves the appropriate level of protection and is significantly less restrictive of trade." NAFTA has similar standards, which are embodied in Article 712. What risk management criterion, if any, is implicit in these standards? The Delaney Clause of the U.S. Food, Drug, and Cosmetic Act bans the use of food additives containing any human or animal carcinogen. Consistent with the SPS Agreement, can the United States invoke the Delaney Clause to ban the importation of food additives containing minute amounts of pesticides?

5. For a discussion of the treatment on environmental issues in the Uruguay Round, see Schultz, "Environmental Reform of the GATT/WTO International Trading System," 18 *World Competition,* Dec. 1994, at 77. For further discussion of the GATT standards, see Daniel C. Esty, *Greening the GATT* (Washington, D.C.: Institute for International Economics, 1994); Petersmann, "International Trade Law and International Environmental Law," 27 *Journal of World Trade,* Feb. 1993, at 43. For theoretical analyses of the interrelationship between environmental regulation and international trade, see the essays, several by members of the GATT Secretariat, in Kym Anderson and Richard Blackhurst, eds., *The Greening of World Trade Issues* (Hertfordshire: Harvester Wheatsheaf, 1992). For environmentalist

perspectives on these issues, see several of the essays in Durwood Zaelke, Robert Housman, and Paul Orbuch, eds., *Trade and the Environment: Law, Policy and Economics* (Washington, D.C.: Island Press, 1993).

6. Consider a restriction on the importation of a product on the ground that it was manufactured by a process that harms the global commons. What inquiry should be undertaken to judge the validity of this process-based restriction? Which nation's valuation of the harm on the global commons should be taken into account? In judging the validity of one nation's restrictions, should one consider the possibility that other countries might follow suit, imposing similar restrictions? If one does not, the cumulative penalty on the country manufacturing the product might eventually become disproportionate to the harm that it imposes on the global commons. Such a rule might also lead to a "race to impose the first restriction." Would such a race be desirable? On the other hand, should the inquiry about the validity of a nation's restrictions be dependent on the decision maker's estimate of the probability that other countries might subsequently impose similar restrictions? Can such an inquiry be plausibly undertaken?

7. Would your answers differ if the process had interjurisdictional, but not global, effects. If nation B is imposing the restriction, should it matter whether the interjurisdictional harms occur in nation B or elsewhere?

8. If the physical effects of nation A's production processes are purely domestic, is nation B justified in adopting a trade restriction on the basis of the existence values to citizens of B of the harmed resources in nation A? Should the contingent valuation methodology (CVM), which is discussed in Chapter IV, be the basis for determining the legality of the restrictions? Whose valuations should be cognizable?

9. What should count as a competitiveness spillover? Consider two scenarios under which the environmental standards in nation A are laxer than those in nation B:

    (a) the environmental standards in nation A are set to maximize social welfare in that country, but they are laxer than those in nation B because the citizens of A value environmental quality less highly.

    (b) the environmental standards in nation A are laxer than those that would maximize social welfare in nation A.

In both cases, there are no interjurisdictional spillovers. If a countervailing duty is appropriate in either instance, how should one determine its magnitude?

10. Assume that nation B's standards for a particular industry are more stringent than nation A's. Table XI.1 defines various scenarios by reference to how the standards in A and B, respectively, compare to those that would maximize social welfare in each of the jurisdictions.

It is useful to illustrate this table by means of an example. The following situation is consistent with the box labeled "g." A's actual standard is 10 parts per million (ppm) of a pollutant,

Table XI.1.

A	B Laxer than optimal	B Optimal	B More stringent than optimal
Laxer than optimal	a	b	c
Optimal	d	e	f
More stringent than optimal	g	h	i

whereas it optimal standard is 12 ppm; thus, A's actual standard is more stringent than the optimal standard. In turn, B's actual standard is 8 ppm (more stringent than A's actual standard) but its optimal standard is 6 ppm (thus, B's standard is less stringent than the optimal standard).

Should B's use of "sticks" be appropriate in all circumstances, merely because A's standards are less stringent? Should it be appropriate only in situations *a, b,* and *c,* in which A's standards are laxer than optimal? Should B be barred from using "sticks" in situations *a, d,* and *g* because its own standards are laxer than optimal? In situations *c, f,* and *i,* where B's standards are more stringent than optimal, should B be permitted to use sticks only if its optimal standards are more stringent than A's standards?

11. The GATT is hostile to process-based regulation. For example, in an influential report, the GATT secretariat stated that "[i]n principle, it is not possible under GATT's rules to make access to one's own market dependent on the domestic environmental policies or practices of the exporting country," adding that this approach "protects trade relations from degenerating into anarchy though unilateral actions in pursuit of unilaterally-defined objectives" (General Agreement on Tariffs and Trade Secretariat, Trade and Environment [Geneva, 1992]). Following the two tuna/dolphin decisions discussed by Chang, can Article XX of the GATT ever be successfully invoked for process-based regulation?

NAFTA is largely silent about process-based regulation. Article 1114 states, however, in language that is probably only precatory, that "it is inappropriate to encourage investment by relaxing domestic health, safety, or environmental measures." Should the failure to adopt standards be treated similarly to the relaxation of standards?

A large literature has developed criticizing the GATT for constraining a nation's ability to impose trade measures on environmental grounds. See Dunoff, "Institutional Misfits: The GATT, the ICJ and Trade-Environment Disputes," 15 *Michigan Journal of International Law* 1043 (1994). In the United States, these views appear to be have had an influence on decision makers. Vice President Gore wrote while he was still a senator: "Just as government subsidies of a particular industry are sometimes considered unfair under the trade laws, weak and ineffectual enforcement of pollution control measures should also be included in the definition of unfair trading practices" (Al Gore, *Earth in the Balance* 343 [Boston: Houghton Mifflin, 1992]). Similarly, the proposed International Pollution Deterrence Act, S. 984, 102d Cong., 1st Sess. (1991), would authorize the imposition of countervailing duties equal to the amount that the foreign firm would have to expend in order to comply with the U.S. standards. Is such legislation desirable?

12. In a part of the article that is not excerpted, Stewart considers the case of multilateral restrictions, imposed as part of international agreements establishing common environmental standards. He argues that such restrictions should be viewed more deferentially than unilateral restrictions. Why should more deference be appropriate? Stewart gives the example of a policy by the European Union, assertedly adopted to combat global warming, that imposes a ban on imports of products produced by processes that yield carbon dioxide emissions greater than those permitted by the EU standards or a tariff on imports of such products. Could one argue that multilateral restrictions ought to be even more suspect because they have greater impact on free trade? If more deference is accorded to such multilateral restrictions, an incentive is created to enter into international agreements establishing common environmental standards. Is such an incentive necessarily desirable? Consider the fact that different nations might have different preferences for environmental protection and that the harm from pollution might also be different. For a useful classification of environmentally based trade restraints, which focuses in part on the unilateral-multilateral distinction, see Charnovitz, "A Taxonomy of Environmental Trade Measures," 6 Georgetown International Environmental Law Review 1 (1993).

13. Several recent international environmental agreements contain multilateral trade restrictions. For example, Article 4.5 of the Basel Convention on the Control of Transboundary Movements of Hazardous Wastes and their Disposal provides that a "[p]arty shall not permit hazardous wastes or other wastes to be exported to a non-Party or to be imported from a non-Party." Article 4.1 of the Montreal Protocol on Substances that Deplete the Ozone Layer provides that "each Party shall ban the import of controlled substances" from nonparties. Article VIII:1(a) of the Convention on International Trade in Endangered Species (CITES) requires parties to "penalize trade in, or possession of" a listed species. For analyses of such agreements, see Baker, "Protection, Not Protectionism: Multilateral Environmental Agreements and the GATT," 26 *Vanderbilt Journal of Transnational Law* 437 (1993); Cameron and Robinson, "The Use of Trade Provisions in International Environmental Agreements and Their Compatibility with the GATT," 2 *Yearbook of International Environmental Law* 3 (1991); McDonald, "Greening the GATT: Harmonizing Free Trade and Environmental Protection in the New World Order," 23 *Environmental Law* 397, 450–62 (1993).

14. How should one assess whether the use of sticks is proportionate to the legitimate interests of the country wielding the sanction? What factors should be relevant?

15. Stewart is more skeptical about the use of sticks than Chang. Who has the better argument?

16. In a part of the article that is not excerpted, Chang recognizes that "sticks" might be subject to abuse, serving the interests of protectionism rather than of genuine environmental concern. He proposes the following presumptive rule:

> [A] trade measure that restricts imports from another country, because that country provides environmental protection that falls below a particular standard, is an illegal disguised restriction on trade if: (1) the environmental protection in question applies to a stock of a natural resource within the foreign country, in the absence of physical spillovers; (2) the foreign producers to be regulated compete with a domestic industry subject to similar regulations to protect a stock of the same resource in the importing country; and (3) use value is an important consideration that supports the regulation of the domestic stock, but not of the foreign stock. (Chang, "An Economic Analysis of Trade Measures to Protect the Global Environment," 83 *Georgetown Law Journal* 2131, 2198 [1995])

Is this rule desirable? Is it too broad? Is it too narrow? Should it ever be appropriate for nation B to rely on the existence values of a resource in A as the basis for trade restraints? Does such reliance effectively turn every domestic environmental effect into an interjurisdictional effect?

17. In the preceding example, should the relative states of development of nations A and B affect the validity of the restrictions? For economic analyses of the impact on the global environment of north–south trade, see Chichilnisky, "North–South Trade and the Global Environment," 84 American Economic Review 851 (1994), and Copeland and Taylor, "North–South Trade and the Environment," 109 Quarterly Journal of Economics 755 (1994). For discussion of the distributional impacts of global environmental policies, see Chapter XII.

18. Assume that nations A and B have the same ambient standards for a particular air pollutant. Nation A has fewer sources of that pollutant, so it meets these ambient standards by means of less stringent emission standards. Should nation B be allowed to restrict the importation from nation A of goods on the ground that they are produced under a less stringent emission standard? Would such restrictions have any valid environmental rationale?

19. The cost of doing business in a country is affected by a number of governmental policies, including expenditures in education, health care, and infrastructure. The stringency of environmental standards is merely one component of this cost. Why should one call for the harmonization of environmental standards if one does not call for the harmonization of other factors affecting the cost of doing business? Should nation B be able to successfully argue that nation A's large investments in education, health care, and infrastructure are equivalent to a subsidy to industry?

20. Empirical studies generally show that differences in environmental standards do not have an important effect on competitiveness. For discussions of these studies, see Stewart, "Environmental Regulation and International Competitiveness," 102 *Yale Law Journal* 2039, 2061–84 (1993); Jaffe, Peterson, Portney, and Stavins, "Environmental Regulation and the Competitiveness of U.S. Manufacturing: What Does the Evidence Tell Us?" 33 *Journal of Economic Literature* 132 (1995). What impact should these studies have on the desirability of trade measures justified on the basis of differences in environmental controls on production processes?

21. Other factors affecting the cost of doing business, such as proximity to raw materials and markets, rainfall, and temperature, are not the product of governmental policies. Why should a nation be allowed to obtain an advantageous trade position as a result of these natural phenomena, but not as a result of the preference of its citizens for less stringent environmental standards?

22. Consider a scenario in which nation A makes large expenditures in health care but has comparatively weak environmental standards. In contrast, nation B, spends little on health care but has far more stringent environmental standards. What would be the effect of the harmonization of environmental standards? Would such harmonization be desirable? For a related discussion, see Revesz, "Rehabilitating Interstate Competition: Rethinking the 'Race-to-the-Bottom' Rationale for Federal Environmental Regulation," 67 *New York University Law Review* 1210, 1244–47 (1992), excerpted in Chapter VII.

XII

International Environmental Law

The readings in this chapter explore issues that are central to the understanding of international environmental law. The articles by Edith Brown Weiss and Robert Solow deal with what is now probably the most widely invoked principle of international environmental law: sustainable development. The principle gained wide notoriety following the publication in 1987 of *Our Common Future*, the report of the World Commission on Environment and Development (often referred to as the Brundtland Report, after its chairperson, who was then prime minister of Norway). This report defined sustainable development as development that "meets the needs of the present without compromising the ability of future generations to meet their own needs." The concept of sustainable development has been featured repeatedly in subsequent international environmental agreements, including the 1992 Rio Declaration on Environment and Development, though little headway has been made in reaching a consensus on a more precise understanding of the concept's scope.

The Weiss article equates sustainable development with intergenerational equity, which she defines by reference to three principles: First, the principle of conservation of options requires each generation to preserve the natural and cultural resource bases so that the options available to future generations are not unduly restricted. Second, the principle of conservation of quality requires each generation to prevent a worsening of the planet's environmental quality. Third, the principle of conservation of access requires each generation to provide its members with equitable rights of access to the legacy of past generations, and to conserve this access for the benefit of future generations.

According to the Solow article, sustainability requires that each future generation have the means to be as well off as its predecessors. He gives content to this principle by proposing a modification to the traditional measure of a nation's economic activity. From Net National Product (NNP)—Gross National Product (GNP) minus the depreciation of fixed capital assets—he would subtract the value of expended nonrenewable resources and environmental assets like clean air and water. Solow argues that each generation must use its nonrenewable and environmental resources in a way that does not detract from the ability of future generations to have a similar standard of living. He admits that certain unique and irreplaceable resources, like certain national parks, should be preserved for their own sake, but maintains that the consumption of nonunique natural and environmental resources ought to be permissible as long as they are replaced by other resources such as equipment or technological knowledge.

The remaining readings explore two of the most salient issues concerning the development of international environmental law. First, unlike the case of national legislation, in the international context there is no central government with authority to impose regulatory solutions. Instead, norms result from agreements among nations. As in a typical free-rider problem, a nonparticipating nation cannot be excluded from the benefits of a global environmental agreement. Because every nation faces the same calculus, the logic of collective action would predict that such agreements will not be reached. The article by Oran R. Young analyzes the conditions that facilitate international environmental agreements. It seeks to explain why the record has been successful in some areas but far less successful in others. For example, the Montreal Protocol on Substances that Deplete the Ozone Layer, adopted in 1987 and significantly strengthened in 1990, is widely regarded as an effective means of protecting stratospheric ozone. In contrast, despite the adoption in 1992 of the Framework Convention on Climate Change, little progress has been made on an agreement concerning actual limitations on the emissions of greenhouse gases, which lead to global warming.

A second salient issue arises as a result of the vast disparity of wealth among members of the international community, the different stages of their industrial development, and their different past contributions to pollution problems. Equity considerations played a large role in the Montreal Protocol and the Rio Declaration and are likely to be central to further discussions concerning climate change. The article by Henry Shue argues that four separate questions of justice must be resolved in connection with the design of a climate change regime: a fair allocation of the costs of preventing global warming that is still avoidable, a fair allocation of the costs of coping with global warming that will not be avoided, a background allocation of wealth that would make bargaining over the prior two issues a fair process, and a fair long-term allocation of emissions of greenhouse gases.

Intergenerational Equity: A Legal Framework for Global Environmental Change

EDITH BROWN WEISS

Sustainable development rests on a commitment to equity with future generations. In 1972 the United Nations Stockholm Conference on the Human Environment recognized that we had a responsibility to "protect and improve" the environment for both present and future generations. . . .

Three principles form the basis of intergenerational equity. First, each generation should be required to conserve the diversity of the natural and cultural resource base, so that it does not unduly restrict the options available to future generations in solving their problems and satisfying their own values, and should also be entitled to diversity comparable to that enjoyed by previous generations. This principle is called "conservation of options." Second, each generation should be required to maintain the quality of the planet so that it is passed on in no worse condition than that in which it was received, and should also be entitled to planetary quality comparable to that enjoyed by previous generations. This is the principle of "conservation of quality." Third, each generation should provide its members with equitable rights of access to the legacy of past generations and should conserve this access for future generations. This is the principle of "conservation of access."

The proposed principles recognize the right of each generation to use the Earth's resources for its own benefit, but constrain the actions of the present generation in doing so. Within these constraints they do not dictate how each generation should manage its resources. They do not require that the present generation predict the preferences of future generations, which would be difficult if not impossible. Rather, they try to ensure a reasonably secure and flexible natural resource base for future generations that they can use to satisfy their own values and preferences. They are generally shared by different cultural traditions and are generally acceptable to different economic and political systems.

While the principle of quality may be viewed as including the principle of diversity, they are separate and complementary. To illustrate this, we can invoke the analogy of a common law trust, whose corpus consists of investments in two different energy companies and a computer company. If the trustee shifts the investments into other energy and computer companies that turn out to be lower in quality as investments, the value of the trust corpus declines, but the diversity of the holdings does not change. By contrast, if the trustee combines all the investments into a single oil company, the value of the holdings may remain the same, but the diversity of the holdings is sharply compromised.

In our planet, environmental quality may decline, but this does not necessarily reduce significantly the diversity of the resource base. Similarly, it may be possible for one generation to sustain the quality of air and water but substantially destroy the diversity of the resource base, as by a significant loss of genetic diversity. Certainly the two principles interact and feed upon each other. It is easier to maintain quality if there are many options available for doing so, and serious water pollution may cause fish to disappear. It is easier to conserve options when there is concern for maintaining quality. Both principles are essential for a robust planet for future generations and must be implemented in tandem.

Conservation of Options

Future generations are more likely to survive and attain their goals if they have a variety of options for addressing their problems. Conserving the diversity of the natural and cultural resource bases is designed to give our descendants a robust and flexible heritage with which to try to achieve a decent and healthy life.

The principle of conserving options rests on the premise that diversity, like quality, contributes to robustness. This can be seen in the contribution of biological diversity to the robustness of ecosystems. If diverse strains and species are present in an ecosystem and the system is perturbed, some strains and species will survive and multiply. While the distribution of the biological population may change, the ecosystem remains viable. By contrast, farmers producing monocultures have to work hard to preserve them, for they are easily eliminated through the invasion of weeds, insects, and other pests. Some theoretical scientific research suggests, however, that as systems become more complex (more species and a richer structure of interdependence), they may become more dynamically fragile. This suggests that we need to understand the special kinds of complexity that promote stability.

Biological diversity as it relates to robustness encompasses change in the species and strains that make up the ecosystem. This point is essential to intergenerational justice, for it means that change, which is essential for economic development, is an integral part of implementing the principle.

The wisdom of conserving options is reflected more broadly in conventional economic practices, such as maintaining diversity in the corpus of a common law trust, portfolios of investments, and national economies. In these latter examples, diversity is primarily viewed as a means of spreading risks to avoid reliance on only one investment or industry. At the same time it offers an effective strategy for improving economic wealth.

The question arises, however, whether conserving options does not disregard the needs of the present generation. It may be argued that the best way to conserve options is to preserve the status quo, which means that poor people in particular will continue to suffer.

This argument applies the principle incorrectly. Conservation of options can be accomplished by new technological developments that create substitutes for existing resources or processes for exploiting them more efficiently, as well as by conservation of existing resources. Certainly any investment in the development of particular resources forecloses other options for that resource. The decision to convert an area to solar panels will foreclose use of the land for crops, at least for that period of time. But the solar panels may help to conserve more scarce energy supplies, such as helium-rich natural gas reserves, or to avoid fossil-fuel emissions that contribute to climate change. To the extent that a hydroelectric dam or mine will destroy a unique natural resource, however, we must proceed extremely cautiously, if at all, because future generations might be willing to pay us handsomely to conserve it for them.

The principle of conservation of options requires that on *balance* the diversity of the resource base be maintained. It acts as an important brake on those who would destroy biological diversity by clear-cutting tropical areas, developing crop mono-cultures to the exclusion of conserving wild cultivars, exhausting all known quantities of essentially non-renewable resources such as oil and helium-bearing natural gas, or discarding the cultural resources of all but a few dominant cultures.

Conservation of Quality

The principle of conservation of quality requires that we leave the quality of the natural and cultural environments in no worse condition than we received it. Recent generations have used resources of air, water, and soils as free resources for dumping their wastes, thereby passing on the costs of their activities to future generations in the form of degraded quality of air and water, with accompanying harms to plant and animal life and to human health.

The principle of conserving quality is consistent with environmentally sustainable growth. It does not mean that the environment must remain unchanged, which would be inconsistent in any event with conserving the present generation's access to the benefits of the planet. In determining whether one generation is conserving quality, trade-offs are inevitable. For example, we may exhaust more reserves of a natural resource and cause modest levels of pollution, but pass on a higher level of income, capital, and knowledge sufficient to enable future generations to develop substitutes for the depleted resource and methods for abating or removing pollutants. A framework must be developed in which such balancing can take place. Necessary components will be predictive indices of resource diversity and resource quality, baseline measurements, and an improved capacity to predict technological change.

It is natural to assume that present trends in natural and social systems will continue. However, breaking-points may exist in key variables beyond which these systems will reorganize and substantially change their properties. Predicting these breaking-points is thus of critical importance, probably more important than predicting specific technological changes, since such breaking-points would indicate the need for deliberate human intervention. . . .

Conservation of Access

Conservation of access gives the members of the present generation a reasonable, non-discriminatory right of access to the natural and cultural resources of our planet.

This means they are entitled to these resources to improve their own economic and social well-being provided that they respect their equitable duties to future genera-tions and do not unreasonably interfere with the access of other members of their generation to these same resources.

This offers a principle of justice between generations and between members of the same generation. The refinement of what conservation of access means as applied to members of the present generation is extremely complex. It implies both that the patrimony itself to which they have access should be comparable in quality and diversity (or robustness) to previous generations and that they should have a minimum level of resources so that they can in fact have access to such a patrimony. Thus, members of the present generation must not degrade the patrimony available, and to the extent that some members are too impoverished to have effective access, must assist them to gain such access.

An Almost Practical Step Toward Sustainability
ROBERT SOLOW

It is a commonplace thought that the national income and product accounts, as cur-rently laid out, give a misleading picture of the value of a nation's economic activ-ity to the people concerned. The conventional totals, gross domestic product (GDP) or gross national product (GNP) or national income, are not so bad for studying fluctuations in employment or analyzing the demand for goods and services. When it comes to measuring the economy's contribution to the well-being of the country's inhabitants, however, the conventional measures are incomplete. The most obvious omission is the depreciation of fixed capital assets. If two economies produce the same real GDP but one of them does so wastefully by wearing out half of its stock of plant and equipment while the other does so thriftily and holds depreciation to 10 percent of its stock of capital, it is pretty obvious which one is doing a better job for its citizens. Of course the national income accounts have always recognized this point, and they construct net aggregates, like net national product (NNP), to give an appropriate answer. Depreciation of fixed capital may be badly measured, and the error affects net product, but the effort is made.

The same principle should hold for stocks of nonrenewable resources and for environmental assets like clean air and water. Suppose two economies produce the same real net national product, with due allowance for depreciation of fixed capital, but one of them is wasteful of natural resources and casually allows its environment to deteriorate, while the other conserves resources and preserves the natural envi-ronment. In such a case we have no trouble seeing that the first is providing less amply for its citizens than the second. So far, however, the proper adjustments needed to measure the stocks and flows of our natural resources and environmental assets are not being made in the published national accounts. . . .

Now I go back to the beginning and make my case in more detail. Suppose we adopt a simplified picture of an economy living in some kind of long run. What I mean by that awkward phrase is that we are going to ignore all those business-cycle problems connected with unemployment and excess capacity or overheating and inflation. From quarter to quarter and year to year this economy fully exploits the resources of labor, plant, and equipment that are available to it.

To take the easiest case—that of natural resources—first, imagine that this economy starts with a fixed stock of nonrenewable resources that are essential for further production. This is an oversimplification, of course. Even apart from the possibility of exploration and discovery, the stock of nonrenewable resources is not a pre-existing lump of given size, but a vast quantity of raw materials of varying grade, location, and ease of extraction. Those complications are not of the essence, so I ignore them.

It is of the essence that production cannot take place without some use of natural resources. But I shall also assume that it is always possible to substitute greater inputs of labor, reproducible capital, and renewable resources for smaller direct inputs of the fixed resource. Substitution can take place on reasonable terms, although we can agree that it gets more and more costly as the process of substitution goes on. Without this minimal degree of optimism, the conclusion might be that this economy is like a watch that can be wound only once: it has only a finite number of ticks, after which it stops. In that case there is no point in talking about sustainability, because it is ruled out by assumption; the only choice is between a short happy life and a longer unhappy one.

Life for this economy consists of using all of its labor and capital and depleting some of its remaining stock of resources in the production of a year's output (GDP approximately). Part of each year's output is consumed, and that gives pleasure to current consumers; the rest is invested in reproducible capital to be used for production in the future. There are various assumptions one could make about the evolution of the population and employment. I will assume them to have stabilized, since I want to talk about the very long run anyway. Next year is a lot like this year, except that there will be more plant and equipment, if net investment was positive this year, and there will be less of the stock of resources left.

Each year there are two new decisions: how much to save and invest, and how much of the remaining stock of nonrenewable resources to use up. There is a sense in which we can say that this year's consumers have made a trade with posterity. They have used up some of the stock of irreplaceable natural resources; in exchange they have saved and invested, so that posterity will inherit a larger stock of reproducible capital.

This intergenerational trade-off can be managed well or badly, equitably or inequitably. I want to suppose that it is done well and equitably. That means two things. First, nothing is simply wasted; production is carried on efficiently. Second, although the notion of intergenerational equity is much more complicated and I cannot hope to explain it fully here, the idea is that each generation is allowed to favor itself over the future, but not too much. Each generation can, in turn, discount the welfare of all future generations, and each successive generation applies the same discount rate to the welfare of its successors. To make conservation an interesting proposition at all, the common discount rate should not be too large.

You may wonder why I allow discounting at all. I wonder, too: no generation 'should' be favored over any other. The usual scholarly excuse—which relies on the idea that there is a small fixed probability that civilization will end during any little interval of time—sounds far-fetched. We can think of intergenerational discounting as a concession to human weakness or as a technical assumption of convenience (which it is). Luckily, very little of what I want to say depends on the rate of discount, which we can just imagine to be very small.

Given this discounting of future consumption, we have to imagine that our toy economy makes its investment and resource-depletion decisions so as to generate the largest possible sum of satisfactions over all future time. The limits to this optimization process are imposed by the pre-existing stock of resources, the initial stock of reproducible capital, the size of the labor force, and the technology of production. . . .

Now I come to the first major analytical step in my argument. If you look carefully at the solution to the problem of intergenerational resource allocation I have just sketched, you see that an excellent approximation of each single period's contribution to social welfare emerges quite naturally from the calculations. It is, in fact, a corrected version of net domestic product. The new feature is precisely a deduction for the net depletion of exhaustible resources. (I use the phrase "net depletion" because it is possible to extend this reasoning to allow for some discovery and development of new resources. In the pure case, where all discovery and development have already taken place, net and gross depletion coincide.)

The correct charge for depletion should value each unit of resource extracted at its net price, namely, its real value as input to production minus the *marginal* cost of extraction. As Hartwick has pointed out, if the marginal cost of mining exceeds average cost, which is what one would expect in an extractive industry, then the simple procedure of deducting the gross margin in mining (that is, the value of sales less the cost of extraction) will overstate the proper deduction and thus understate net product in the economy. . . .

Now I want to start down an apparently quite different path, but I promise that it will eventually link up with the unromantic measurement issues I have discussed so far, and will even reinforce the argument I have made.

I do not have to remind you that "sustainability" has become a hot topic in the last few years, beginning, I suppose, with the publication of the Brundtland Commission's report, *Our Common Future*, in 1987. As far as I can tell, however, discussion of sustainability has been mainly an occasion for the expression of emotions and attitudes. There has been very little analysis of sustainable paths for a modern industrial economy so that we have little idea of what would be required in the way of policy and what sorts of outcomes could be expected. As things stand, if I express a commitment to sustainability, all that tells you is that I am unhappy with the modern consumerist life-style. If I pooh-pooh the whole thing, on the other hand, all you can deduce is that I am for business as usual. . . . not a very satisfactory state of affairs. . . .

If sustainability means anything more than a vague emotional commitment, it must require that something be conserved for the very long run. It is very important to understand what that something is: I think it has to be a generalized capacity to produce economic well-being.

It makes perfectly good sense to insist that certain unique and irreplaceable assets should he preserved for their own sake; nearly everyone would feel that way about Yosemite or, for that matter, about the Lincoln Memorial, I imagine. But that sort of situation cannot be universalized: it would be neither possible nor desirable to "leave the world as we found it" in every particular.

Most routine natural resources are desirable for what they do, not for what they are. It is their capacity to provide usable goods and services that we value. Once that principle is accepted, we are in the everyday world of substitutions and trade-offs.

For the rest of this talk, I will assume that a sustainable path for the national economy is one that allows every future generation the option of being as well off as its predecessors. The duty imposed by sustainability is to bequeath to posterity not any particular thing—with the sort of rare exception I have mentioned—but rather to endow them with whatever it takes to achieve a standard of living at least as good as our own and to look after their next generation similarly. We are not to consume humanity's capital, in the broadest sense. Sustainability is not always compatible with discounting the well-being of future generations if there is no continuing technological progress. But I will slide over this potential contradiction because discount rates should be small and, after all, there is technological progress.

All that sounds bland, but it has some content. The standard of living achievable in the future depends on a bundle of endowments, in principle on everything that could limit the economy's capacity to produce economic well-being. That includes nonrenewable resources, of course, but it also includes the stock of plant and equipment, the inventory of technological knowledge, and even the general level of education and supply of skills. A sustainable path for the economy is thus not necessarily one that conserves every single thing or any single thing. It is one that replaces whatever it takes from its inherited natural and produced endowment, its material and intellectual endowment. What matters is not the particular form that the replacement takes, but only its capacity to produce the things that posterity will enjoy.

The Politics of International Regime Formation: Managing Natural Resources and the Environment
ORAN R. YOUNG

The record clearly shows that institutional bargaining results in the formation of new international regimes under some conditions. Consider, just to name some recent examples, the development of institutional arrangements pertaining to transboundary

Reprinted with permission of the author and publisher from Oran R. Young, "The Politics of International Regime Formation," *International Organization*, Vol. 43 #3, pp. 349—375. Copyright © 1989 by the World Peace Foundation and the Massachusetts Institute of Technology.

radioactive fallout, stratospheric ozone, and Antarctic minerals. Yet success is far from ensured in such endeavors. Like self-interested actors in all social arenas, those attempting to work out the terms of international regimes are often stymied by bargaining impediments that prolong negotiations over institutional arrangements and can easily result in deadlocks. It is not surprising, for instance, that the negotiations over a regime for deep seabed mining took so long and had such an ambiguous outcome. Nor is it hard to understand why interested parties have so far failed to devise workable arrangements to protect the biological diversity of the earth's ecosphere or to cope with the global climate change expected to result from the emission of greenhouse gases.

The . . . task, then, is to make use of the model of institutional bargaining to pinpoint the determinants of success and failure in efforts to form institutional arrangements in international society. . . . I initiate this process by deriving some hypotheses about factors governing the likelihood of success in efforts to form international regimes. . . .

Institutional Bargaining Can Succeed Only When the Issues at Stake Lend Themselves to Contractarian Interactions

Those engaged in efforts to form international regimes experience incentives to approach this process as a problem-solving exercise aimed at reaching agreement on the terms of a social contract when the absence of a fully specified zone of agreement encourages integrative bargaining and the presence of imperfect information ensures that a veil of uncertainty prevails. In situations governed by unanimity rules, a contractarian environment of this sort is necessary to avoid the positional deadlocks that commonly arise in connection with distributive bargaining. It is therefore critical to observe that collective-action problems in international society of the sort engendering an interest in devising arrangements to institutionalize cooperation vary in the degree to which they lend themselves to treatment in contractarian terms. And it is worth noting that those involved in efforts to form international regimes often differ markedly in terms of the skill they display in presenting problems of regime formation in contractarian terms.

To see the relevance of this factor to the success of institutional bargaining, consider, to begin with, some extreme cases. It is exceedingly difficult, for instance, to portray the problem of controlling acid precipitation in North America in contractarian terms both because the producers of the relevant emissions and the victims of acid deposition are so clearly identified at the outset and because there is not much overlap in the membership of the two groups. The problem of controlling transboundary radioactive fallout resulting from nuclear accidents, by contrast, is comparatively easy to treat in contractarian terms. Although a good deal is known about the dangers of radioactive fallout, individual members of international society ordinarily cannot know in advance whether they will occupy the role of site of an accident, victim state, or unharmed bystander with respect to specific accidents. This is exactly the sort of situation that gives rise to incentives to consider the common good in devising institutional arrangements.

Although they are less extreme, other cases add to our understanding of the importance of this proposition about the significance of contractarianism. There are important differences, for example, between the problems of ozone depletion and global climate change that affect the extent to which they lend themselves to formulation in contractarian terms. While the impact may vary somewhat on the basis of latitude, human populations in every part of the world will be harmed if the depletion of stratospheric ozone continues at its present rate. In the case of global warming, on the other hand, there will almost certainly be winners and losers who are comparatively easy to differentiate. Significant increases in sea level will cause severe damage to certain low-lying coastal areas (for example, one-half to two-thirds of Bangladesh could easily be inundated) while bestowing benefits on other regions. Global warming is expected to make some areas increasingly hospitable to large-scale agriculture even as other areas lose their current role in agricultural production. And the impact of these differences on processes of regime formation is heightened by the fact that the sources of greenhouse gas emissions are numerous and widely dispersed, whereas the producers of CFCs [chlorofluorocarbons] are few in number and located in a relatively small number of states. Under the circumstances, it is no cause for surprise that the foundations for an international regime designed to protect the ozone layer are now in place, whereas a regime to deal with global climate change is not yet in sight.

The Availability of Arrangements That All Participants Can Accept as Equitable (Rather than Efficient) Is Necessary for Institutional Bargaining to Succeed

Economists and others who approach the issue of regime formation as a problem in comparative statics generally place primary emphasis on the achievement of allocative efficiency in discussing the formation of new institutions as well as in evaluating the performance of existing arrangements. Such analysts are apt to be highly critical of arrangements that encourage misallocation of scarce resources or that seem likely to produce outcomes lying inside the relevant welfare frontier. To be more concrete, they find much to criticize in arrangements allocating some of the choicest deep seabed mining sites to the Enterprise,[1] imposing across-the-board percentage cuts on the production of CFCs by current producers, or reserving at least one orbital slot for each state that may become interested in satellite broadcasting.

Yet those who negotiate the terms of international regimes seldom focus on these questions of allocative efficiency. In a negotiating environment featuring the rule of unanimity, they must occupy themselves, for the most part, with considerations of equity on the understanding that institutional bargaining in international society can succeed only when all of the major parties and interest groups come away with a sense that their primary concerns have been treated fairly. Allocative efficiency is an abstract concept. No one can determine whether the outcomes flowing from a given regime are in fact efficient until much later. And even then, economists often disagree vigorously in their assessments of the efficiency of observable

---

[1][An international authority that would have conducted seabed mining, in a parallel system with private miners. ED.]

outcomes. Equity, by contrast, is an immediate concern that evokes strong feelings on all sides. To return to the previous examples, no reasonable observer could have expected the less developed countries participating in the law of the sea negotiations to accept an arrangement that explicitly excluded the Enterprise from mining operations, thereby ensuring that a few highly industrialized states would dominate this commercial activity. There is a sense of fairness that everyone can relate to in across-the-board percentage cuts which is hard to match in more complex arrangements featuring charges or transferable production permits. And it is surely easy to understand why the less developed countries regard as unjust any system that features the allocation of orbital slots on a first come, first served basis. While it is important to recognize that there are no objective standards of equity which can be applied to human affairs, it is also worth noting that identifiable community standards regarding equity do exist in specific social settings. And there is much to be said for the proposition that satisfying these standards is a necessary condition for international regime formation, whatever outside observers may think of the long-term consequences of the resultant arrangements with respect to allocative efficiency.

The Existence of Salient Solutions (or Focal Points Describable in Simple Terms) Increases the Probability of Success in Institutional Bargaining

Those endeavoring to craft statutes in municipal legislatures sometimes proceed, for tactical reasons, to construct formulas that are so complex or obscure that interest groups actually or potentially opposed to the relevant provisions have difficulty comprehending what is being put to a vote. No doubt, such tactics can prove useful in the efforts to form winning coalitions that dominate legislative bargaining. For the most part, by contrast, salience based on simplicity and clarity contributes to success in institutional bargaining involving numerous parties operating under unanimity rules. The idea of a simple ban or prohibition on pelagic sealing, for example, was a key factor in the success of the negotiations that produced the original regime for the conservation of northern fur seals in 1911. In more contemporary terms, the salience of the formula of across-the-board percentage cuts in the production and consumption of CFCs certainly played a role in the successful effort to reach agreement on the 1987 protocol regarding ozone depletion. And the fact that early warning procedures are markedly simpler than provisions covering compensation for damages surely has much to do with the ease of achieving agreement on the early notification convention of 1986 as well as with the failure to incorporate compensation provisions into the two 1986 conventions relating to nuclear accidents.

Conversely, it is hard to avoid the conclusion that the complexity of arrangements encompassing permits or licenses, production controls, technology transfers, the role of the Enterprise, and so forth bedeviled the effort to negotiate a regime for the deep seabed (known as the Area) in the law of the sea negotiations and played a significant role in accounting for the ambiguity of the final outcome. And similar problems might well plague efforts to come to terms on the provisions of a regime relating to global climate change. The power of salience can become a serious constraint on efforts to devise appropriate institutional arrangements. It constitutes a barrier to the introduction of some clever and attractive devices that students of institutional design have come up with to handle collective-action problems. But this in no way detracts from the role of salience as a determinant of success in the formation of regimes in international society.

The Probability of Success in Institutional Bargaining Rises When Clear-cut and Effective Compliance Mechanisms Are Available

It is common knowledge among those who study collective-action problems that negotiators can fail to reach agreement on arrangements capable of yielding benefits for all parties concerned because they do not trust each other to comply with the terms of the arrangements once they are established. This places a premium on the development of requirements that are easy to verify, as in the case of cuts in the production of CFCs by a small number of clearly identified producers. It also accounts for the attractions of arrangements that are comparatively easy to police, such as the licensing system for deep seabed mining contemplated under the regime for the Area. Presumably, some such reasoning played a role, as well, in the decision to orient the regime for endangered species toward the regulation of international trade in contrast to the control of habitat destruction within individual nations. While trade restrictions are hard enough to verify and police, it is not easy even to imagine how to implement a regime requiring the individual members to take effective steps to control the forces causing habitat destruction within their jurisdictions. Similar problems would undoubtedly afflict any effort to devise, a workable regime to protect biological diversity.

At the same time, the lack of well-entrenched and properly financed supranational organizations in international society ensures that international regimes must rely heavily on the ability and willingness of individual members to elicit compliance with key provisions within their own jurisdictions. A problem that has dogged the regime for endangered species, for example, is the sheer inability of many states to control the activities of poachers and others involved in the illegal trade in furs, skins, and animal parts within their jurisdictions. Contrast this with the case of the fur seal regime, under which any harvest of seals was either closely regulated or actually carried out by state agencies, thereby enabling municipal governments to exercise effective control over the relevant activities whenever they chose to do so. Under the circumstances, it is easy enough to understand why regime formation in international society is most apt to succeed when the participants can rely on relatively simple, nonintrusive compliance mechanisms that municipal governments can operate without undue effort or the need to expend scarce political resources. The 1987 protocol on ozone depletion, which has a remarkably straightforward formula coupled with explicit delegation of implementation to individual participants, offers a clear illustration of this proposition.

For the Most Part, Exogenous Shocks or Crises Increase the
Probability of Success in Efforts to Negotiate the Terms of
International Regimes

Even in negotiations that allow considerable scope for integrative bargaining under
a veil of uncertainty, institutional bargaining exhibits a natural tendency to bog down
into a kind of sparring match in which participants jockey for positional advantages
and lose track of their common interest in solving the relevant collective-action
problems. All too often, the net result is a failure to reach agreement regarding
feasible arrangements that would prove mutually beneficial. Given this background,
it will come as no surprise that exogenous shocks or crises frequently play a significant role in breaking these logjams and propelling the parties toward agreement on
the terms of institutional arrangements. The precipitous decline in the northern fur
seal population in the early years of this century and the extraordinary drop in blue
whale stocks in the 1930s clearly played major roles in inducing the relevant parties
to drop their bargaining ploys in the interests of reaching agreement on the provisions
of regulatory regimes before it was too late. It is hard to overstate the shock value of
the 1986 Chernobyl accident in motivating the parties to come to terms on at least
some of the provisions of a regime for nuclear accidents within six months of this
dramatic event. And the 1985 discovery and subsequent publicization of an ozone
"hole" over Antarctica emerged clearly as a driving force behind the efforts which
produced the 1987 protocol on stratospheric ozone and which may well lead to additional regulatory arrangements in the near future, despite the fact that ozone depletion
over Antarctica is not an immediate threat to major centers of human population.
      Compare these cases with the problem of global climate change. There is a good
case to be made for the proposition that the disruptive impacts of nuclear accidents and
ozone depletion are likely to pale by comparison with the consequences of the global
warming trend over the next century. To date, however, we have not experienced an
exogenous shock or crisis in this realm that can compare with the Chernobyl accident
or the ozone hole in capturing and galvanizing the attention of policymakers and
broader publics alike. Talk of a creeping crisis with regard to global warming simply
cannot produce the impact of the exogenous shocks mentioned previously as a force
in breaking the logjams that commonly arise in institutional bargaining. This is no
doubt frustrating to those working on a number of important collective-action problems. It is hard to contrive credible crises, and there is no reason to suppose that the
occurrence of exogenous shocks will correlate well with the ultimate importance of
the problems at hand. But none of this can detract from the role of exogenous shocks
or crises as a determinant of success in efforts to build regimes in international society.

Institutional Bargaining Is Likely to Succeed
When Effective Leadership Emerges;
It Will Fail in the Absence of Such Leadership

We come back, in the end, to the role of leadership in determining outcomes arising
from institutional bargaining in international society. It is no exaggeration to say that

efforts to negotiate the terms of international regimes are apt to succeed when one or more effective leaders emerge. In the absence of such leadership, they will fail. Those engaged in institutional bargaining must strive to invent options capable of solving major problems in a straight-forward fashion and to fashion deals that are acceptable to all. To the extent that the participants have incentives to engage in integrative bargaining and to the extent that a veil of uncertainty prevails and linkages among problems allow for logrolling, the task of those negotiating the terms of international regimes will be made easier. But such considerations cannot eliminate the crucial role of entrepreneurship at the international level.

Entrepreneurial leaders in institutional bargaining are neither hegemons who can impose their will on others nor ethically motivated actors who seek to fashion workable institutional arrangements as a contribution to the common good or the supply of public goods in international society. Rather, international entrepreneurs are actors who are skilled in inventing new institutional arrangements and brokering the overlapping interests of parties concerned with a particular issue-area. . . . [N]ongovernmental organizations or even individuals can become leaders in efforts to form international regimes. The role of the Comité Spécial de l'Année Geophysique Internationale in establishing SCAR [Scientific Committee on Antarctic Research] in 1958 and, through SCAR, in forming the regime for Antarctica in 1959 is comparatively well known. But the role of IUCN [International Union for the Conservation of Nature and Natural Resources] in promoting the regimes governing trade in endangered species and conservation of polar bears as well as the role of UNEP [United Nations Environment Programme] in creating the regime for controlling pollution in the Mediterranean Basin are also striking examples of success in international entrepreneurship. And the remarkable role of Mustafa Tolba. UNEP's executive director, in shepherding the negotiations regarding the protection of stratospheric ozone to a successful conclusion is worthy of much more systematic examination. None of this means, of course, that states cannot assume leadership roles in negotiating international regimes: far from it. The activities of the United States in connection with the 1987 protocol on ozone, of France in the ease of Mediterranean pollution control, and of several developing countries in the context of deep seabed mining standout, to name just a few examples.

Neither the mainstream utilitarians nor the power theorists work with constructs capable of offering significant analytic leverage on the type of entrepreneurial leadership under consideration here. Yet there is ample evidence to demonstrate convincingly that such entrepreneurial activities are necessary for success in regime formation at the international level. It follows, therefore, that an enhanced effort to understand entrepreneurial leadership must loom large in any research program directed toward the study of institutional bargaining in international society.

Subsistence Emissions and Luxury Emissions
HENRY SHUE

Leaving aside the many important questions about justice that do not have to be raised in order to decide how to tackle threats to the global environment, we will find four questions that are deeply involved in every choice of a plan for action. (1) What is a fair allocation of the costs of preventing the global warming that is still avoidable?; (2) What is a fair allocation of the costs of coping with the social consequences of the global warming that will not in fact be avoided?; (3) What background allocation of wealth would allow international bargaining (about issues like (1) and (2)) to be a fair process?; (4) What is a fair allocation of emissions of greenhouse gases (over the long-term and during the transition to the long-term allocation)? . . .

Allocating the Costs of Prevention

Whatever sums are spent in the attempt to prevent additional warming of the climate must somehow be divided up among those who are trying to deal with the problem. The one question of justice that most people readily see is this one: who should pay for whatever is done to keep global warming from becoming any worse than necessary?

One is tempted to say: "to keep it from becoming any worse than it is already going to be as a result of gases that are already in the air." Tragically, we will in fact continue to make it worse for some time, no matter how urgently we act. Because of the Industrial Revolution, the earth's atmosphere now contains far more accumulated CO_2 than it was normal for it to contain during previous centuries of human history. This is not speculation: bubbles of air from earlier centuries have been extracted from deep in the polar ice, and the CO_2 in these bubbles has been directly measured. Every day we continue to make large net additions to the total concentration of CO_2.

Several industrial nations have unilaterally committed themselves to reducing their emissions of CO_2 by the year 2000 to the level of their emissions in 1990. This may sound good, and it is obviously better than allowing emissions levels to grow in a totally uncontrolled manner, as the United States and many other industrial nations are doing. The 1990 level of emissions, however, was making a net addition to the total every day, because it was far in excess of the capacity of the planet to recycle CO_2 without raising the surface temperature of the planet. A reduction to the 1990 level of emissions means *reducing the rate* at which we are adding to the atmospheric total to a rate below the current rate of addition, but it also means *continuing to add to the total*.

Stabilizing emissions at a level as high as the 1990 level will not stabilize temperature—it will continue the pressure to drive it up. In order to stabilize temperature, emissions must be reduced to a level at which the accumulated *concentration* of CO_2 in the atmosphere is stabilized. CO_2 must not be added by human processes faster than natural processes can handle it by means that do not raise the surface temperature. Natural processes will, of course, have to "handle" whatever concentration

Reprinted by permission from 15 *Law & Policy* 39, copyright © 1993, Blackwell Publishers Ltd.

of CO_2 we choose to produce, one way or another; some of those ways involve adjustments in parameters like surface temperature that *we* will have a hard time handling. There is, therefore, nothing magic about the 1990 level of emissions. On the contrary, at that historically unprecedented level of emissions, the atmospheric concentration would continue to expand rapidly—it merely would not expand as quickly as it will at present levels or at the higher business-as-usual future levels now to be expected.

Emissions must be stabilized at a much lower level than the 1990 level, which means that emissions must be sharply reduced. The most authoritative scientific consensus said that in order to stabilize the atmospheric concentration of CO_2, emissions would have to be reduced below 1990 levels by more than 60 percent! Even if this international scientific consensus somehow were a wild exaggeration and the reduction needed to be, say, a reduction of only 20 percent from 1990 levels, we would still face a major challenge. Every day that we continue to add to the growing concentration, we increase the size of the reduction from current emissions necessary to stabilize the concentration at an acceptable total.

The need to reduce emissions, not merely to stabilize them at an already historically high level, is only part of the bad news for the industrial countries. The other part is that the CO_2 emissions of most countries that contain large percentages of the human population will be rising for some time. I believe that the emissions from these poor, economically less-developed countries also ought to rise insofar as this rise is necessary to provide a minimally decent standard of living for their now impoverished people. This is, of course, already a (very weak) judgment about what is fair: namely, that those living in desperate poverty ought not to be required to restrain their emissions, thereby remaining in poverty, in order that those living in luxury should not have to restrain their emissions. Anyone who cannot see that it would be unfair to require sacrifices by the desperately poor in order to help the affluent avoid sacrifices will not find anything else said in this article convincing, because I rely throughout on a common sense of elementary fairness. Any strategy of maintaining affluence for some people by keeping other people at or below subsistence is, I take it, patently unfair because so extraordinarily unequal—intolerably unequal.

Be the fairness as it may, the poor countries of the globe are in fact not voluntarily going to refrain from taking the measures necessary to create a decent standard of living for themselves in order that the affluent can avoid discomfort. For instance, the Chinese government, presiding over more than 22 percent of humanity, is not about to adopt an economic policy of no-growth for the convenience of Europeans and North Americans already living much better than the vast majority of Chinese, whatever others think about fairness. Economic growth means growth in energy consumption, because economic activity uses energy. And growth in energy consumption, in the foreseeable future, means growth in CO_2 emissions.

In theory, economic growth could be fueled entirely by forms of energy that produce no greenhouse gases (solar, wind, geothermal, nuclear [fission or fusion] and hydroelectric). In practice, these forms of energy are not now economically viable (which is not to say that none of them would be if public subsidies, including government-funded research and development, were re-structured). China specifically has vast domestic coal reserves, *the* dirtiest fuel of all in CO_2 emissions, and no economically viable way in the short run of switching to completely clean technologies or importing the cleaner fossil fuels, like natural gas, or even the cleaner technologies for burning its own coal, which do exist in wealthier countries. In May 1992 Chen Wang-xiang, general secretary of China's Electricity Council, said that coal-fired plants would account for 71 to 74.5 percent of the 240,000 megawatts of generating capacity planned for China by the year 2000 (1992). So, until other arrangements are made and financed, China will most likely be burning vast and rapidly increasing quantities of coal with, for the most part, neither the best available coal-burning technology, nor the best energy technology overall. The only alternative China actually has with its current resources is to choose to restrain its economic growth, which it will surely not do, rightly or wrongly. (I think rightly.)

Fundamentally, then, the challenge of preventing additional avoidable global warming takes this shape: how does one reduce emissions for the world as a whole while accommodating increased emissions by some parts of the world? The only possible answer is: by *reducing* the emissions by one part of the world by an amount *greater than* the increase by the other parts that are increasing their emissions.

The battle to reduce total emissions should be fought on two fronts. First, the increase in emissions by the poor nations should be held to the minimum necessary for the economic development that they are entitled to. From the point of view of the rich nations, this would serve to minimize the increase that their own reductions must exceed. Nevertheless, the rich nations must, second, also reduce their own emissions somewhat, however small the increase in emissions by the poor, if the global total of emissions is to come down while the contribution of the poor nations to that total is rising. The smaller the increase in emissions necessary for the poor nations to rise out of poverty, the smaller the reduction in emissions necessary for the rich nations—environmentally sound development by the poor is in the interest of all.

Consequently, two complementary challenges must be met—and paid for—which is where the less obvious issues of justice come in. First, the economic development of the poor nations must be as "clean" as possible—maximally efficient in the specific sense of creating no unnecessary CO_2 emissions. Second, the CO_2 emissions of the wealthy nations must be reduced by more than the amount by which the emissions of the poor nations increase. The bills for both must be paid: someone must pay to make the economic development of the poor as clean as possible, and someone must pay to reduce the emissions of the wealthy. These are the two components of the first issue of justice: allocating the costs of prevention.

Allocating the Costs of Coping

No matter what we do for the sake of prevention from this moment forward, it is highly unlikely that all global warming can be prevented, for two reasons. First, what the atmospheric scientists call a "commitment to warming" is already in place simply because of all the additional greenhouse gases that have been thrust into the atmosphere by human activities since around 1860. . . . We have done whatever we have done, and now its consequences, both those we understand and those we do not understand, will play themselves out, if not this month, some later month. Temperature at the surface level—at our level—may or may not already have begun to rise. But the best theoretical understanding of what would make it rise tells us that it will sooner or later rise because of what we have already done—and are unavoidably going to continue doing in the short-and medium-term. Unless the theory is terribly wrong, the rise will begin sooner rather than later. In the century and a quarter between the beginnings of the Industrial Revolution and 1993, and especially in the half century since World War II, the industrializing nations have pumped CO_2 into the atmosphere with galloping vigor. As of today the concentration has already ballooned. . . .

The second issue of justice, then, is: how should the costs of coping with the unprevented human consequences of global warming be allocated? The two thoughts that immediately spring to mind are, I believe, profoundly misguided; they are, crudely put, to-each-his-own and wait-and-see. The first thought is: let each nation that suffers negative consequences deal with "its own" problems, since this is how the world generally operates. The second is: since we cannot be sure what negative consequences will occur, it is only reasonable to wait and see what they are before becoming embroiled in arguments about who should pay for dealing with which of them. However sensible these two strategic suggestions may seem, I believe that they are quite wrong and that this issue of paying for coping is both far more immediate and much more complex than it seems. This brief overview is not the place to pursue the arguments in any depth, but I would like to telegraph why I think these two obvious-seeming solutions need at the very least to be argued for.

To Each-His-Own

Instantly adopting this solution depends upon assuming without question a highly debatable description of the nature of the problem, namely, as it was put just above, "let each nation that suffers negative consequences deal with 'its own' problems." The fateful and contentious assumption here is that whatever problems arise within one's nation's territory are *its own*, in some sense that entails that it can and ought to deal with them on its own, with (only) its own resources. This assumption depends in turn upon both of two implicit and dubious premises.

First, it is taken for granted that every nation now has all its own resources under its control. Stating the same point negatively, one can say that it is assumed that no significant proportion of any nation's own resources are physically, legally, or in any other way outside its own control. This assumes, in effect, that the international distribution of wealth is perfectly just, requiring no adjustments whatsoever across national boundaries! . . .

Second is an entirely independent question that is also too quickly assumed to be closed: it is taken for granted that no responsibility for problems resulting within one nation's territory could fall upon another nation or upon other actors or institutions outside the territory. Tackling this question seriously means attempting to wrestle with slippery issues about the causation of global warming and about the connection, if any, between causal responsibility and moral responsibility, issues to be discussed more fully later. Once the issues are raised, however, it is certainly not a foregone conclusion, for instance, that coastal flooding in Bangladesh (or the total submersion of, for example, the Maldives and Vanuatu) would be entirely the responsibility of, in effect, its victims and not at least partly the responsibility of those who produced, or profited from, the greenhouse gases that led to the warming that made the ocean water expand and advance inland. On quite a few readings of the widely accepted principle of the "polluter pays" those who caused the change in natural processes that resulted in the human harm would be expected to bear the costs of making the victims whole. Once again, I am not trying to settle the question here, but merely to establish that it is indeed open until the various arguments are heard and considered.

Wait-and-See

The other tactic that is supposed to be readily apparent and eminently sensible is: stay out of messy arguments about the allocation of responsibility for potential problems until we see which problems actually arise—we can then restrict our arguments to real problems and avoid imagined ones. Unfortunately, this too is less commonsensical than it may sound. To see why, one must step back and look at the whole picture.

The potential costs of any initiative to deal comprehensively with global warming can be divided into two separate accounts, corresponding to two possible components of the initiative. The first component, introduced in the previous section of this article, is the attempted prevention of as much warming as possible, the costs of which can be thought of as falling into the prevention account. The second component, briefly sketched in this section, is the attempted correction of, or adjustment to—what I have generally called "coping with"—the damage done by the warming that for whatever reasons goes unprevented.

It may seem that if costs can be separated into prevention costs and coping costs, the two kinds of costs could then be allocated separately, and perhaps even according to unrelated principles. Indeed, the advice to wait-and-see about any coping problems assumes that precisely such independent handling is acceptable. It assumes in effect that prevention costs can be allocated—or that the principles according to which they will be allocated, once they are known, can be agreed upon—and prevention efforts put in motion, before the possibly unrelated principles for allocating coping costs need to be agreed upon. What is wrong with this picture of two basically independent operations is that what is either a reasonable or a fair allocation of the one set of costs may—I will argue, does—depend upon how the other set of costs is allocated. The respective principles for the two allocations must not merely not be unrelated but be complementary.

In particular, the allocation of the costs of prevention will directly affect the ability to cope later of those who abide by their agreed-upon allocation. To take an extreme case, suppose that what a nation was being asked to do for the sake of prevention could be expected to leave it much less able to cope with "its own" unprevented problems, on its own, than it would be if it refused to contribute to the prevention efforts—or refused to contribute on the specific terms proposed—and instead invested all or some of whatever it might have contributed to prevention in its own preparations for its own coping. For example, suppose that in the end more of Shanghai could be saved from the actual eventual rise in sea-level due to global warming if China simply began work immediately on an elaborate and massive, Dutch-style system of sea-walls, dikes, canals, and sophisticated flood-gates—a kind of Great Sea-Wall of China—rather than spending its severely constrained resources on, say, purification technologies for its new coal-fueled electricity generating plants and other prevention measures. From a strictly Chinese point of view, the Great Sea-Wall might be preferable even if China's refusal to contribute to the prevention efforts resulted in a higher sea-level at Shanghai than would result if the Chinese did cooperate with prevention (but then did not have time or resources to build the Sea-Wall fast enough or high enough).

This fact that the same resources that might be contributed to a multilateral effort at prevention might alternatively be invested in a unilateral effort at coping raises two different questions, one primarily ethical and one primarily non-ethical (although these two questions are not unrelated either). First, would it be fair to expect cooperation with a multilateral initiative on prevention, given one particular allocation of those costs, if the costs of coping are to be allocated in a specific other way (which may or may not be cooperative)? Second, would it be reasonable for a nation to agree to the one set of terms, given the other set of terms—or, most relevantly given that the other set of terms remained unspecified? Doing your part under one set now while the other set is up for grabs later leaves you vulnerable to the possibility of the second set's being stacked against you in spite of, or because of, your cooperation with the first set. It is because the fairness and the reasonableness of any way of allocating the costs of prevention depends partly upon the way of allocating the costs of coping that it is both unfair and unreasonable to propose that binding agreement should be reached now concerning prevention, while regarding coping we should wait-and-see.

The Background Allocation of Resources and Fair Bargaining

This last point about potential vulnerability in bargaining about the coping terms, for those who have already complied with the prevention terms, is a specific instance of a general problem so fundamental that it lies beneath the surface of the more obvious questions, even though it constitutes a third issue of justice requiring explicit discussion. The outcome of bargaining among two or more parties, such as various nations, can be binding upon those parties that would have preferred a different outcome only if the bargaining situation satisfies minimal standards of fairness. An unfair process does not yield an outcome that anyone ought to feel bound to abide by if she can in fact do better. A process of bargaining about coping in which the positions of some parties were too weak precisely because they had invested so much of their resources in prevention would be unfair in the precise sense that those parties that had already benefited from the invested resources of the consequently weakened parties were exploiting that very weakness for further advantage in the terms on which coping would be handled.

In general, of course, if several parties (individuals, groups, or institutions) are in contact with each other and have conflicting preferences, they obviously would do well to talk with each other and simply work out some mutually acceptable arrangement. They do not need to have and apply a complete theory of justice before they can arrive at a limited plan of action. If parties are more or less equally situated, the method by which they should explore the terms on which different parties could agree upon a division of resources or sacrifices (or a process for allocating the resources or sacrifices) is actual direct bargaining. Other things being equal, it may be best if parties can simply work out among themselves the terms of any dealings they will have with each other.

Even lawyers, however, have the concept of an unconscionable agreement; and ordinary non-lawyers have no difficulty seeing that voluntarily entered agreements can have objectionable terms if some parties were subject, through excessive weakness, to undue influence by other parties. Parties can be unacceptably vulnerable to other parties in more than one way, naturally, but perhaps the clearest case is extreme inequality in initial positions. This means that morally acceptable bargains depend upon initial holdings that are not morally unacceptable—not, for one thing, so outrageously unequal that some parties are at the mercy of others.

Obviously this entails in turn that the recognition of acceptable bargaining presupposes knowledge of standards for fair shares, which are one kind of standard of justice. If we do not know whether the actual shares that parties currently hold are fair, we do not know whether any actual agreement they might reach would be morally unconscionable. The simple fact that they all agreed is never enough. The judgment that an outcome ought to be binding presupposes a judgment that the process that produced it was minimally fair. . . .

Allocating Emissions: Transition and Coal

The third kind of standard of justice is general but minimal: general in that it concerns all the resources and wealth that contribute to the distribution of bargaining strength and weakness, and minimal in that it specifies, not thoroughly fair distributions, but distributions not so unfair as to undermine the bargaining process. The fourth kind of standard is neither so general nor so minimal. It is far less general because its subject is not the international distribution of all wealth and resources, but the international distribution only of greenhouse gas emissions in particular. And rather than identifying a minimal standard, it identifies an ultimate goal: what distribution of emissions should we be trying to end up with? How should shares of the limited global total of emissions of a greenhouse gas like CO_2 be allocated among nations and among individual humans? Once the efforts at prevention of avoidable warming are complete, and once the tasks of coping with unprevented harms are dealt with, how should the scarce capacity of the globe to recycle the net emissions be divided?

So far, of course, nations and firms have behaved as if each of them had an unlimited and unshakable entitlement to discharge any amount of greenhouse gases that it was convenient to release. Everyone has simply thrust greenhouse gases into the atmosphere at will. The danger of global warming requires that a ceiling—probably a progressively declining ceiling—be placed upon total net emissions. This total must somehow be shared among the nations and individuals of the world. By what process and according to what standards should the allocation be done?

I noted above the contrast between the minimal and general third kind of standard and this fourth challenge of specifying a particular (to greenhouse emissions) final goal. I should also indicate a contrast between this fourth issue and the first two. Both of the first two issues are about the allocation of costs: who pays for various undertakings (preventing warming and coping with unprevented warming)? The fourth issue is about the allocation of the emissions themselves: of the total emissions of CO_2 compatible with preventing global warming, what percentage may, say, China and India use—and, more fundamentally, by what standard do we decide? Crudely put, issues one and two are about money, and issue four is about CO_2. We need separate answers to, who pays? and to, who emits? because of the distinct possibility that one nation should, for any of a number of reasons, pay so that another nation can emit more. The right answer about emissions will not simply fall out of the right answer about costs, or vice versa.

We will be trying to delineate a goal: a just pattern of allocation of something scarce and valuable, namely greenhouse-gas emissions capacities. However, a transition period during which the pattern of allocation does not satisfy the ultimate standard may well be necessary because of political or economic obstacles to an immediate switch away from the status quo. For instance, current emissions of CO_2 are very nearly as unequal as they could possibly be: a few rich countries with small populations are generating the vast bulk of the emissions, while the majority of humanity, living in poor countries with large populations, produces less altogether than the rich minority. It seems reasonable to assume that, whatever exactly will be the content of the standard of justice for allocating emissions, the emissions should be divided somewhat more equally than they currently are. Especially if the total cannot be allowed to keep rising, or must even be reduced, the per capita emissions of the rich few will have to decline so that the per capita emissions of the poor majority can rise.

Nevertheless, members of the rich minority who do not care about justice will almost certainly veto any change they consider too great an infringement upon their comfort and convenience, and they may well have the power and wealth to enforce their veto. The choice at that point for people who are committed to justice might be between vainly trying to resist an almost certainly irresistible veto and temporarily acquiescing in a far-from ideal but significant improvement over the status quo. In short, the question would be: which compromises, if any, are ethically tolerable? To answer this question responsibly, one needs guidelines for transitions as well as ultimate goals: not, however, guidelines for transitions instead of ultimate goals, but guidelines for transitions *in addition to* ultimate goals. For, one central consideration in judging what is presented as a transitional move in the direction of a certain goal is the distance travelled toward the goal. The goal must have been specified in order for this assessment to be made.

Notes and Questions

1. What differences are there between the notions of sustainable development espoused by Weiss and Solow? Which do you find more attractive? For further discussion of the problems in defining sustainable development, see Hahn, "Toward a New Environmental Paradigm," 102 *Yale Law Journal* 1719 (1993); Lele, "Sustainable Development: A Critical Review," 19 *World Development* 607 (1991).

2. How categorical are Weiss' commands? How would she figure out whether the options available to future generations are "unduly restricted?" How would she judge whether the resource base passed on to future generations is "reasonably secure and flexible?" What does it mean to "proceed extremely cautiously" with the destruction of unique natural resources? How should one determine whether "on balance" the diversity of the resource base has been maintained? For a fuller version of Weiss' argument, see *In Fairness to Future Generations: International Law, Common Patrimony, and Intergenerational Equity* (New York: United Nations University, 1989).

3. Assume that the probability of the extinction of a species rises somewhat each time that a member of the species is killed. When the rise in this probability is above some threshold, the species is deemed to be endangered. At that point, the principle of conservation of options presumably would ban additional killings of members of the species. How should the principle deal with probability rises below this threshold? How should this threshold be determined?

4. Weiss states that future generations might be willing to "pay us handsomely" to conserve unique natural resources. But how should we determine whether resources that we wish to expend are ones that future generations would find unique? Can such an inquiry be performed without violating Weiss' maxim that the present generation should not predict the preferences of future generations?

5. Similarly, Solow would preserve irreplaceable assets. How is this command consistent with his central argument concerning the substitutability of natural and other resources? How would he determine whether a resource is irreplaceable?

6. How should one determine, for the purposes of Weiss' argument, whether environmental quality has declined? Should improvements in the quality of certain environmental resources be allowed to make up for deterioration in other areas? If so, how should one determine whether the improvements are sufficient to counteract the deterioration?

7. Weiss states that "we may exhaust more reserves of a natural resource and cause modest levels of pollution, but pass on a higher level of income, capital, and knowledge." To what extent is this concession consistent with her principle that "we leave the quality of the natural and cultural environments in no worse condition than we received it?" To what extent does this concession make her position similar to Solow's?

8. Does the principle of intergenerational equity logically imply a requirement of intragenerational equity? If so, what principle of intragenerational equity is implied by Weiss' argument? Should trade-offs between intergenerational and intragenerational equity be permissible? More specifically, should it be permissible to sacrifice principles of intergenerational equity in order to secure a more equitable current, and future, distribution? To what extent is Weiss willing to compromise the principles of conservation of options and quality in order to further the principle of conservation of access?

9. What role is the discounting of the welfare of future generations playing in Solow's argument? Would the argument be significantly affected if such discounting were deemed inappropriate?

10. In addition to the principle of sustainable development, other central maxims of international environmental law include the precautionary principle and the "polluter pays" principle. The former prescribes that scientific uncertainties be resolved in favor of environmental controls. Is this principle helpful in determining the content of specific environmental policies? Can all uncertainties ever be resolved in accordance with the precautionary principle? For further discussion, see Bodansky, "Scientific Uncertainty and the Precautionary

Principle," 33 *Environment* 4 (1991); Cameron and Abouchar, "The Precautionary Principle: A Fundamental Principle of Law and Policy for the Protection of the Environment," 14 *Boston College International and Comparative Law Review* 1 (1991); Hey, "The Precautionary Concept in Environmental Law and Policy: Institutionalizing Caution," 4 *Georgetown International Environmental Law Review* 303 (1992). For example, Bodansky (p. 5) states that the precautionary principle "represents a rejection of 'risk neutrality,' which measures risk purely as the product of the magnitude and probability of harm." In contrast, Hey states (p. 305) that the concept "rejects environmental policies based on the assimilative capacity of the environment, i.e., it rejects a policy based on the assumptions that science can accurately determine the assimilative capacity of the environment and that, once determined, sufficient time for preventive action will remain." Which definition is more attractive? The Rio Declaration provides in Principle 15 that "[w]here there are threats of serious or irreversible damage, lack of full scientific certainty shall not be used as a reason for postponing cost-effective measures to prevent environmental degradation." How does this formulation compare with Bodansky's and Hey's statements? Is the precautionary principle consistent with the maximization of social welfare?

11. A typical formulation of the polluter pays principle contained in international agreements states that "by virtue of [this principle,] costs of pollution prevention, control and reduction measures shall be borne by the polluter." This formulation does not specify whether the polluter must pay for the damages caused by pollution. As long as this issue is not firmly resolved, what is the practical effect of the principle? Does it provide guidance for the setting of environmental standards? On its face, the principle would appear to bar government subsidies, but even with respect to this issue it has been the subject of exceptions. For further discussion, see Smets, "The Polluter Pays Principle in the Early 1990s," in Luigi Campiglio, Laura Pineschi, Domenico Siniscalco, and Tullio Treves, eds., *The Environment After Rio: International Law and Economics* (London: Graham & Trotman, 1994), at 131. Does the principle prescribe whether the initial allocation of marketable permits should be done by auction or, instead, given for free to current polluters. For further discussion of this issue, see Chapter VI.

12. In Principle 16 of the Rio Declaration, the polluter pays principle is subject to several qualifications: "National authorities should endeavor to promote the internalization of environmental costs and the use of economic instruments, taking into account the approach that the polluter should, in principle, bear the cost of pollution, with due regard to the public interest and without distorting international trade and investment." What are possible reasons for this formulation?

13. Does Young make a persuasive distinction between issues that lend themselves to contractarian interactions and ones that do not? It is likely that every environmental agreement will produce an unequal distribution of benefits and burdens. Consider two possible agreements. One produces net benefits for all nations but large disparities in the distribution of these benefits. Another produces some net gainers and some net losers but smaller distributional disparities. Which agreement would Young say is more likely? Which agreement is, in fact, more likely?

14. Young suggests that lack of equity is more likely to derail an agreement than lack of efficiency. Consider, however, ways in which lack of efficiency might interfere with the existence of contractarian interactions.

15. Young also suggests that there is a trade-off between the goals of efficiency and equity. Need there be such a trade-off? For example, could an equitable allocation of permits to emit greenhouse gases, coupled with rights to trade such permits, satisfy both goals?

16. The existence of clear cut and effective compliance mechanisms is another factor adduced by Young as conducive to the formation of international environmental agreements. From this perspective, how would you rate the likelihood of limitations on the emissions of greenhouse gases? For further discussion of the factors favoring international environmental agreements, see Barrett, "The Problem of Global Environmental Protection," in Dieter Helm, *Economic Policy Towards the Environment* (Oxford: Blackwell, 1991); Richard Elliot Benedick, *Ozone Diplomacy: New Directions in Safeguarding the Planet* (Cambridge, Mass.: Harvard University Press, 1991), chap. 14; Hahn and Richards, "The Internationalization of Environmental Regulation," 30 *Harvard International Law Journal* 421 (1989); Keohane, Haas, and Levy, "The Effectiveness of International Environmental Institutions," in Peter M. Haas, Robert O. Keohane, and Marc A. Levy, eds., *Institutions for the Earth* (Cambridge, Mass.: MIT Press, 1993); Keohane and Ostrom, "Introduction," in *Local Commons and Global Interdependence: Heterogeneity and Cooperation in Two Domains* (London: Sage, 1995); Sebenius, "Designing Negotiations Towards a New Regime: The Case of Global Warming," 15 *International Security* 110 (1991).

17. What answers should be given to Shue's four questions of justice? For further analyses of equity issues between developed and developing countries, see "Magraw, Legal Treatment of Developing Countries: Differential, Contextual, and Absolute Norms," 1 *Colorado Journal of International Environmental Law and Policy* 69 (1990); Young and Wolf, "Global Warming Negotiations: Does Fairness Matter?" *Brookings Review,* Spring 1992, at 46.

18. Principle 6 of the Rio Declaration states that "[t]he special situation and needs of developing countries, particularly the least developed and those most environmentally vulnerable, shall be given special priority." Principle 7 states: "In view of the different contributions to environmental degradation, States have common but differentiated responsibilities. The developed countries acknowledge the responsibility that they bear in the international pursuit of sustainable development in view of the pressures that their societies place on the global environment and of the technologies and financial resources they command." Principle 11 states: "Environmental standards . . . should reflect the environmental and developmental context to which they apply. Standards applied by some countries may be inappropriate and of unwarranted economic and social cost to other countries, in particular developing countries." What equity standards are embedded in these principles? Are differential standards undesirable from a trade perspective? In this connection, recall the discussions of Chapter XI. For further discussion of equity issues raised by the Rio Declaration, see Porras, "The Rio Declaration: A New Basis for International Cooperation," 1 *Review of Comparative and International Environmental Law* 245 (1992).

19. The conflicting interests of developing and developed countries in connection with the Montreal Protocol are set forth in Benedick, supra, chapters 12, 13. Under Article 5, developing countries with a consumption level of controlled substances of less that 0.3 kilograms per capita were entitled to delay compliance with the provisions of the protocol by ten years. At the time, the interest of the developing countries was to expand their use of chlorofluorocarbons (CFCs) and halons during the transitional period. By 1989, two years after the adoption of this provision, the developing countries had a more complex agenda. In light of the successful efforts toward phaseout of these substances by the developed countries, the developing countries became concerned that, as the production of CFCs fell, their cost would increase and substitute technologies would be more expensive. As a result, developing countries sought financial assistance to meet the incremental costs imposed on them by the Montreal Protocol, as well as the transfer of substitute technologies under favorable conditions. Articles 10 and 10a, the London Revisions to the Montreal Protocol, adopted in

1990, are responsive to these concerns. Is this a plausible model for dealing with global warming? For further discussion of the Montreal Protocol, see Barratt-Brown, "Building a Monitoring and Compliance Regime Under the Montreal Protocol," 16 *Yale Journal of International Law* 519 (1991).

20. There is probably a consensus that the emissions of greenhouse gases in developed countries must be reduced, but that in developing countries these emissions ought to continue to rise at least in the short term. For example, the Framework Convention on Climate Change states in its preamble "that the largest share of historical and current global emissions of greenhouse gases has originated in developed countries, that per capita emissions in developing countries are still relative low and that the share of global emissions originating in developing countries will grow to meet their social and development needs." What measures of equity might be useful in determining appropriate increases and reductions, respectively?

21. What role should past emissions of greenhouse gases play in designing measures of equity? Is it appropriate to penalize nations that emitted such gases before there was a consensus about the global warming problem? How does the equity of such an outcome compare with retroactive liability under Superfund—an issue that is discussed in Chapter X?

22. Evaluate the following bases for allocating permits for the emission of greenhouse gases:
   (a) past practice
   (b) land area
   (c) GNP
   (d) population
Does your evaluation depend on whether the permits are tradeable? In this connection, it is useful to compare, for example, the 1990 energy-related emissions of carbon dioxide of the United States, France, Mexico, and China. Table XII.1 shows their emissions per capita and per unit of GNP. See International Energy Agency, Climate Change Policy Initiatives [Paris: OECD, 1992], at 28, table 3. For further discussion of the equity issues involved in the allocation of permits, see Barrett, "Acceptable Allocations of Tradeable Carbon Emissions Entitlements in a Global Warming Treaty," in United Nations Conference on Trade and Development, *Combating Global Warming: Study on a Global System of Tradeable Carbon Emission Entitlements* (New York: United Nations, 1992), at 85; Grubb and Sebenius, "Participation, Allocation and Adaptability in International Tradeable Emission Permit Systems for Greenhouse Gas Control," in Organisation for Economic Co-operation and Development, *Climate Change: Designing a Tradeable Permit System* (Paris: OECD, 1992), at 185.

23. The Framework Convention on Climate Change contains, in Article 4.2, a "joint implementation" provision that states that parties may implement required emission limitations jointly with other parties (the Montreal Protocol also allows limited trading, under Article 2.5). This provision is thought to authorize trades among nations in emission permits, enabling a country with high abatement costs to meet its responsibility under the convention

Table XII.1.

	Emissions of CO_2 Per Capita	Emissions of CO_2 Per Unit of GNP
United States	19.97	1.09
France	6.80	0.64
Mexico	3.72	1.63
China	2.11	5.78

by financing abatement in a country in which abatement costs are lower. What countries should one expect to find in the latter category? Is "joint implementation" desirable for developed countries? Is it desirable for developing countries? Is "joint implementation" suspect if it produces disproportionate reductions in developing countries? For general discussion of the convention, see Bodansky, "The United Nations Framework Convention on Climate Change: A Commentary," 18 *Yale Journal of International Law* 451 (1993); Stone, "Beyond Rio: 'Insuring' Against Global Warming," 86 *American Journal of International Law* 445 (1992).

24. Consider the following criticisms of joint implementation advanced by developing countries:
   (a) that it will enable developed countries to continue to increase greenhouse gas emissions by financing reductions in developing countries, consequently retarding industrial development in these countries
   (b) that cheap options for reducing emissions in developing countries will be exploited by developed countries, with developing countries later having to face more costly abatement options

25. Consider two central issues for the design of a trading regime in greenhouse permits:
   (a) Should trading be limited to carbon dioxide or should it also include other greenhouse gases?
   (b) Should trading be limited to sources of carbon dioxide or should it also include sinks, such as expanding forests, which remove carbon from the atmosphere?

What are the advantages and disadvantages of the competing resolutions of these issues? What difficulties arise if sinks are made part of the trading system? For discussion of the first issue, see Stewart and Wiener, "The Comprehensive Approach to Global Climate Policy," 9 *Arizona Journal of International and Comparative Law* 83 (1992). In a portion of his article that is not excerpted, Henry Shue argues, in opposition of a comprehensive market in greenhouse gas emissions:

> To suggest simply that it is a good thing to calculate cost-effectiveness across all sources of all GHGs [greenhouse gases] is to suggest that we ignore the fact that some sources are essential and even urgent for the fulfillment of vital needs and other sources are inessential or even frivolous. What if, as is surely in fact the case, some of the sources that it would cost least to eliminate are essential and reflect needs that are urgent to satisfy, while some of the sources that it would cost most to eliminate are inessential and reflect frivolous whims?

How strong is this argument? How would one design a market to take account of his concerns?